THEORIES OF EVOLUTION

THEORIES OF EVOLUTION

By

H. JAMES BIRX

B.S., M.S., M.A., Ph.D.

Chairman
Anthropology/Sociology Department
Canisius College, Buffalo, New York

CHARLES C THOMAS • PUBLISHER
Springfield • Illinois • U.S.A.

Published and Distributed Throughout the World by
CHARLES C THOMAS • PUBLISHER
2600 South First Street
Springfield, Illinois 62717

© *1984 by* CHARLES C THOMAS • PUBLISHER
ISBN 0-398-04902-5
Library of Congress Catalog Card Number: 83-12046

With THOMAS BOOKS *careful attention is given to all details of manufacturing
and design. It is the Publisher's desire to present books that are satisfactory as to
their physical qualities and artistic possibilities and appropriate for their
particular use.* THOMAS BOOKS *will be true to those laws of quality that
assure a good name and good will.*

Library of Congress Cataloging in Publication Data

Birx, H. James.
  Theories of evolution.

  Bibliography: p.
  Includes index.
  1. Human evolution—Philosophy. I. Title.
GN281.B59   1983   573.2   83-12046
ISBN 0-398-04902-5

Printed in the United States of America
PS-R-3

SPECIAL ACKNOWLEDGMENT

Although (unless otherwise explicitly indicated and/or clear from the context), I, H. James Birx, alone bear the sole and exclusive responsibility for the accuracy of all factual and theoretical statements in this book, as well as all opinions expressed herein, I wish to recognize my special indebtedness to Dr. George V. Tomashevich, Professor of Anthropology, State University of New York College at Buffalo, for his manifold intellectual and stylistic contributions to various parts of this work, including the Prologue and especially Chapter Twelve on Knežević, as well as several additional chapters.

I am particularly obligated to this senior colleague and eminent American natural and social scientist, scholar, and poet for placing at my disposal, over a period of several months of intensive discussion and close cooperation, the rich reservoir of his unusually broad and versatile erudition embracing, *inter alia*, not only general anthropology, world history, history of philosophy, and Slavic and East European Studies but also his extensive knowledge of the history and philosophy of science. He has called to my attention the work of Petr Kropotkin and several other leading Russian biologists of the pre-Soviet period, Božidar Knežević, Ksenija Atanasijević, and other outstanding philosophers and scientists of his native Yugoslavia, Jean Rostand of France, as well as interesting glimpses of protoevolutionary thought in the early philosophies of ancient India and China. For all these and many other helpful contributions I thank him.

FURTHER ACKNOWLEDGMENTS

I am especially grateful to my friend Gary R. Clark for both his encouragement and inspiration from the conception to the completion of this volume; his interest in my project contributed in a very large measure to the success of this undertaking.

The author also appreciates the help offered from his following colleagues and friends during the writing of this manuscript: Dr. Kenneth S. Balmas, David A. Bosworth, II, Dr. Allan L. Canfield, Anthony J. DeBlasi, Dr. Eugene P. Finnegan, S.J., Dr. Arthur A. Hitchcock, Thomas M. Koehler, Robert A. Lorenz, Dr. Michael P. Penetar, David L. Smith, Dr. Victor A. Tomovich, Angelo M. Turco, George J. Turco, Louis D. Turco, Dr. Marilyn S. Watt, and Henry E. Winters, Jr.

The writer is very thankful for the endorsements of the following scholars: Dr. Philip Appleman, Frederick Edwords, Dr. John Kraus, Dr. Paul Kurtz, Mary Lukas, Dr. Sol Tax, and Dr. Oleg Zinam.

It is both a distinct honor and pleasure to include the appropriate and eloquent poem "The Skeletons of Dreams" by Dr. Philip Appleman, Distinguished Professor of English at Indiana University (this poem first appeared in *The New York Times*, Sunday, November 29, 1981).

Profound gratitude goes to Mr. Payne E. L. Thomas of Charles C Thomas, Publisher for his gracious accommodations and lasting concern throughout the preparation and production of this book.

Finally, the author extends his heartfelt compliments to both Marie E. Coleman and Rosemary A. Sortino for their conscientious and expert secretarial assistance.

To the immortal memory of
CHARLES ROBERT DARWIN
(1809-1882)

Charles Robert Darwin (1809-1882). Courtesy of American Museum of Natural History.

This preservation of favourable individual differences and variations, and the destruction of those which are injurious, I have called Natural Selection, or the Survival of the Fittest. ... There is grandeur in this view of life, with its several powers, having been originally breathed into a few forms or into one; and that, whilst this planet has gone cycling on according to the fixed law of gravity, from so simple a beginning endless forms most beautiful and most wonderful have been, and are being, evolved.

Charles Darwin
On the Origin of Species
Chapters IV, XV

If the anthropomorphous apes be admitted to form a natural sub-group, than as man agrees with them, not only in all those characters which he possesses in common with the whole Catarhine group, but in other peculiar characters, such as the absence of a tail and of callosities, and in general appearance, we may infer that some ancient member of the anthropomorphous sub-group gave birth to man. . . . In a series of forms graduating insensibly from some ape-like creature to man as he now exists, it would be impossible to fix on any definite point when the term "man" ought to be used. But this is a matter of very little importance. . . . Man still bears in his bodily frame the indelible stamp of his lowly origin.

Charles Darwin
The Descent of Man
Chapters VI, VII, XXI

Whenever we abandon nature, and give ourselves up to the fantastic flights of our imagination, we become lost in vagueness, and our efforts culminate only in errors. The only knowledge that it is possible for us to acquire is and always will be confined to what we have derived from a continued study of nature's laws; beyond nature all is bewilderment and delusion: such is my belief.

Jean-Baptiste Lamarck, *Zoological Philosophy (1809)*

Evolution is an integration of matter and concomitant dissipation of motion; during which the matter passes from an indefinite, incoherent homogeneity to a definite, coherent heterogeneity; and during which the retained motion undergoes a parallel transformation.

Herbert Spencer, *First Principles (1862)*

The question of questions for mankind (the problem which underlies all others, and is more deeply interesting than any other) is the ascertainment of the place which Man occupies in nature and of his relations to the universe of things.

Thomas H. Huxley, *Evidence as to Man's Place in Nature (1863)*

First labour, after it and then with it speech: these were the two most essential stimuli under the influence of which the brain of the ape gradually changed into that of man, which for all its similarity is far larger and more perfect.

Friedrich Engels, *The Part Played by Labour in the Transition from Ape to Man (1876)*

What is the ape to man? A laughingstock or a painful embarrasment. And man shall be just that for the overman: a laughingstock or a painful embarrasment. You have made your way from worm to man, and much in you is still worm. Once you were apes, and even now, too, man is more ape than any ape. Whoever is the wisest among you is also a mere conflict and cross between plant and ghost. But do I bid you become ghosts or plants? Behold, I teach you the overman. The overman is the meaning of the earth. Let your will say: the overman shall be the meaning of the earth! I beseech you, my brothers, remain faithful to the earth, and do not believe those who speak to you of otherworldly hopes!

Friedrich Nietzsche, *Thus Spake Zarathustra (1883)*

Humanity is but a transitory phase of the evolution of an eternal substance, a particular phenomenal form of matter and energy, the true proportion of which we soon perceive when we set it on the background of infinite space and eternal time.

Ernst Haeckel, *The Riddle of the Universe (1900)*

Error creates the great, the lasting, the warm; truth crushes and grinds to powder, makes ephemeral and extinguishes the flame of everything. Error creates eternity, truth turns it into moments. Error creates worlds, suns, the universe; truth turns the worlds into atoms. Error creates oceans, truth breaks oceans into drops. Error is warm, truth is cold. Error makes of man a God-chosen creature, truth makes of him but an ape. Error puts the earth in the center of the universe, truth makes of her a grain in the cosmos. Error elevates, truth lowers. Error makes of man an entire book, truth makes of him but a single letter. Hence, truth is needed by only a few who are strong enough to endure it. This is why it is so scarce among men.

Božidar Knežević, *Aphorism 533 (1902)*

The evolution of life, looked at from this point, receives a clearer meaning, although it cannot be subsumed under any actual idea. It is as if a broad current of consciousness had penetrated matter, loaded, as all of consciousness is, with an enormous multiplicity of interwoven potentialities. It has carried matter along to organization, but its movement has been at once infinitely retarded and infinitely divided.... From this point of view, not only does consciousness appear as the motive principle of evolution, but also, among conscious beings themselves, man comes to occupy a privileged place. Between him and the animals the difference is no longer one of degree, but of kind.

Henri Bergson, *Creative Evolution (1907)*

Is evolution a theory, a system or a hypothesis? It is much more: it is a general condition to which all theories, all hypotheses, all systems must bow and which they must satisfy henceforward if they are to be thinkable and true. Evolution is a light illuminating all facts, a curve that all lines must follow.

Pierre Teilhard de Chardin, *The Phenomenon of Man (1940)*

With the rise of the evolutionary movement and the various sciences contributing to it, the idealistic philosophers and fideists were deprived of their concepts of soul, mind, consciousness, will, and other traditional props.... Man is incidental to existence; he occupies a very minor position in the universe. This basic fact of the nondependence of the field of existence on man, or on any mind, makes philosophical idealism impossible.

Marvin Farber, *Basic Issues of Philosophy (1968)*

Evolution is purposeless, nonprogressive, and materialistic.... I predict the triumph of Darwinian pluralism. Natural selection will turn out to be far more important than some molecular evolutionists imagine, but it will not be omnipotent, as some sociobiologists seem to maintain.

Stephen Jay Gould, *Ever Since Darwin (1973)*

Perhaps the origin and evolution of life is, given enough time, a cosmic inevitability.... Evolution is a fact, not a theory.... The origin and evolution of life are connected in the most intimate way with the origin and evolution of the stars.... As long as our inquiries are limited to one or two evolutionary lines on a single planet, we will remain forever ignorant of the possible range and brilliance of other intelligences and other civilizations.

Carl Sagan, *Cosmos (1980)*

The Skeletons of Dreams

He found giants
in the earth: Mastodon,
Mylodon, thigh bones
like tree trunks, Megatherium, skulls
big as boulders—once,
in this savage country, treetops
trembled at their passing.
But their passing was silent as snails,
silent as rabbits: nothing at all recorded
the day when the last of them came
crashing through creepers and ferns,
shaking the earth a final time,
leaving behind them crickets,
monkeys, and mice.
For think: at last it is nothing
to be a giant—the dream
of an ending haunts tortoise and Toxodon,
troubles the sleep of the woodchuck
and the bear.

Back home in his English garden,
Darwin paused in his pacing,
writing it down in italics
in the book at the back of his mind,
>   *When a species has vanished*
>   *from the face of the earth,*
>   *the same form never reappears...*
So after our millions of years
of inventing a thumb and a cortex,
and after the long pain
of writing our clumsy epic,
we know we are mortal as mammoths,
we know the last lines of our poem.
And somewhere in curving space
beyond our constellations,
nebulae burn in their universal law:
nothing out there ever knew
that on one sky-blue planet
we dreamed that terrible dream.
Blazing along through black nothing
to nowhere at all, Mastodons of heaven,
the stars do not need our small ruin.

                                          Philip Appleman

CONTENTS

xiii

THEORIES OF EVOLUTION

REFLECTIONS ON EVOLUTION

E volution is a fact.
While whirling through seemingly eternal time and infinite space, our planet has undergone eons of change: it is neither at the center of this universe nor a fixed orb, as it was believed to be by most of the ancient thinkers. The material earth is not merely six thousand years old, and all living things on its surface were not suddenly created by a divine power to exist as static types or kinds for all ages to come. Even symboling man as the human animal and sociocultural being is not a unique form of life isolated from organic history, biological principles, or the essential unity of life.

Rocks, fossils, artifacts, and genes tell an incredibly different story: this planet with its web of life is unmistakably just another fragment of the endless flux of sidereal evolution. Our earth is nearly five billion years old, life first emerged about four billion years ago, and humanlike forms have walked upright on land for at least four million years. The documented evidence is clear and cannot be ignored.

On our terrestrial world, which is merely a cosmic speck of matter and energy, life as we know it emerged at least once, with some organisms (both plants and animals) adapting and thriving over vast periods of geological time, although others, in fact most,

---

Some of the material in this Prologue was first presented by the author in two papers delivered in public symposiums held at Canisius College: "Man, Science, and Nature: Toward a Comprehensive and Critical Approach to Philosophical Anthropology" (April 26, 1981) and "Darwin and Philosophy: Critical Reflections" (April 19, 1982).

have eventually perished as evidenced in the paleontological record. Throughout the complex process of biological evolution, survival has been the exception while extinction is the rule, and process does not guarantee progress.

This present volume marks the centennial of the death of Charles Robert Darwin (1809-1882), the greatest and most significant author on the subject of biological evolution since this conceptual view of organic history began to emerge long before the appearance of the earth sciences (not to mention modern social theory, dynamic philosophy, and process theology). During the global voyage of H.M.S. *Beagle* in the third decade of the last century, the crew referred to the young Darwin as "the philosopher": Darwin, however, claimed that even Lamarck and Cuvier were only schoolboys compared to the towering genius of the great Aristotle.

Darwin's own achievement represented a crucial turning point in critical thought that resulted in a conceptual revolution: before Darwin, the idea of evolution had been simply a rational speculation on the nature of reality; with Darwin, the evolutionary framework became a scientific explanation for the history of life itself; and after Darwin, all of the sciences (from astronomy to psychology) as well as philosophy and even theology have had to accept both its truth and its implications for the place of humankind within this universe.

The topic of evolution is so vast and intricate in terms of its rich and diverse range of facts and concepts that this book must focus on only those dramatic highlights (e.g. the thoughts of Darwin and Teilhard de Chardin) and major consequences (e.g. the ongoing creation and/or evolution controversy). In fact, one may speak of the evolution of the theory of evolution.

In the Mediterranean world during the Presocratic period in western intellectual history (600-400 BC), the rational speculations of several naturalist cosmologists resulted in ideas anticipating the scientific theory of biological evolution. There were the insights of Thales, Anaximander, Xenophanes, Heraclitus, and Empedocles: life originated in water, the makeup of man resembles that of a fish, fossils in rock strata are the remains of past but different organisms, cyclical change is the pervasive essential characteristic of this dynamic cosmos, and all living things must adapt to changing environments if they are to survive and reproduce (this last crucial point even anticipated Darwin's major explanatory principle of "natural selection" and Spencer's controversial phrase "the survival of the fittest").

The early atomism of Leucippus of Miletus, Democritus of Abdera, and Epicurus of Samos foreshadowed the modern materialist interpretation of an evolving universe. Also, the ancient thought-systems of both India and China recognized the flux of nature and man's direct relationship to all living things.

The systematic Aristotle (384-322 BC), the ingenious founder of the science of biology who made pioneering contributions to both embryology and taxonomy as well as grounded his own process metaphysics in change and development, was not an evolutionist. The Stagirite regarded organic nature as an eternally fixed "Great Chain of Being" of immutable plant and animal forms: a hierarchy of life referred to as the ladder of nature, a static scale of terrestrial existence from the simple minerals to the rational man at its apex.

Aristotle's safe and secure worldview was one of pervasive order and harmony that, unfortunately, blinded subsequent thinkers to the conceptual contributions of Lucretius (96-55 BC) and Plotinus (205-270 AD): the former gave a general outline of human sociocultural development from prehistoric times to the Roman Empire against the background of his own atomist/materialist evolutionary cosmology, while the latter offered a cyclical view of reality as the everlasting one and the ephemeral many rooted in Neoplatonism and mysticism.

After the so-called Dark Ages and Medieval Period, which saw no significant advances in science or natural philosophy, critical thinkers of the Italian Renaissance boldly challenged the dogmatic Aristotelian/Ptolemaic/Thomistic worldview: one recalls the significant breakthroughs of Cusa, Copernicus, Bruno, Kepler, and Galileo. Advances in astronomy, physics, and natural philosophy paved the way for later important discoveries in the earth sciences. In fact, during this time, it was Leonardo da Vinci who seriously speculated on the age of rocks and the significance of fossils within his own dynamic view of the material world.

Following the Renaissance, first Leeuwenhoek (1632-1723) and then Linnaeus (1707-1778) gave a tremendous impetus to the emerging science of biology. Just as Galileo's scientific use of the telescope had brought the celestial objects of the vast macrocosm into the range of human vision, so now Leeuwenhoek's scientific use of the microscope brought the biological objects of the living microcosm before the eyes of the curious observer. Linnaeus's descriptions and classifications of plants and animals revealed the similarities among organisms, although he held to the fixity of types despite his growing bewilderment over the finding of varieties within the assumed immutable species.

Some enlightened philosophers of the eighteenth century even glimpsed the theory of evolution. Although the process views of Condorcet, Comte, and Hegel centered on social history, their historical perspective established an intellectual atmosphere that stimulated the coming of the subsequent earth sciences in the nineteenth century.

Among the early naturalist writings, Hutton's *Theory of the Earth* (1785, 1795) followed by Lyell's *Principles of Geology* (1830-1833) clearly placed historical geology on a firm foundation. Erasmus Darwin's *Zoonomia* (1796) offered an account for the origin of life on earth from a single living filament that existed millions of years before the history of humankind. Lamarck and Cuvier contributed to invertebrate and vertebrate paleontology, respectively. Jacques Boucher de Perthes fathered the science of prehistoric archaeology after he found human artifacts associated with the fossils of now extinct mammals deep in rock strata. These advances in science (along with continued progress in embryology and taxonomy) occurred before the naturalist Charles Darwin met the fundamentalist Robert FitzRoy, which was an ironic event that set the stage for both a scientific revolution and the creation/evolution controversy.

Darwin is the father of the scientific theory of biological evolution. He presented a view of humankind, organic development, and the history of our planet that seriously challenged all previous interpretations and evaluations of man, life, and the earth. His powerful theory of evolution, with its illuminating ideas and awesome perspective, remains a disturbing worldview for many (if not even an anathema to some groups as is dramatically demonstrated in the present revival of the creation/evolution controversy).

It is both ironic and tragic that this debate between fundamentalist creationists and scientific evolutionists should still continue in the very year that commemorates the centennial anniversary of Charles Darwin's death. Of course, there is a wide range of interpretation and perspective in the world's literature on evolution. This volume explores a part of this literature.

In the last century, the natural philosopher and mathematician Auguste Comte (1798-1857) called for a social physics: the serious empirical study of human thoughts and behavior patterns within a historical perspective. He had rightly been impressed with the rigorous methods, both scientific and mathematical, and amazing discoveries of the natural scientists (one thinks of Kepler, Galileo, and Newton), as well as the technological advances of the industrial revolution. His lofty aim was to humanize science through the empirical understanding and rational appreciation of mankind by elevating the study of humanity to the level of objective and critical inquiry.

Despite his pervasive optimism, Comte maintained that there would be limits to empirical inquiry; for example, he claimed that humankind would never be able to determine the chemical composition of our own sun (much less ever know the makeup of distant stars). Yet, thanks to methods of indirect observation and resultant inference unimaginable in Comte's time, modern astrophysics and astrochemistry are able to determine the size, distance, and molecular basis of stars both unseen and unknown to Comte and the naturalists of his century.

This point clearly indicates that one should not be too quick in setting unwarranted limits to the possibilities of empirical discoveries as yet unmade and to their unforeseeable consequences. No closure should be dogmatically placed on further technological progress and scientific inquiry within the rational guidelines of a free, open, and responsible society. To be sure, a distinction among the natural sciences, social sciences, and humanities is always necessary. Likewise, in the human world, ethical principles and preferential judgments are not reducible to mere scientific statements (although this does not preclude the need for empirical facts and rational conceptual frameworks to guide human thought and action, especially in matters concerning morality). The enlightened Comte saw a new frontier and boldly paved the way for the emergence of the social sciences as we know them today, e.g. sociology, anthropology, and psychology.

Also in the last century, sometimes referred to as the Age of Ideology, Ludwig Feuerbach (1804-1872), a representative of the so-called Hegelian left, fathered a conceptual framework that may be called scientific philosophical anthropology. Although he studied theology and philosophy, Feuerbach became disillusioned with the religious worldview as well as Hegelian idealism. His own critical study of humankind within nature advocated a rigorous interpretation of man as both an evolved animal and a morally concerned sociocultural being (a view grounded in the special sciences and the canons of logic).

Therefore, Feuerbach represents a significant philosophical break from the earlier German idealists (e.g. Leibniz, Kant, and Hegel) and a starting point for a naturalist humanism. He embraced the scientific/rational attitude, acknowledged the value of both the natural and social sciences, and incorporated the theory of biological evolution into his own conceptual framework of man within nature. In essence, Feuerbach saw humanity within sociocultural development and our species as a product of organic history.

It is now over one hundred years since Comte founded sociology as the empirical science of human thought and behavior as well as positivism as the objective approach of philosophy (although, unfortunately, his position is a narrowly conceived view of human methodology). It is also over one hundred years since Feuerbach called for a scientific philosophical anthropology as the new conceptual framework of the future.

During the twentieth century, the academic discipline of general anthropology has considerably altered our view of humankind within natural history and sociocultural development. Being both interdisciplinary and intradisciplinary in its approach to the human world, general anthropology is now a truly scientific and holistic study of mankind within nature. As such, modern anthropology is simultaneously a natural science as well as a social science and one of the humanities: man, its very subject matter, partakes of and belongs to all three realms. Man began in nature, lives in nature, and will end surrounded by nature as the all-encompassing material reality.

Yet, despite the remarkable advances of both the natural and social sciences (not to mention the incredibly rapid progress in communication and technology), there still remains a wide gap between the conceptual framework of the informed and expert minority and the interpretations held by the uninformed and unexpert majority. Too often personal biases, prejudices, and mere opinions as well as traditional beliefs have predetermined how facts and relationships (ideas, concepts, hypotheses, and theories) are incorporated into one's own worldview or even if they are to be taken seriously at all. The continuing quest for a sound understanding and sober appreciation of humankind within this universe requires inquisitive minds, an open society, and free inquiry within an ethical framework motivated by the ethos of plurality, responsibility, intellectual modesty, and mutual tolerance for all.

To put scientific philosophical anthropology in a proper perspective, an examination of some major achievements in science (especially cosmology, astronomy, and evolutionary biology), and their far-reaching implications and consequences, is both necessary and highly desirable.

Rejecting mere opinions and traditional beliefs, as well as legends and myths, the Presocratics (600-400 BC) as natural cosmologists encouraged the critical observation of nature itself and rigorous rational deliberation: intellectually, they turned chaos into cosmos, i.e. the universe of chaos became a cosmos of order. Without scientific instruments, but with the aid of the acute intellect, they investigated the heavens through the power of logical and disciplined imagination. Some of these early thinkers anticipated biological evolution, envisioned other worlds, claimed the existence of extraterrestial life (including extraplanetary intelligences), and intuited the flux and essential unity of all reality. Even today, one may envy their bold curiosity, relative freedom of expression, and sweeping cosmic perspective. These intuitions of the Presocratic intellect occurred more than two millenia before Darwin, Einstein, and Sagan!

These incredible speculations were later superceded, not altogether happily, by the Aristotelian worldview: an ordered and secure but also dogmatic and closed encyclopedic interpretation of an alleged finite nature grounded in a geocentric and geostatic cosmology as well as an anthropocentric view of the terrestrial world. A privileged position was given to man and our earth, and this conceptual framework was later supported by Ptolemy and Aquinas. As a result, humankind and this planet both enjoyed their special place in the cosmic scheme of things until the Italian Renaissance centuries later.

Like Aristarchus in antiquity, Nicolas Copernicus proposed and defended a heliocentric universe (his astronomy was founded less on mathematics than on natural philosophy). Taken seriously, this revised ancient viewpoint shifted the physical center of the universe away from our earth by regarding the sun as the true center of the cosmos. As such, man's habitat was now held to be merely one of several planets that revolve around the nearest star.

Kepler's mathematical discoveries of the eliptical orbits of the planets and his three laws of planetary motion, Galileo's use of the telescope for celestial inquiry resulting in his fathering descriptive astronomy, and the subsequent work in astronomy/ cosmology from Newton to Einstein all attest to the diminished status of our planet earth within a universe seemingly eternal in time, infinite in space, and endlessly changing. This revolutionary viewpoint had already been boldly suggested and rationally argued by the brilliant but tragic genius from Nola, Giordano Bruno (1548-1600), in whose farsighted opinion the center of the eternal and infinite cosmos is everywhere and its circumference is nowhere. Based on this visionary breakthrough, modern astronomy and cosmology humble man before the awesome immensity of this universe.

If both an earth-centered and sun-centered model of the universe are no longer sound, can a man-centered cosmology still be admissible as a true interpretation of physical reality? Here we must turn our eyes away from the starry heavens above and to the rocks of terra firma below with their fossils and artifacts, empirical evidence that factually demonstrates the creative evolution of life on this planet (including the origin and organic history of the human species with all its works: physical, mental, and symbolic).

The natural origin, historical development, and essential unity of life on earth was anticipated by several Presocratics of antiquity (although it would be, of course, misleading to refer to them as evolutionists in the modern sense). Likewise, several Oriental thinkers of antiquity also anticipated the evolutionary perspective. Yet, Aristotle's view of nature argued that every plant and animal species has its unique natural place in the eternally fixed "Great Chain of Being": a continuous but static terrestrial hierarchy of biological types from simple plants through organic forms of ever-increasing complexity and sensitivity to intelligent man at its apex.

This great thinker and logician from Stagira rejected the possibility that any species of plant or animal had ever become extinct, interpreted fossils as being merely chance aberrations in rock strata, and held that like begets like or kind begets kind. Although Aristotle made undeniable contributions to taxonomy and embryology (as well as studied the growth, development, and reproductive activities of organisms), the father of biology was clearly not an evolutionist. However, the Roman philosopher and poet Lucretius did, in fact, anticipate the modern cosmology and wrote in general about the biological and sociocultural evolution of the human animal.

The long interval of the Dark Ages and Medieval Period was not sufficient time to discredit the Aristotelian worldview. And Linnaeus, the first significant biologist after Aristotle, also held to the eternal fixity of species (however, he was bewildered by the discovery of varieties within the allegedly immutable species of plants that he studied).

The evolution theory was reawakened from its long historical slumber by some thinkers of the French Enlightenment, although they lacked the empirical evidence necessary to factually document the mutability of plant and animal forms throughout the flow of organic history. It was the natural philosopher Lamarck who first wrote a book solely for the purpose of presenting the theory of evolution as a working conceptual framework to account for the creative diversity yet essential unity of all life on this planet. His major book, *Zoological Philosophy* (1809), was a speculation on the origin and evolution of animals due to spontaneous generation, the evolutionary laws of use/disuse and the inheritance of acquired characteristics, and a vitalist interpretation of life. Lamarck's pioneering work deserves to be remembered, although his explanations to account for the exact mechanisms of biological evolution are now no longer acceptable to the scientific community.

Charles Robert Darwin did not develop his scientific theory of biological evolution in a vacuum. Like all creative intellects, he too was a product of history and a particular sociocultural milieu. It has been suggested that the idea of evolution was "in the air," but theories do not exist as ontological entities independent of human knowers and doers.

The so-called Age of Enlightenment had been an intellectual period that paved the way for the nineteenth century, with its rich ideas and emerging sciences. The old world-system was crumbling. Naturalists were seriously challenging the traditional views of space, time, history, and even the nature of change.

Rocks, fossils, and artifacts were now the subject matter of scientific inquiry. It became generally realized that man has a history, and that everything in nature (including the earth and the cosmos) is a part of the ongoing flux of physical reality. Naturalists critically questioned the hitherto accepted age of the earth, explanation for the origin and history of life, and even the antiquity and development of the human animal. Historical geology, comparative paleontology, and prehistoric archaeology became sciences in their own right (not to mention the continued advances in botany, taxonomy, and embryology). The growing empirical evidence dramatically argued for a worldview incredibly different from the conceptual framework that had dominated western thought for about twenty centuries.

The earth seemed to be millions if not even billions of years old, the creativity of biological development seemed to be grounded in natural causes, and the human species seemed to be a relatively recent product of organic history totally within the material world.

It fell to the young naturalist Charles Darwin to bring together all these growing empirical facts, new concepts, and challenging perspectives into a comprehensive and intelligible view of earth history (especially the development of life and humankind in terms of science and reason). Like Haeckel before him, Knežević rightly refers to Darwin as the Newton of biology: although the concept of biological evolution was "in the air" so to speak, nevertheless there was the need both to document empirically this theoretical framework with sufficient evidence as well as to argue its validity with implacable logic (not to mention acknowledging its far-reaching implications for the special sciences, philosophy, and even theology).

After studying both medicine and theology, the would-be Anglican minister Charles Darwin was remarkably free from the dogmas of science, philosophy, and theology. His genius was open to scientific discoveries, widening experiences, and challenging ideas. He took the Hutton/Lyell theory of uniformitarianism or geological gradualism seriously. He journeyed for five years around the globe (1831-1836) as naturalist and traveling companion aboard the *Beagle* under the eccentric aristocrat and fundamentalist creationist Captain FitzRoy, and later read with great reward the Malthusian principle of population in 1838.

Darwin tells us that it was his visit to the Galapagos Islands that, in retrospect, convinced him that biological species are in fact dynamic and mutable (not immutable as was the general consensus among biologists of his own time). On this isolated prehistoric-like archipelago, the naturalist experienced the creative results of evolutionary forces: giant tortoises that varied in the size and shape of their shells from island to island, land and marine iguanas that reminded him of those mighty dinosaurs of an earlier primeval era, and the many species of finches with their beaks that differed in size and shape from habitat to habitat among the islands. Darwin's primary interest had shifted from geology to biology and would later even incorporate anthropology and psychology. After circumnavigating the earth, he was convinced that species are subject to change and referred to biological evolution as "descent with modification" (he even wrote an essay in 1842, which he expanded into a sizable manuscript in 1844, but, despite being encouraged by his learned friends, published neither of the two writings). As is commonly known, Darwin's major work on the scientific theory of biological evolution primarily by means of slight variation and natural selection is his book entitled *On the Origin of Species* (1859). What is not so commonly known is that in this volume he did not include the human animal within his own process view of organic history. Not until the publication of *The Descent of Man* (1871) did Darwin extend his theory of slow and gradual evolution to account also for the origin and natural development of the human species. Huxley and Haeckel had already done so.

Today, any serious study of man and consideration of his place in nature cannot ignore the Darwinian framework (although recent advances in genetics and our modern understanding of selective principles have somewhat modified the early explanation for biological evolution). It must be pointed out, however, that those promoting the fact of evolution may differ in the emphasis they place on the various explanatory mechanisms of this process. Evolutionists also represent a wide range of interpretations, evaluations, and philosophies found in the literature (from the writings of Spencer, Huxley, and Haeckel in the last century to the worldviews of Bergson, Whitehead, and Teilhard de Chardin in this century).

Copernicus had removed the earth from its assumed privileged position as the fixed center of the universe. In turn, Darwin now removed man from his alleged special place in organic nature. Just as the earth is one of a possibly countless number of planets among the billions of galaxies in the known cosmos, so is the human being merely one of the millions of species that have emerged in this world (most have survived only to later vanish within the ongoing flux of biological history). Darwin recognized man's close relationship to the higher primates (especially the three great apes). Like Huxley and Haeckel, he too held that the human animal biologically differs from the apes merely in degree rather than in kind. From recent discoveries in immunology, neoteny, and genetics to comparative studies in embryology, morphology, and ethology, modern science clearly demonstrates beyond dispute that the human animal is in fact closer to the three great apes than even Huxley, Haeckel, or Darwin could have imagined in the last century.

Scientific philosophical anthropology can neither ignore the advances in astronomy/cosmology nor blindly disregard the overwhelming facts supporting the evolutionary framework.

On 30 June 1860, less than a year after the publication of Charles Darwin's *On the Origin of Species*, organized science and organized religion openly collided at a public meeting of the British Association for the Advancement of Science held in Oxford's University Museum Library. The clash between these two institutions occurred in the persons of naturalist Thomas Henry Huxley and Bishop Samuel Wilberforce, respectively. This infamous debate gave public attention to the creation/evolution controversy, which is essentially an ongoing debate between the scientific attitude and the religious attitude over the meaning, purpose, and place of man within nature (recall that this controversy was initiated by the personal dispute between Darwin the naturalist and FitzRoy the fundamentalist during the global voyage of the *Beagle*). The learned and eloquent Huxley successfully defeated the naive and pompous Wilberforce: it was a major victory for science in general, and a significant turn of events for Darwin in particular. The scientific theory of biological evolution had successfully overcome its first major challenge.

Astonishing as it may seem, this very same struggle between science and religion continues in some quarters unabated and essentially unmodified (if not even intensified) more than a century later. Although, as rightly stressed by Tomashevich and others, Darwin never claimed in public or print that humankind descended from the now living higher or later apes (let alone from monkeys) but only that we and they had once shared a very distant common ancestry, his position nevertheless continues to be grossly misrepresented either through ignorance or malice or as a result of both. These attacks on the theory of evolution go unabated despite the overwhelming amount of established evidence to the contrary. In fact, for some, it is contradictory to refer to scientific creationism or religious evolutionism.

Views of nature have both changed and varied throughout the history of science and philosophy. One may argue that there are essentially two contrasting interpretations of physical reality: on the one hand, a static and closed model of a finite cosmos as represented by the Aristotelian framework, and on the other, a dynamic and open view of an eternal and infinite universe as presented by Giordano Bruno. The author advocates the Brunian model of the cosmos in light of empirical inferences, scientific discoveries, and bold rational speculations.

In our century, Alfred North Whitehead wrote about process as well as reality. Einstein eventually acknowledged the empirical evidence supporting an expanding universe (although he had to abandon his own earlier steady-state model of reality). Teilhard de Chardin envisioned creation as a cosmogenesis: briefly, one may view evolution as creation or creation as evolution. Among many other thinkers of our age, these three giants of thought saw man as an integral part of the flux of nature. In such a worldview, no ultimate philosophical system or definitive statement on humankind is ever possible. Since the universe is continuously changing, nothing is ever finished or complete. Creative possibilities in nature may be infinite, with every end implying a new beginning. Likewise, the material universe seems utterly indifferent to the fleeting existence of the human species or even life.

The modern view of man and the heavens is primarily due to scientific inquiry and rational speculation. What is science? In the author's opinion, it is an ongoing search for truth based on an empirical method of formulating and testing hypotheses and theories confirmed wherever possible by experience, experimentation, critical reasoning, and other rigorous means of verification. Science seeks systematically related bodies of factual knowledge and causal relationships that lead to predictions in terms of degrees of statistical probabilities. Science is objective, rational, self-critical, self-correcting, self-revising, and self-improving (it is essentially and irreconcilably antidogmatic, selectively cumulative, and always open-ended). Of course, one must distinguish between science and scientific practitioners (see George V. Tomashevich, "Reflections on Science and Religion" in *Free Inquiry*, 1 (3): 34-36).

The present human predicament is that rational as well as irrational man is conscious of his humble status and material finitude between the infinitely great and the infinitely small in the face of eternity and certain death. In spite of Sagan's intergalactic perspective and O'Neill's space colonies (and despite the awesomely wondrous as well as potentially dangerous promises of genetic engineering and solar energies) man, who likes to refer to his own species as the wise animal, must come to grips with his own existential imperative: adapt, survive, and thrive with unsettling and sobering truths or fail and vanish from this universe in a euphoria of self-deluding and self-flattering error and perspective.

Perhaps there is life, including intelligent beings and civilizations, among the plurality of other worlds across the light-years of stellar evolution. Perhaps some superior intelligences elsewhere have even, in fact, achieved that relative practical immortality through science and technology foreseen by the natural philosopher Condorcet in the French Enlightenment. Even so, it seems reasonable to assume that they too are subject to the same laws and inevitable fate of this material universe.

The assumption that the universe is eternal in time gives the human species almost infinite hope. That the cosmos seems infinite in space puts before humankind an almost eternal series of challenges to be met and overcome. That reality is creative flux gives one a perpetual sense of awe and wonder. Aristotle taught that philosophy begins in wonder. Whitehead once remarked that philosophy ends in wonder. The author suggests that the task of scientific philosophical anthropology is to explore the awesome possibilities of the creative and evolutionary process between these two points of wonderment.

While the idea of evolution did not originate with Charles Darwin, he does remain the central figure in the historical development of this conceptual framework: his persuasive writings established the scientific theory of biological evolution on a firm foundation in terms of empirical evidence and rational deliberation. Unlike Lamarck and Wallace, Darwin's interpretation of organic history is free from both metaphysical aspects and supernatural elements. With profound insight, he gave the theory of evolution an irresistible force with his principle of natural selection (yet, his explanation for organic history was neither complete in perspective nor totally accurate in terms of mechanisms).

Once Darwin had published his two volumes on evolution, it is remarkable how rapidly this view of life was accepted by most naturalists before the end of the last century (thereby demonstrating the inescapable power of science and reason). The major early supporters of the Darwinian theory were Thomas Henry Huxley in England, Ernst Haeckel in Germany, Asa Gray in the United States, and Vladimir O. Kovalevskii in Russia, not to forget the contributions of geopaleontologist and zoologist K. F. Rul'e, embryologist Aleksandr O. Kovalevskii, geographer and social theorist Petr A. Kropotkin (*Mutual Aid: A Factor of Evolution*, 1902 and rev. ed. 1904), bacteriologist Élie Metchnikoff (*The Nature of Man*, 1903), and biochemist A. I. Oparin (*The Origin of Life*, 1923 and 1936).

It may be safely argued that Darwin was, in fact, the major intellectual figure of the last century. In that iconoclastic age of ideology, his work gave a lasting impetus to scientific naturalism and rational humanism: man, life, our planet, and this cosmos will never again be viewed within the framework of eternal fixity. Darwin had grasped the two supporting pillars of the traditional worldview in western intellectual history (Aristotle and Aquinas) in his creative but lethal grip of scientific analysis and rational synthesis and, as a direct result of his thoughts and writings, brought the whole edifice of earth-bound and human-centered ideas crashing to the ground; only the lingering dust remains to blind some to the fact of evolution. However, one must endure the facts of reality; the material universe has a way of continuously asserting itself on the side of science and reason rather than traditional beliefs and personal opinions.

Marx was greatly indebted to Darwin (as well as Feuerbach and Hegel) but claimed that progressive sociocultural development has passed through several major qualitative leaps of advancement. Gone was the continuously slow and gradual development of Darwinian evolution (Darwin, of course, never joined the ranks of the Marxists). Engels wrote about the metaphysical foundation of dialectical materialism; of particular interest is his unfinished manuscript entitled *The Part Played by Labour in the Transition from Ape to Man* (1876). Lenin grounded his worldview in pervasive materialism. It may be argued that dialectical materialism and scientific evolutionism are mutually inclusive explanations for dynamic reality. This is a rich area for critical inquiry.

Influenced by the materialist theory of biological evolution, Sigmund Freud grounded the human psyche and sexuality in the lowly animal origin of our species. Like Marx and Nietzsche, he was an atheist (Darwin, at least an agnostic, had remained silent on the theological implications of his naturalist interpretation of organic history).

Darwin had clearly upset cherished ideas and religious beliefs about the significance of man in nature. The theory of evolution places this planet within cosmic history, life within the biochemical development of our earth, and the human species within the emergence of the primates. The naturalist interpretation of world history, based on the rigorous study of rocks and fossils, dealt a fatal blow to the fundamentalist creationist belief in a rigid acceptance of the Mosaic cosmogony as related in the story of *Genesis* found in the Old Testament of the Holy Bible (even to challenge Aristotle had been to attack the Roman Catholic Church). An evolutionist may be a theist, panentheist, deist, pantheist, agnostic, or atheist. Be that as it may, the biblical account of divine creation is not admissible as empirical evidence, and a courtroom is not the final place to determine the truth or falsity of a scientific doctrine: as always, nature is the ultimate arbitrator.

Primitive man relied upon magical forces to explain the then unknown (e.g. birth, growth and development, disease, and death). In the recent past, Kant claimed that science would never understand the growth and development of even a blade of grass (much less its origin in organic history), while Comte held that science would never be able to determine the chemical composition of our sun (not to mention the makeup of other stars in this universe). Modern man turns to nature as well as science and reason (especially medicine and technology) to understand as well as predict and perhaps even control the same phenomena mentioned above.

Evolution is the sidereal thread in the ongoing material continuum, which encompasses everything from quasars (those cosmic fossils), pulsars, and black holes to all organic human phenomena on our planet. The seemingly permanent crust of the earth is, actually, a changing surface of geological structures and natural processes. The entire evolution of life on earth (from the primeval DNA/RNA codes and the first amino acids, through unicellular and then multicellular organisms, including invertebrates and later vertebrates such as fishes, amphibians, reptiles, birds, and mammals, all the way to human beings) represents only a fleeting instant in the endless eons of cosmic time and change. Evolutionists are now interested in the seemingly explosive adaptive radiation of Cambrian life at the beginning of the Paleozoic era, the extensive extinction of life at the end of the Mesozoic era, and the emergence of hominids during the Plio-Pleistocene junction of the Cenozoic era.

The scientific theory of organic evolution requires a dispassionate and objective recognition of the empirical, experiential, and experimental evidence available at the time. As aptly observed by George V. Tomashevich, although factually documented almost beyond belief by the special sciences (geology, paleontology, taxonomy, biogeography, comparative anatomy and physiology, embryology, biochemistry, genetics, ecology, nonhuman and human psychology and behavior, anthropology, sociology, astrophysics, and astrochemistry as well as other empirical disciplines), the idea of evolution continues to be treated as if it were indeed a matter of mere hypothesis that makes too many people needlessly uncomfortable.*

In modern science, there is the gradualist/saltationist debate with the Neodarwinists defending the slow and continuous evolution of life, while a few paleontologists (notably Eldredge and Gould) are favoring the relatively sudden appearance of new species within small isolated populations.

One is free to disagree with, but not to ignore, the writings of the evolutionists. Today, rigorous spokesmen in favor of the evolutionary theory include astronomer Carl Sagan (*Cosmos*, 1980) and paleontologist Steven Jay Gould (*Ontogeny and Phylogeny*, 1977): they stress the overwhelming empirical evidence

---

*George V. Tomashevich in a public lecture on "The Evolution of the Idea of Evolution" delivered at Canisius College on the occasion of the Darwin Centennial Celebration, organized and chaired by Dr. Birx on April 19, 1982.

that supports the theory of evolution on both the celestial and terrestrial levels of inquiry, respectively. Also, naturalist David Attenborough (*Life on Earth*, 1979) presents a panoramic view of the origin, history, and proliferation of life on this planet.

Evolution is a theory, a fact, and a framework. It has both explanatory and predictive power. There are different evolutionary rates (modes and tempos) and distances as well as various models, interpretations, and perspectives (from the celestial to the terrestrial, from the macrocosm to the microcosm). Evolutionists distinguish between macroevolution (e.g. the dynamic history of the cosmos or the phyletic development of life on earth) and microevolution (e.g. the origin of a new variety of species): one sees the results of macroevolution in the fossil records of rock strata and experiences microevolution in the speciation of bacteria and polyploidy in higher organisms.

A naturalist humanist must consider the scientific ramifications, philosophical implications, and theological consequences of the fact of evolution. One must distinguish between the scientific evidence for evolution and philosophies of evolution. All interpretations of organic history are subject to verification, modification, or falsification in light of new facts and widening experiences.

In philosophy, the evolutionary framework is crucial: it sheds important light on metaphysics (ontology and cosmology) and epistemology as well as places ethics, logic, and aesthetics in a process perspective. One may ask about the 'how' and the 'why' of evolution (as well as its implications for human freedom and responsibility). The impact of Darwinian thought echoes and reverberates through the writings of the emergent evolutionists (Morgan, Alexander, Smuts, and Sellars) as well as the leading American pragmatists (Peirce, James, and Dewey). This impact is also present in the materialist humanist writings of Marvin Farber. Within the broad range of plausible interpretations of evolution, from pervasive materialism (mechanism) at the one extreme to mystical spiritualism (finalism) at the other, a naturalist explanation for the universe encompassing both cosmology and anthropology can incorporate all the documented facts and relationships of the special sciences and human experience.

The modern worldview is neither earth-bound nor human-centered, but it clearly reflects the relative insignificance of this planet and our own species within the awesome vastness of this dynamic universe. One may even imagine the convergence of evolutionary science and process theology. Yet, it is necessary to recall that no single individual has the final word on reality, experience, and human values.

Evolution remains an impelling and compelling theory whose scope ranges from sidereal galaxies to human ethics. The human realm of ethics, morals, and values is distinct within (but not separated from) physical nature. Plato's ideal forms, Aristotle's unmoved mover, and Kant's categorical imperative ultimately removed the essence of man, his rational faculty, from the material world of change and evolution. Darwin recognized that our species's moral potential distinguishes it from the other animals (including the apes and monkeys), while Nietzsche and Freud pointed out that culture is a mask that covers humankind's animality.

Among the important critical works in the area of ethical principles and moral conduct from the evolutionary perspective are Herbert Spencer's *The Principles of Ethics* (1893), Thomas Henry Huxley's *Evolution and Ethics* (1894), Petr A. Kropotkin's *Mutual Aid: A Factor of Evolution* (1902) and *Ethics: Origin and Development* (1924), Pierre Teilhard de Chardin's "Building the Earth" (1931), Ashley Montagu's *Darwin: Competition and Cooperation* (1952), Julian Huxley's *Religion Without Revelation* (1957), C.H. Waddington's *The Ethical Animal* (1960), Antony Flew's *Evolutionary Ethics* (1967), and Philip Appleman's "Darwin: On Changing the Mind" and "Darwin Among the Moralists" in *Darwin* (1979).

Since humankind is becoming more and more capable of guiding and controlling its future evolution on earth and elsewhere, there is an obvious need and growing urgency for enlightened science and wise judgments. The materialist naturalist as evolutionist humanist may anticipate with critical optimism the convergence of science and religion, thought and action, duty and inclination, and the integrity of the individual with the collective desires of society as a whole.

The theory of evolution is the conceptual foundation of all modern biology, and it remains an exciting and rewarding area of scientific and philosophical inquiry. The key concepts for organic evolution are variation, adaptation, survival, reproduction, and extinction. Darwinism (the explanatory *principle* of natural selection) and evolutionism (the scientific *fact* of organic history on this planet) are not synonymous, and Neodarwinism (the *synthetic theory* of biological evolution that incorporates all the findings of modern genetics) seems to be an incomplete account for the total development of all life on earth. There are even a few adherents to Neolamarckism, although the alleged mechanism of the inheritance of acquired characteristics by offspring as the result of use or disuse modifying organs in their parents has yet to be empirically demonstrated beyond any reasonable doubt. Neither Darwinism nor Neodarwinism is synonymous with evolutionism: in terms of mechanisms, the fact of biological evolution very likely extends beyond the explanatory powers of genetic variability and natural selection. As such, one may safely anticipate further breakthroughs in understanding the process of organic evolution and appreciating its results.

What does evolution have to say about man's relationship to life within earth history? Any open-minded critical thinker cannot ignore the irrefutable truth and the indubitable consequences of the evolutionary perspective (particularly as it sheds light on man's relationship to the apes and life's link to the cosmos). Man is cosmic evolution aware of itself (at least once), and our species is completely immersed within both the majesty and travail of this material universe.

In modern evolutionary biology, promising areas of further inquiry include cladistics, heredity (e.g. selfish genes and neutral alleles), cosmic influences, sociobiology, genetic engineering, altruistic and group behavior, and the emerging science of exobiology. Mathematics will play a major role in understanding and appreciating life in this universe, although there is a crucial need to distinguish between reality (ontology) and formal inquiry (epistemology).

At this time, one can only investigate the single process of evolution that has been taking place on our earth. However, at a future date, one can imagine a comparative study of evolutionary processes taking place among numerous worlds (an area of scientific inquiry that will grow out of the emerging science of exobiology).

Did life actually come from somewhere else in the cosmos? Recently, several renowned scientists have seriously revived the panspermia hypothesis for the presence of life on earth: Francis Crick in *Life Itself: Its Origin and Nature* (1981), and Fred Hoyle with Chandra Wickramasinghe in *Evolution From Space: A Theory of Cosmic Creationism* (1981). Somewhat reminiscent of the philosophy of Anaxagorous in antiquity, this hypothesis claims that genes or spores are drifting across the vastness of space (or have been sent throughout the universe by a higher civilization) and those that reached our planet gave rise to biological history. Yet, directed or nondirected panspermia and terrestrial evolution are not necessarily mutually exclusive explanatory frameworks: one way or another, both human life and organic matter owe their origin to the stars.

As appropriately observed by Professor Tomashevich, particularly important places in the intellectual history of the theory of evolution are occupied by the centennial observances of two of its most significant dates. In 1959, the first of these, masterfully organized by Dr. Sol Tax at the University of Chicago, marked the publication of Darwin's epoch-making volume *On the Origin of Species* of 1859 (cf. Sol Tax, ed., *Evolution After Darwin,* three volumes, Chicago: The University of Chicago Press, 1960). In 1982, the second of these, suggested by this author and ably arranged by Dr. Paul Kurtz at the State University of New York at Buffalo, commemorated the death of Darwin on April 19, 1882 (cf. Paul Kurtz, ed., *Free Inquiry,* special issue entitled "Science, The Bible, and Darwin: An International Symposium," Vol. 2, No. 3, Summer 1982).

Likewise, this book was researched and written during two relevant centennial years: 1981 commemorated the birth of Pierre Teilhard de Chardin (the Jesuit priest and geopaleontologist whose mystical vision of human destiny within the framework of a spiritual cosmos seriously introduced the theory of evolution into theology), while this past year observed the death of Charles

Robert Darwin (the naturalist whose analytic and synthetic powers as well as rational insights into the workings of nature founded the scientific theory of biological evolution on a firm foundation). The author wishes to stress that this book deliberately gives special emphasis to Darwin, Teilhard de Chardin, and the ongoing creation and/or evolution controversy.

In the centennial year of his death, we recall with deep respect and profound gratitude the enormous debt modern thought owes to the scientific genius of Charles Robert Darwin. His remarkable capacity for analytical penetration and synthetic integration launched a comprehensive and intelligible theory of the history of life on earth. It is most appropriate that this volume on evolution is dedicated to his immortal memory.

Theories of Evolution is the first volume in a series of four works on the place of man and life in this universe. The other three titles are *Human Evolution, Life in the Cosmos,* and *Eternity and Infinity: Speculations in Cosmology.* All four books will be available from Charles C Thomas, Publisher.

Chapter 1

THE EMERGENCE OF HUMANKIND

Since the work of Linnaeus, the human animal has been classified as a member of the primate order (this same taxon also includes the prosimians, new world monkeys, old world monkeys, and the five apes).[1] The first primates emerged from Mesozoic insectivores. In fact, since that time, the primate order has been undergoing successful evolutionary adaptive radiation throughout the entire seventy million years of the Cenozoic era. As a result of biological evolution, our own species is directly related to, although not descended from, the three great apes (orangutan, chimpanzee, and gorilla) and, to a lesser degree, the two hylobates (gibbon and siamang). Although all the other primates are primarily arboreal creatures, the human being has adapted to a terrestrial existence. This has been accomplished through its acquiring an erect posture, bipedal locomotion, the ability to use and subsequently make as well as own stone implements, and (what is essential) a superior central nervous system, including the cerebral cortex, which allows for symboling with the resultant emergence of verbal language along with all other sociocultural developments.

In his major work, *On the Origin of Species* (1859), the ever-cautious Charles Darwin was reluctant to extend his own theory of biological evolution to account for the emergence and natural development of the human species. In the conclusion of the book's first edition, the brilliant scientist merely wrote, "Light will be thrown on the origin of man and his history."[2] It is puzzling that Darwin refrained from including in this, his principle scientific statement, a section on the evolution of the human animal. Those who actually read this pivotal volume must easily have seen that the very same evolutionary principles used to account for the appearance and organic history of the plants and other animals could also be extended to explain the origin and development of our zoological group as well. Briefly, one may safely argue that the controversy over the Darwinian worldview was essentially due to the theoretical implications and physical consequences it held for a factual understanding and rational appreciation of the place of humankind within the process of nature as a whole.

In his significant work, *Zoological Philosophy* (1809), published in the year of Darwin's birth and exactly fifty years before the appearance of *On the Origin of Species*, the French naturalist Lamarck offered the theory of animal evolution to the scientific community. His book was the first written solely for the purpose of presenting and defending the evolutionary view of life. Within its pages, the author argues that the human species originated and slowly descended from an orangutanlike hominoid form. Although a simplistic and specifically incorrect interpretation of hominoid evolution, it was nevertheless a bold and, in a general way, intuitively insightful hunch anticipating later and better-documented views on this still supremely delicate subject.

In Germany, Arthur Schopenhauer developed a process metaphysics that saw man as the apex of a pyramidally structured terrestrial evolution. He held that our species had a simian origin in the old world tropics: the first human beings were suddenly born from the chimpanzee in Africa and the orangutan in Asia. Nevertheless, Schopenhauer's outlandish speculations on this subject did not have a tangible effect on the emergence of the scientific theory of biological evolution.

Correctly imagining how controversial his theory of evolution would be to most philosophers and theologians as well as scientists, Darwin hesitated about writing an explicit account of the emergence and evolution of the human animal in *On the Origin of Species* (his first book devoted specifically to the general organic history of life on earth). Certainly the human fossil record at that time was so meager that one could almost excuse Darwin's avoidance of the whole subject. Only a few Neanderthal and Cro-Magnon skeletons with some paleolithic tools and weapons had been found up to that time to suggest the possibility of a prehistoric origin for our species from some earlier anthropoid-like form.

As "Darwin's bulldog" in England, the courageous Thomas Henry Huxley not only defended the theory of evolution against the misguided and unfortunate attacks of Bishop Samuel Wilber-force ("Soapy Sam") but also was the first to write a book for the specific purpose of extending the Darwinian framework to include the human zoological group. Huxley's first book, surprisingly also entitled *On the Origin of Species* (1863) in the American edition of 1881, defended Darwin's theory of "descent with modification" primarily due to natural selection. His *Evidence as to Man's Place in Nature* (1863) argued not only that the human animal evolved from a common ancestry shared by the great apes, but also that man's difference from these pongids was merely quantitative rather than qualitative (i.e. man is closer to the higher apes than the apes themselves are to the lower monkeys). In short, it was Huxley, not Darwin, who first placed humankind squarely within organic history.

In 1863, Sir Charles Lyell also wrote about the antiquity of man. Unlike Huxley, however, the father of modern geology was reluctant to embrace the evolutionary framework and certainly never accepted the truth of human evolution. Darwin had been at least generally familiar with the geological speculation of James Hutton as well as greatly influenced by the scientific work of Charles Lyell. Yet, he could not convince Lyell of the truth of his biological theory of organic history and its implications for our own species.

Finally, in 1871, Darwin published his still more provocative and controversial volume, *The Descent of Man*. By this time, however, both Huxley and Haeckel had written about the evolution of the human animal. Darwin himself now wrote that man and the apes share a common prehistoric ancestry, that this hominoid group would be found on the continent of Africa, and that man differs from the pongids merely in (admittedly significant) degree.[3] He saw the essential distinction between man and the great apes in the moral potential of the former, which simply does not appear in the latter (recall that Immanuel Kant had been impressed by the starry heavens above and the moral law within).[4] As a rigorous evolutionist, Darwin held that all the present complex aspects of the human animal had evolved from earlier, simpler beginnings (including not only the biological but also the psychological and sociocultural ones).

Unlike Huxley and Haeckel, however, Darwin was not eager to consider the philosophical and theological implications of his mechanist/materialist interpretation of organic evolution. Despite passages dealing with these subjects in his early metaphysical notebooks, Darwin kept his personal thoughts on such sensitive topics almost exclusively to himself.[5] Huxley coined the term "agnostic" to refer to his own position, which echoes Herbert Spencer's epistemological conclusion about an unknown and a supposed unknowable as elements of his process metaphysics of cosmic evolution. As a monist, Haeckel adopted the position of a materialist pantheism. The shy, sensitive, and always tactful Darwin was at least an agnostic, although certainly never an aggressive or militant atheist.

Regrettably enough, most of the major fossil evidence to support human evolution was discovered after Darwin's death: Java Man (*Pithecanthropus erectus*) was found in the early 1890s, the Taung juvenile skull (*Australopithecus africanus*) in 1924, and Peking Man (*Sinanthropus pekinensis*) in 1926. However, even the scientific community was at first reluctant to accept the implications of the hominid status correctly assigned to these very important paleoanthropological finds.

The year 1959 not only marked the centennial celebration of the appearance of Charles Darwin's *On the Origin of Species*, a major intellectual event organized by the eminent anthropologist Sol Tax and held at the University of Chicago[6] but also represented a crucial turning point in paleoanthropology. In July of that same year, Mary D. Leakey had discovered a hominid skull in Bed 1 of Olduvai Gorge in Tanzania, central East Africa, thus ending a thirty-year search for early man in this area of the world. This find, at that time the oldest known human fossil remain, was analyzed, described, and interpreted by the late Louis S. B. Leakey and classified as *Zinjanthropus boisei* but mistakenly held to be the maker of the Oldowan pebble tools/weapons found in the same rock strata. The use of the potassium/argon radiometric dating technique determined the prehistoric discovery to be about 1.75 million years old. The Leakeys had, in fact, made an enormously significant breakthrough. The worldwide publicity and scientific attention given to the *Zinjanthropus* specimen resulted in increasing those funds being made available for the continued research by these and other professional anthropologists in quest for more empirical evidence capable of factually documenting the beginning and early evolution of these then apparent ancestors of our own species.

Only two short years later, in 1961, Dr. Leakey discovered an even more hominid fossil skull from Olduvai Gorge that he classified as *Homo habilis*. This specimen is remarkably hominid (far more humanlike than all the previously discovered australopithecine material), had a cranial capacity of about 750 cc, and was the actual maker of the paleolithic artifacts associated with the Oldowan culture. The significance of the *Zinjanthropus* find, now adjudged to be only a specimen of the *Paranthropus robustus* form, faded into the background but remains a reminder to prehistorians not to make hasty and premature decisions about the immediate taxonomic and biosocial importance and ultimate scientific implications of a particular fossil hominid discovery before sufficient evidence is in.

Following in his parents' footsteps and rewarded with early success, Richard E. Leakey took the ongoing quest for the origins of humankind in central East Africa to the Koobi Fora site at Lake Turkana.[7] There, in 1972, he discovered fossil evidence of *Homo habilis* (including the now famous Skull No. KNM-ER-1470, recently dated to be 1.9 million years old, although first thought by Leakey to be considerably earlier). Despite some honest mistakes in specific interpretations of this find that have been reputably exploited and capitalized upon by certain academic competitors in a rather puerile manner, Leakey's overall contribution to human paleontology remains impressive and highly respectable.[8]

In light of all the cumulative fossil evidence, the early Pleistocene hominids of the Villafranchian time represented at least three different genera: the robust *Paranthropus* form, the gracile *Australopithecus* type, and *Homo habilis,* the most hominid of the group. Yet there still remained the need to unearth the skeletal fragments and concomitant other traces of the ancestors of even these hominids, the earliest known at that time.

In 1974, paleoanthropologist Donald C. Johanson discovered "Lucy" (*Australopithecus afarensis*) at the Hadar site in the Afar Triangle of Ethiopia.[9] This partial late Plio-Pleistocene fossil skeleton of a single hominidlike female is about 2.9 million years old and, according to its discoverer, representative of the small-brained but erect-walking ancestors of humankind. In fact, the Lucy skeleton is hominidlike in every general anatomical feature except for its primitive skull and dental characteristics as well as small cranial capacity (these aspects are more hominoidlike than hominidlike). Dr. Johanson's team has also found a "First Family" as old as Lucy, as well as related Plio-Pleistocene hominid materials (including jaws and a knee joint) close to four million years old. This evidence from Ethiopia is helping to shed light on the provenience of the whole human zoological group.

As if Johanson's awesome luck was in itself not enough to increase excitement in the science of physical anthropology, in 1978 Mary D. Leakey had the astounding good fortune to unearth three tracks of hominid footprints about 3.7 million years old from Site G at Laetoli in Tanzania, near Lake Eyasi south of Olduvai Gorge.[10] The Pliocene Laetolil Beds have also yielded numerous hominid fossils, chiefly lower jaws and teeth, as well as a wide variety of nonhuman animal remains (as yet no stone tools/weapons have been found at Laetoli). Johanson suggests that both these footprints and his Pliocene findings from the Hadar site belong to the same distinctive hominidlike form *Australopithecus afarensis* (although this model and interpretation of early hominid phylogeny is not shared by the Leakey family). Yet one significant generalization can safely be made: these incipient hominids stood erect and walked fully upright with a free-striding human bipedal gait long before using, and later making, paleolithic tools/weapons and subsequently acquiring a very large brain concurrent with speech.[11]

In the fall of 1981, over a period of ten days, scientists unearthed eight bone fragments from two different individuals in the desolate and remote Afar Triangle desert of north-central Ethiopia. These fragments are very probably the fossil remains of man's thus far most ancient known ancestors. They were found in the Awash River Valley, only forty-five miles south of the site that had yielded in 1974 the now famous Lucy skeleton. This crucial hominidlike material was spotted by two members of an international paleoanthropological research team headed by Dr. J. Desmond Clark of the University of California at Berkeley: the anatomist Dr. Timothy D. White discovered the upper part of a left femur about six inches long that belonged to an upright-walking male about sixteen years old, while Leonard Krishtalka came upon seven skull fragments that are the oldest humanlike fossils ever found.

Geobiological evidence and the radiometric dating of a lava flow of cindery volcanic debris at the site show these petrified bones to be about four million years old. As a result, this site may prove to be the richest and most important fossil area in the world. All the facts suggest that the two Lucy-type specimens belong to the *Australopithecus afarensis* stage of early hominid evolution and, if this is true, argue for biological stability over one million years (this would give added support to extending the Eldredge/ Gould hypothesis of punctuated equilibria to account for the origin and early phase of humankind).

Although admittedly meager, this impressive new evidence clearly substantiates the claim that our oldest ancestors were, indeed, walking erect on two legs with a small cranial capacity of about 400 cc for several million years before the modern cranial capacity of 1500 cc was first obtained by *Homo sapiens neanderthalensis* about 200,000 years ago. Of course, more fossil evidence will have to be found before those remaining questions concerning the emergence of humankind are finally answered and conflicting interpretations are settled once and for all.

In the final chapters of *The Making of Mankind* (1981), Richard E. Leakey considers the implications of human sociocultural evolution from the Ice Age cave art to the achievements of modern science and technology.[12] He stresses the survival value of cooperation (recall the mutual aid theory of Kropotkin) and claims human aggression to be primarily a learned phenomenon (contrast with the innate aggression theory of Lorenz).[13] The Kalahari tribesmen are viewed as "living fossils" representing our prehistoric ancestors in their thoughts and behavior patterns.

During the fall of 1982, Richard E. Leakey and his scientific colleagues discovered humanoid fossils in the Samburu Hills north of Nairobi, Kenya, which should shed considerable light on the origin of man: those fossils, about fourteen million years old, were deposited at the time when the dryopithecines and ramapithecines (now referred to as the kenyapithecines) probably began diverging from each other, while those fossils about eight million years old may represent the emergence of the earliest true hominids as scavengers with incipient bipedalism long before meat eating and encephalization.

One may safely assume that the paleoanthropologists of the future will discover additional hominid fossil evidence that could very likely clarify those issues still surrounding the emergence and early history of our species. Such finds would greatly help to fill in the admitted gaps now inherent in the current scientific overview of human development within the naturalist framework of organic evolution. Likewise, it is absurd to maintain that the present (or even lasting) incompleteness of the geopaleontological record in regards to any empirical knowledge of man's place within nature warrants a theological leap of blind faith. The symboling animal as tool user, tool maker, and tool owner is merely one of those millions of species that have inhabited this planet earth, whose biological history is one of speciation and extinction. What is clearly needed is more science.

In light of our growing fossil evidence, Darwin was prophetically right to select Africa as the cradle of humankind. (Haeckel had thought Asia to be the birthplace of our species.) Comparative studies of the Miocene hominoid crania and teeth of the African dryopithecinae complex point to a major separation that occurred more than 12 million years ago within this large and diversified group of higher primates: one line represents the pongidlike dryopithecines, which are ancestral to the present living apes, while another line consists of the hominidlike kenyapithecines, directly antecedent to the later true hominids and eventually conducive to *Homo sapiens sapiens* of today.[14] Some anthropologists claim a much later date for the pongid-hominid divergence based on a comparative study of the postcranial anatomy and biochemistry of modern human populations and the two now living African apes (especially the chimpanzee).[15] Of course, in this complex history of the hominoids, there were probably numerous unsuccessful major lines of evolutionary descent as represented by the extinction of the pongids *Oreopithecus* and *Gigantopithecus*. *Homo erectus* appeared about 1.5 million years ago and may have been responsible for the disappearance of certain ground-dwelling pongid and possibly even hominid forms.[16]

Yet, as in the case of other great thinkers, Darwin's speculations on human evolution were not always either theoretically or factually correct in every detail. In light of subsequent discoveries, he turned out to be in error in holding that the modern cranial capacity of about 1500 cc was reached in our prehistoric ancestors before they acquired an erect posture and made stone implements for defense. As one can see today, the fossil evidence clearly indicates that just the opposite was the case: the human cranial capacity did not start to expand greatly until about half a million years ago. Although the neurophysiological complexity of the early hominid brain must be taken into consideration as well as its sheer size, a sufficient explanation for the relatively rapid expansion of the human cranium roughly between 200,000 and 50,000 years ago is still forthcoming (a considerably smaller human cranial capacity had obviously been sufficient for mere adaptation to and survival in the early natural environment).

Despite the fact that there is still no universal consensus among anthropologists as to the specific classification of all the hominid fossil material (not to mention the lingering so-called Neanderthal problem), a general outline of human evolution does, nevertheless, emerge. With the extinction of the lesser hominid forms like *Australopithecus* and *Paranthropus*, it is the australopithecine *Homo habilis* that seems ancestral to the pithecanthropine or *Homo erectus* stage (e.g. Java Man and Peking Man), to be followed by the Neanderthal phase of *Homo sapiens neanderthalensis*.[17] The subsequent emergence of Cro-Magnon Man preceded the appearance of modern man (both referred to, as

Homo sapiens sapiens), with a hunting/gathering life-style, which was soon to be replaced by the coming of agriculture (with the cultivation of plants and the domestication of animals), metallurgy, and finally civilization.[18]

In his theory of biological evolution, Darwin was unable to give a scientific explanation for the sudden chance appearance of those beneficial, because they were useful, physical variations that represent the raw material for the process of natural selection. To all appearances, he never read Mendel's monograph, *Experiments in Plant Hybridization* (1866), and therefore unfortunately missed the wondrously helpful implications for his own incomplete theory of this already available pioneering introduction to the science of genetics free from Lamarck's use/disuse and the inheritance of acquired characteristics. Instead, for whatever reason, Darwin did not appreciate the mathematical significance of his experiments with plants and animals in order to understand the transmission of physical traits from generation to generation. His own particulate theory of pangenesis was a reluctant attempt to see inheritance in Neolamarckian terms (Darwin had unsuccessfully attempted to account for all of organic evolution within a geological context that greatly underestimated the true age of our earth).

In modern science, one speaks of the synthetic theory of biological evolution grounded in the Darwin/Wallace explanatory mechanism of natural selection as amended by the Mendelian principles of heredity and supplemented by our increasing comprehension of the mutation theory, DNA/RNA molecules, and population genetics. The ongoing biochemical research into the origin of life on earth is an area of scientific speculation that Darwin chose not to delve into seriously. His paternal grandfather, Erasmus Darwin, had held that life appeared from a single original organic filament in the primeval oceans of our planet, thus strangely and inadvertently anticipating the current "primeval soup" theory associated with Oparin and others in this century.

If Darwin were with us today, it is tempting to speculate what the scientific father of biological evolution would think of the several new areas of empirical research: comparative primate ethology, altruistic and cooperative group behavior (first studied by Petr Kropotkin in his *Mutual Aid: A Factor of Evolution* of 1902), sociobiology and genetic engineering,[19] apparently and at least temporally, neutral genes, and at last the emerging science of exobiology.[20] His open-minded and open-ended naturalist stand-point is broad and tolerant enough to incorporate all the positive findings of the physical and social sciences within an evolutionary context both terrestrial and celestial (including the most recent hominid fossil evidence).

Recently, the Eldredge/Gould hypothesis of punctuated equilibria has introduced into the ever more synthetic theory of biological evolution the possibility that new plant and animal species appear more or less 'suddenly' and dramatically throughout organic history, as is alleged to be documented in the admittedly incomplete fossil record.[21] However, Darwin argued in favor of a slow, incremental, and continuous evolution of life on earth. Thus, he maintained that a new species emerges as the result of the gradual accumulation of slight, favorable chance variations over long periods of time. His interpretation of biological evolution did not allow for the sudden appearance of major alterations or saltations in the assumed historical continuity of living things.

Of course, it must be pointed out that Darwinian phyletic gradualism and the Eldredge/Gould hypothesis of punctuated equilibria are not necessarily mutually exclusive conceptual frameworks to account for the historical emergence of new flora and fauna throughout organic evolution. Similarly, the appeal to sudden leaps in biological history does not necessarily argue for a creationist or supernatural interpretation of the process of organic evolution. Clearly, there is a profound difference between the postulated material leaps of punctuated equilibria and the traditional metaphysical leaps of religious faith (the geopaleontologist and Jesuit priest Pierre Teilhard de Chardin argued for transitional evolutionary "critical thresholds" that account for the sudden major qualitative leaps among levels of upward moving planetary emergence in order to justify his assumption of the uniqueness of man).[22]

For the paleontologists Eldredge and Gould, the presumed

equilibrium or relative stasis of a small population is followed by a more or less dramatically punctuated jump resulting in an event of rapid speciation (it is curious to note that while Teilhard's position is motivated by Catholic theology, Gould's position seems to be motivated by Marxist ideology). In a manner reminiscent of the once heated dispute between the catastrophists and the uniformitarians in historical geology (not to forget the still earlier conflict between the Neptunists and the Vulcanists), the truth in the above-mentioned controversy most probably also lies somewhere in a dynamic and synthetic middle position, bridging, reconciling, and transcending gradualism as the thesis and saltationism as the antithesis. As such, the author proposes that this more comprehensive view be called punctuated gradualism, partaking of the partial truth of each of the opposed positions.

The current revival of the deeply regrettable creation/evolution controversy and the persistent racial, ethnic, and religious prejudices throughout the world demand a rigorous adherence to morally responsible science and ethically guided reason for a clearer understanding and proper appreciation of the place of our species within the process of nature. Geologically speaking, humankind is a relatively recent new species that is inextricably a part of the evolving continuity and essential unity of organic history on earth. Human uniqueness resides essentially in a superior central nervous system, particularly the cerebral cortex with its marvellous symbolic functions and resultant sociocultural consequences. Recent rigorous ethological studies of the great apes (orangutan, chimpanzee, and gorilla) in their wild habitats clearly demonstrate that these three pongids are psychosocially closer to the human animal than was thought by Huxley, Haeckel, and even Darwin in the last century (not to mention the astonishing genetic, biochemical, and behavioral similarities).[23]

More and more, the evolving human animal is able to direct and control the process of biological evolution (including its own unfolding course, character, and destiny). This is becoming more clearly possible as a result of the increasing transition from biological chance and necessity to sociocultural freedom and responsibility. With a strange twist of ambiguous irony, the ongoing development of science and technology is providing a deeper understanding of our lowly origins in the remote past, threatening the present existence of life itself, including our own species, as a result of the apparently geometric advance in knowledge and only arithmetic growth in the wisdom required to use it well,[24] and at the same time promising the wondrous possibility of solving our most pedestrian problems on earth in the future and eventually propelling us toward the distant stars if not beyond.

NOTES AND SELECTED REFERENCES

[1]The author first presented many of the ideas expressed in this chapter in his paper "Charles Darwin and Fossil Man" delivered at the State University of New York at Buffalo (April 16, 1982) as part of an international symposium on Science, the Bible, and Darwin: they then appeared in an article published in a special issue of *Free Inquiry*, ed. by Dr. Paul Kurtz, 2(3):34-37.

[2]Cf. Charles Darwin, *The Origin of Species* (New York: Pelican Classics, 1981, p. 458). The sixth and last edition of Darwin's *Origin* appeared in 1872.

[3]Cf. Charles Darwin, *The Origin of Species* and *The Descent of Man* (New York: Modern Library, 1936, esp. chapter six of *Descent*, pp. 512-528). Also refer to Russell Tuttle, "Darwin's Apes, Dental Apes, and the Descent of Man: Normal Science in Evolutionary Anthropology" in *Current Anthropology*, 15(4):389-426.

[4]*Ibid.*, p. 495. Also refer to Michael T. Ghiselin, "Darwin and Evolutionary Psychology" in *Science*, 179(4077):964-968, and Stephen Jay Gould, "Nonmoral Nature" in *Natural History*, 91(2):19-20, 22, 26.

[5]Cf. Paul H. Barrett, ed., *Metaphysics, Materialism, and the Evolution of Mind: Early Writings of Charles Darwin* (Chicago: The University of Chicago Press, Phoenix Edition 1980). Also refer to Neal C. Gillespie, *Charles Darwin and the Problem of Creation* (Chicago: The University of Chicago Press, Phoenix Edition 1979), and Eugenia Shanklin, "Darwin vs. Religion" in *Science Digest*, 90(4):64-69, 116.

[6]Cf. Sol Tax, ed., *Evolution After Darwin* (Chicago: The University of Chicago Press, 3 vols., 1960).

[7]Cf. Richard E. Leakey and Roger Lewin, *Origins* (New York: E. P. Dutton, 1977), and Richard E. Leakey and Roger Lewin, *People of the Lake: Mankind and Its Beginnings* (Garden City, New York: Anchor Press/Doubleday, 1978). Also refer to Richard E. Leakey, ed., *The Illustrated Origin of Species* by Charles Darwin (New York: Hill and Wang, 1979, esp. pp. 14-15, 52, 90, 114, 217, 222).

[8]Cf. Constance Holden, "The Politics of Paleoanthropology" in *Science*, 213(4509):737-740, and Edward P. H. Kern with Donna Haupt, "Battle of the

Bones: A Fresh Dispute Over the Origins of Man" in *Life*, 4(12):109-116, 118, 120.

[9]Cf. Donald C. Johanson and Timothy D. White, "A Systematic Assessment of Early African Hominids" in *Science*, 202(4378): 321-330, and Donald C. Johanson and Maitland A. Edey, *Lucy: The Beginnings of Humankind* (New York: Warner Books, 1982).

[10]Cf. Richard L. Hay and Mary D. Leakey, "Footprints of Laetoli" in *Scientific American*, 246(2):50-57, 170.

[11]Cf. C. Owen Lovejoy, "The Origin of Man" in *Science*, 211(4480):341-350.

[12]Cf. Richard E. Leakey, *The Making of Mankind* (New York: E. P. Dutton, 1981, esp. pp. 160-247).

[13]Cf. Petr Kropotkin, *Mutual Aid: A Factor of Evolution* (Boston: Extending Horizons Books, 1914), and Konrad Lorenz, *On Aggression* (New York: Bantum Books, 1967, esp. pp. 228-265).

[14]Cf. Sherwood L. Washburn, "The Evolution of Man" in *Scientific American*, 239(3):194198, 201-202, 204, 206, 208, 242; Eric Delson, "Paleoanthropology: Pliocene and Pleistocene Human Evolution" in *Paleobiology*, 7(3):298-305; and Boyce Rensberger, "Ancestors: A Family Album" in *Science Digest*, 89(3):34-43.

[15]Cf. Jorge J. Yunis and Om Prakash, "The Origin of Man: A Chromosomal Pictorial Legacy" in *Science*, 215(4539):1525-1530.

[16]Cf. G. Philip Rightmire, "Patterns in the Evolution of *Homo erectus*" in *Paleobiology*, 7(2):241-246.

[17]Cf. C. Loring Brace, "The Fate of the 'Classic' Neanderthals: A Consideration of Hominid Catastrophism" in *Current Anthropology*, 5(1):3-43, C. Loring Brace, "Neanderthal: Ridiculed, Rejected, But Still Our Ancestor" in *Readings in Physical Anthropology and Archaeology* edited by David E. Hunter and Phillip Whitten (New York: Harper & Row, 1978, pp. 130-136) and Ralph S. Solecki, "Neanderthal is Not an Epithet But a Worthy Ancestor" in *Ibid.*, pp. 114-120, and Erik Trinkaus and William W. Howells, "The Neanderthals" in *Scientific American*, 241(6):118, 122, 125-133.

[18]Cf. Robert J. Wenke, *Patterns in Prehistory: Mankind's First Three Million Years* (New York: Oxford University Press, 1980).

[19]Cf. Edward O. Wilson, *Sociobiology: The Abridged Edition* (Cambridge, Massachusetts: The Belknap Press of Harvard University Press, 1980), and Charles J. Lumsden and Edward O. Wilson, *Genes, Mind, and Culture: The Coevolutionary Process* (Cambridge, Massachusetts: Harvard University Press, 1981).

[20]Cf. H. James Birx and Gary R. Clark, "The Cosmic Quest" in *Cosmic Search*, 4(1):28, 38.

[21]Cf. Stephen Jay Gould and Niles Eldredge, "Punctuated Equilibria: The Tempo and Mode of Evolution Reconsidered" in *Paleobiology*, 3(2):115-151.

[22]Cf. Pierre Teilhard de Chardin, *The Phenomenon of Man* (New York: Harper Torchbooks, 2nd ed., 1965, p. 272).

[23]Cf. H. James Birx, "The Great Apes: A Close Encounter with Extinction" in *Collections*, 60(2):16-25.

[24]Important works on this subject range from the early book of Fairfield Osborn, *Our Plundered Planet* (Boston: Little/Brown, 1948) to the recent volume by Paul Ehrlich and Anne Ehrlich, *Extinction: The Causes and Consequences of the Disappearance of Species* (New York: Random House, 1981).

Chapter 2

EARLY THOUGHTS ON EVOLUTION

I n ancient India, among the orthodox texts of Hinduism, the *Upanishads* (c. 800 BC) are the most indigenous of the Vedic writings.[1] Unfortunately, they are open to several conflicting interpretations. In general, however, these religious and philosophical speculations attempted to deal with the alleged sacred power implicit in sacrificial rites and ritual performances. One tradition maintains that this immutable and neutral force is, in fact, the single principle that pervades the entire cosmos and underlies its multiplicity of phenomena: as such, the human individual as a microcosm mirrors the whole world process as the macrocosm (there is also the later added belief in moral progress through reincarnation or metempsychosis). According to the *Upanishads*, the eternal element within finite man is claimed to be identical with the divine power sustaining the infinite universe itself. Acknowledging the limitations of the human perspective, this statement of an illusive process metaphysics clearly regards man, life, and "material" objects as the products of endless finite but serial transformations of the same unknowable stuff in space grounded in a transcendent temporal reality. It is no surprise, but of serious result, that this worldview did not engender the scientific attitude in India. Although the *Upanishads* did recognize both time and change, the seemingly "material" cosmos is held to be an illusion, and therefore human inquiry into the nature of things is forever inconsequential and of no value.

In ancient China, an individual or a group of wise elders referred to as Lao Tzu (c. sixth century BC) is said to have founded

41

the Taoist school of thought. This philosophy, known as Taoism and recorded in the classic work of two books entitled *Lao tzu* or *Tao te ching*, takes both the material world and its human situation seriously.[2] Tao is the central concept identical to the One or the totality of Nature itself: it is the simple, eternal, absolute, mystical, indescribable, spontaneous, and forever unknowable beginning and way of all things. Tao is the independent and single creative source and sustaining support of the myriad things in the inexhaustible real universe as it goes through endless cycles of slow growth and rapid decay. The supreme goal of all life, especially human existence, is survival (unlike Darwinism, it is the submissive and weak in Taoism that survives and thrives in the changing world). In the first century BC, Chuang Tzu interpreted the Tao as the dynamic and equalizing principle in the universal flux of all particular things. As such, the neutral Tao has now taken on a moral significance favoring the good. As such, in an obviously crude form, Chuang Tzu seemed to have anticipated natural evolution in conceiving of the transformation of species in a cyclical reality that is ever-changing and developing from the simple to the complex: "All things come from the originative process of Nature and return to the originative process of Nature."[3]

Later, in the cryptic *I Ching*, or *Book of Changes*, which was influenced in part by Taoism, one finds a more mechanical statement of the process law of cyclic change through the interaction of the passive yin and active yang (the two natural forces engendered by the Great Ultimate).[4] Be that as it may, the Neotaoists devoted themselves primarily to the chemical sciences rather than to the biological sciences (except for descriptive, but not comparative, anatomy). It is interesting to note that, in the introduction of western thought to China in this century (which included the writings of Spencer, Huxley, Kropotkin, and Bergson along with those of other process thinkers), glaringly absent seem to be the major works of Charles Darwin![5]

The evolutionist perspective did not have its origin with the writings of Charles Darwin. In western antiquity, several Presocratic natural cosmologists (600-400 BC) had presented concepts that directly or indirectly anticipated the scientific theory of biological evolution.[6] They rejected earlier myths, legends, personal opinions, and religious beliefs. Instead, their specula- tions on man in the universe were grounded in the observation of nature as well as the critical reflection on human experience and the rigorous use of reason. Man was usually seen to be a product of, and totally within, a dynamic physical reality.

Thales claimed that life first appeared in water, the basic stuff of the cosmos, and then unfolded to eventually produce land- dwelling creatures. His student Anaximander went further in maintaining that in the movement of living things from primordial slime to land the line leading to man had once passed through a fishlike stage of development. It was Xenophanes who first seriously recognized both the biological and historical signifi- cance of fossils as the remains of once living but different flora and fauna (even today, the paleontological record is the single most convincing body of evidence to support the theory of evolution). Heraclitus dealt with both the being and becoming of nature, teaching that cyclical change is the fundamental characteristic of reality. Yet, it was Empedocles who came closest to anticipating the modern explanation of organic evolution. In an attempt to explain the origin of plants and then animals, he envisioned the primordial surface of this world covered with free-floating organs that haphazardly came together by sheer chance forming organ- isms: most of the resulting animals (including plants) were monstrosities that perished in the struggle for existence. Never- theless, some organisms did have the necessary structures and functions that enabled them to successfully adapt to their environments and thereby survive and reproduce. In his bizarre attempt to account for the origin and history of life on earth, Empedocles had actually glimpsed the Darwin/Wallace explana- tory principle of natural selection (which was extended by Herbert Spencer in his own biosocial concept of the survival of the fittest).

Suffice it to say that over 2000 years ago there were those bold thinkers who, in their philosophical reflections on the structures and functions of nature, recognized both the historical continuity and essential unit of all living things on our planet. However, this creative advance in human thought, with its emphasis on free inquiry and a naturalist overview, was to end with the emergence of the encyclopedic Aristotelian synthesis.

In retrospect, it is ironic that Aristotle (384-322 BC), the father of biology and several other areas, should have come so close to discovering the principle of evolution yet preferred to believe in the eternal fixity of all plant and animal forms.[7] Despite his concern with reproductive methods as well as the growth and development of organisms, Aristotle's general concept of the terrestrial continuum or "Great Chain of Being"[8] held to the immutability of biological types. He argued that each organic form has its eternally fixed natural place in the hierarchical order of the living world. Although the Stagirite founded the sciences of embryology and taxonomy, providing much valid anatomical and physiological knowledge, he never contributed to the formulation even of an anticipatory foreshadowing of the theory of evolution. For him, man as the sociopolitical and rational animal occupies the apex of this allegedly static ladder of nature (his undeniable claim to immortal fame and gratitude by posterity is undoubtedly his towering contribution to logic in general and syllogistic reasoning in particular). Unlike Anaximander and especially Xenophanes, Aristotle ignored the significance of the paleontological record by claiming fossils to be merely chance aberrations in rock strata. He even claimed that no form of life had ever become extinct and supported spontaneous generation to account for the origin of at least some low forms of life (e.g. worms, frogs, mice, and flies). Although Darwin did refer to Aristotle as "the philosopher" and saw himself as merely a schoolboy when compared to the Greek's comprehensive and unquestionably profound genius, it is obvious that (in terms of theory) the English naturalist is far closer to the Presocratic outlook than to the subsequent Aristotelian worldview. For the Stagirite, nature is merely the monotonous and eternal recurrence of the same (there is no history as such). So, in the final analysis, Darwinian science and Aristotelian philosophy do not mix.

In the epic work *On the Nature of the Universe*, inspired by Epicurus, the Roman poet and mechanist/materialist/atomist philosopher Titus Lucretius Carus (c. 99-55 BC) outlined the natural origin and history of life on earth as well as the biological and sociocultural development of humankind from a prehistoric caveman stage through hunting/gathering to civilization with language and metallurgy.[9] Lucretius' general insights are remarkably sound, considering he recorded them over two thousand years ago! In the *Enneads*, the dynamic idealist and pantheist mystic Plotinus (205-270 AD) as the founder of Neoplatonism presented a cyclical process metaphysics: the three immanent spheres of matter or void, life or soul, and mind or intelligence necessarily fulgurate from the divine One as the transcendent Good itself.[10] Unfortunately, the significant views of both Lucretius and Plotinus were overshadowed by the authority given to Aristotelianism: the later Aristotelians were so blind in their intellectual loyalty to their Stagirite master that they failed to appreciate Lucretius's interpretation of the biocultural history of mankind and Plotinus's process view of reality. As echoed even in our own time, this failure clearly illustrates the danger of an uncritical appeal to even the highest authority (i.e. to any kind of cult of personality).

As the preeminent figure in the golden age of rational studies in medieval Islam, Avicenna of Afghanistan (980-1037) was most probably the leading scientist and humanist of his time.[11] Integrating his own work as a physician, logician, and poet, he advocated a holistic view of humankind within nature and a worldview grounded in a pantheist synthesis of science, philosophy, and religion. Besides a devotion to medicine, his encyclopedic interests included physics, astronomy, geology, paleontology, zoology, anatomy, sociology (he taught that love is a driving force in society), and geometry. Avicenna even introduced the idea that the origin of the human species was in an earlier animal form. His mystical interpretation of the universe as a whole argued for the emanation of multiplicity from unity and the return of multiplicity to unity: an overview reminiscent of cosmological speculations from Plotinus and Cusa to Spencer and Teilhard de Chardin.

During the so-called western Dark Ages and Medieval Period, thinkers were more concerned with the problems of Christian theology than with free inquiry into the natural history of life, the human animal, and the development of the cosmos itself. In general, man was viewed as the final and special product of a limited series of divine acts of creation: priority was given to religious beliefs and mysticism rather than to an empirical investigation of the material world and the use of reason. There was very little scientific inquiry and naturalist philosophy during this metaphysically preoccupied millennium. Throughout this time, the human being, with his rational faculty and presumably immortal soul, was believed to represent a link between all the lower animals on the one hand and the whole hierarchy of higher spiritual entities on the other.

It is not generally known that Leonardo da Vinci (1452-1519) of the Italian Renaissance, the greatest genius ever to walk the surface of our earth, was interested in geology and paleontology besides his primary devotion to both art and anatomy (the naturalist humanist rejected superstition and authority, never taking theology seriously).[12] His insatiable curiosity boldly questioned and bravely explored the very workings of the material world: What is the age of our planet? Why are birds able to fly? How does the human eye work? His intellectual strength lay in an exceptionally acute perception of concrete material reality. His universal mind wanted to understand the human body, animals and plants, and the forces of the physical world; it could analyze bones and muscles, synthesize art and science, and even envision some of the technological marvels to come (e.g. the airplane, armored tank, and submarine).

A true natural philosopher, Leonardo was always open to new facts and experiences. As such, he made original investigations into earth history. Wandering through the Swiss Alps and the mountains of Tuscany, he became fascinated with the beauty and detail of the geological and paleontological evidences about him and their far-reaching implications. To him, the rock strata containing marine fossils suggested a story considerably different from the interpretation given by most thinkers in antiquity. Since marine fossils of many sizes and various species were found in disparate layers above the sea level, he knew that the surface of our planet had not been eternally fixed throughout all time.

Leonardo sought a scientific explanation for the phenomena, rejecting all religious interpretations of creation and destruction. He saw time as the evil destroyer of all things and became obsessed with catastrophic events involving wind, water, and fire (like the later Neptunists in geology, he too thought water to be the basic modifier of the crust of the earth). Consequently, he envisioned the end of our world as a result of the disappearance of all water into the interior of our planet, followed by fire destroying all terrestrial life (including the human species). Clearly, this view of things is a far cry from the Aristotelian model of an eternally fixed and therefore safe and secure universe as the enduring static home of human existence.

In these layers of mountain stone, Leonardo studied the fossilized marine evidence: seashells, sea snails, oysters, corals, scallops, cockles, crabs, cuttlefish, traces of worms, and the bones and teeth of fishes. To account for these objects at the tops of high mountains far removed from the sea, he took time and change seriously. To explain rationally the historical origin of these things, his penetrating intellect leaped ahead of contemporary thought to embrace conceptions of both geological catastrophism and uniformitarianism, as well as perhaps biological history (ideas that are still hard for some thinkers to accept today). Holding to the plasticity of the earth, he claimed that there had been periodic upheavals of the mud layers from the bottom of the salt waters: as such, mud layers from the ancient floors of the seas had emerged to form the mountains, and erosion by rivers later uncovered the strata of marine fossils.

Leonardo recognized both the biological and historical significance of the fossil shells and marine animals he found imbedded in these mountain rocks at high elevations (recall the insights of Xenophanes, in sharp contrast to the views of Aristotle). He rightly held them to be the remains of creatures that had once lived on the beaches or in the seas, and had been later lifted up. He remained fascinated by the destructive force of water on geological structures and its far-reaching consequences over vast periods of time, claimed the earth to be at least 200,000 years old, and may have even intuited the natural theory of biological evolution (if only in a most general way). These were incredible discernments on his part, since most thinkers at that time held fast to the biblical account of a relatively recent divine Creation and simply did not take change and history seriously. Yet, Leonardo had speculated on the processes of sedimentation, fossilization (mineralization), and natural erosion. In doing so, he pushed back the horizon of time. He even acknowledged the similarities among the monkeys, apes, and man over two centuries before the works of Linnaeus and Lamarck.[13]

In his dynamic cosmology, Leonardo rejected an earth-centered model of the universe and even anticipated both the universality of gravitation and the scientific use of the telescope. Reminiscent of Plato, he had once written, "Let no one who is not a mathematician read my works." In later years, however, he changed his view of the cosmos from a mathematico-mechanist interpretation to that of an organismic model of the universe. In his mature vision of things, he vitalized all of reality: the planet earth itself is a living organism. He wrote that animals are the image of the world: man is the microcosm while the universe is the macrocosm, the two being identical in their nature. Like Heraclitus, Leonardo eventually wrote that reality is in a state of continuous flux (the ultimate basis of all phenomena is not mathematics but life, movement, and change). In fact, his overview supported a plurality of worlds within the eternality and infinity of the universe. He shed theology for a naturalist philosophy grounded in pervasive necessity, simplicity, monistic change, and areligious pantheism.

In brief, the ideas of Leonardo da Vinci clearly represent a major break from the Classic and Medieval interpretations of nature. Even today, his vision and accomplishments remain an inspiration to the soaring intellect and creative imagination of each free seeker for truth and wisdom.

Giordano Bruno (1548-1600), the greatest philosopher of the Italian Renaissance, may be said to have at least glimpsed in an *a priori* fashion the natural theory of organic evolution.[14] In maintaining the essential unity of nature and implying the development of lower or simple life forms into higher or complex organisms, he apparently recognized the historical transformation of all life on earth. The incomparable process cosmologist of his age, Bruno courageously taught the eternity and infinity of our creatively unfolding universe as well as boldly claimed the existence of a plurality of inhabited worlds scattered throughout an unbounded space of spirited matter in endless time. Among others, not until Darwin and Sagan would Bruno's unorthodox perspective be given the needed empirical justification.

Following the Renaissance, one has the emergence of both modern science (Kepler, Galileo, and Newton) and modern philosophy (Bacon, Descartes, and Spinoza): a mechanist world-view grounded in mathematics, especially geometry and the calculus, now dominated western thought. The dynamic universe was generally represented as a vast machine running on divinely given celestial laws by a deity independent of and indifferent to the ongoing physical world and human existence as well. Astronomers openly acknowledged material changes in the imperfect cosmos, e.g. the sudden appearance of comets and novae along with those irregularities in the seemingly circular orbits of the wandering stars or planets. Nevertheless, it still remained for a "Newton of biology" to synthesize the terrestrial sciences within a historical framework that took planetary time and organic transformations seriously.

In sharp contrast to Descartes's mechanist interpretation of animals as merely complex machines, it was Leibniz's process view of the history of life as a slowly emerging monadic continuum that helped to pave the intellectual way for the theory of organic evolution. Briefly, man first studied motion in the inorganic objects most remote from our earth and only gradually in those organic structures nearer to him until, finally, he examined himself: it is an incredible example of parallel development that the advance of scientific knowledge and philosophical understanding has more or less repeated the acutal order of cosmic history as seen from the planetary perspective.

Again, the science of biology had its origin in Presocratic curiosity (e.g. Empedocles's speculations on the emergence of first asexual and then sexual creatures and his claim that the world of life is still incomplete) and the pioneering studies of the great Aristotle at the Lyceum (the Stagirite classified and described at least five hundred species of mammals, birds, and fishes as well as mollusks, polyps, sponges, and other marine forms of life). There was also the early work in anatomy and medicine by such researchers as Galen, Avicenna, da Vinci, and Versalius.

Yet, after Aristotle, the first significant biologist is Carolus Linnaeus (1707-1778).[15] This Swedish botanist and physician was particularly interested in the reproductive methods and sexual characteristics of flowering plants in Lapland, which resulted in his devising a new system of artificial classification by means of binary designation called binomial nomenclature (each plant and animal type or kind is assigned its own Latinized scientific name consisting of two terms: a general first term indicating the genus, and a specific second term indicating the species). As such, he is rightly considered to be the founder of modern taxonomy.

Linnaeus revived the science of biology, both systematized and expanded the descriptive classification of living species, referred to the human animal as *Homo sapiens* (the designation still used today), placed our own zoological group along with the apes, monkeys, and lemurs into the same primate order (he claimed man's closest living relative to be the orangutan), and even discerned the adaptive harmony between different animals and their various environments established through adequate senses. He published his findings and ideas in *The System of Nature* (1735, 1766), a significant contribution to science.

However, as a Christian and Aristotelian, Linnaeus believed himself to be divinely chosen and inspired to devote his career to taxonomy in order to help fill in those numerous gaps in the assumed hierarchical continuum of the "Great Chain of Being" and thereby prove the existence of God by empirically demonstrating the order, design, and intelligence pervasive throughout static nature. Having accepted the Stagirite's doctrine of the immutability of species, he was bewildered when he did find varieties among the so-called eternally fixed plant forms. With an ironic twist, modern biology uses comparative taxonomy as one of the several major evidences to empirically support the scientific truth of organic evolution.[16] Since the myth of Adam, the naming and study of plants and animals continues.

One hundred years separates the appearance of Linnaeus's *The System of Nature* and Darwin's visit to the Galapagos Islands; for ten decades, the scientific pendulum of biology was swinging away from the rigid idea of the eternal fixity of forms and toward the growing awareness of the ongoing emergence of new species over vast periods of time and change as well as birth and death.

In biology, the French scientist and philosopher Pierre Louis Moreau de Maupertuis (1698-1759) rejected the hereditary theory of embryonic preformation popular at that time in favor of epigenesis. Unable to give a mechanist account for the origin and nature of life, he developed a metabiological stance grounded in corpuscular psychism, or atomistic dualism: all matter has sensitivity as well as intelligence and memory to some degree. Before Mendel, Maupertuis had applied the mathematical theory of probability to his own particulate theory of heredity in a serious empirical investigation of albinism and polydactyly. Contributing to evolutionist speculation, he maintained that a process of random chance mutation along with the survival of the fittest organisms in changing environments (especially geographically isolated areas) could explain the structural transformations of various species emerging throughout time. Although Maupertuis died one hundred years before the publication of Darwin's *On the Origin of Species*, he had actually at least suggested the major explanatory elements of Neodarwinism or the synthetic theory of biological evolution in modern science.[17]

The French naturalist Georges Louis Leclerc, Comte de Buffon, (1707-1788) asserted early in his monumental forty-four volume work *Natural History: General and Particular* (1749-1804) that species may undergo modification and, consequently, produce varieties if not even new species.[18] All the existing forms of life could be the descendants by true generation of different forms as common ancestors in the past. Although drawing attention to vestigial organs, he did not offer an explanation for the transformation of species. Buffon's later opinions on the controversial subject of organic history fluctuated and (perhaps reluctantly but intentionally) became obscure. In fact, his early ideas on transformism having been both condemned and censured in 1751 by the ecclesiastical authorities at the Sorbonne, Buffon all but withdrew his support for the hypothesis of biological evolution (recall the similar fate of Galileo and Teilhard de Chardin).

Within the framework of continuity and plenitude, Buffon clearly distinguished between true species as entities of nature and the limited mutability of varieties. He wrote about the sterility of hybrids, the heritability of acquired characters, the unity of type demonstrated in comparative anatomy studies, and the origin and natural history of birds within an implied evolutionary overview.

Buffon the scientist placed great emphasis on rigorous experimentation as well as the critical observation of nature: he insisted that science and religion should be strictly separated. His own views, however, often illustrate the power of rational speculation and the need for compromise. In cosmogony, he boldly proposed the nebular theory to account for the origin of the earth: this planetary system had resulted when the glancing blow of a comet against the molten surface of our sun freed a mass of matter which eventually separated into various solar objects. In geology, he recognized seven distinct epochs of natural history spanning over vast periods of time involving millions of years (thereby arguing for the formation and destruction of mountains by the flux and reflux of tides and currents and explaining fossil evidence as the remains from the organic development of plants and animals in remote ages). He is also remembered for his biological theory of organic molecules that, since they are pervasive throughout nature, are a source of living organisms (perhaps through spontaneous generation). Briefly, it is deeply regrettable that Buffon's pioneering insights as offered in his sweeping view of both cosmic evolution and earth history were silenced in their development by some of the myopic and dogmatic religionists who were very influential during his own lifetime. Ironically, then, Buffon was both an asset and a liability to the emerging theory of evolution.

In Germany, influenced by Linnaeus and Buffon, Johann

Friedrich Blumenbach (1752-1840) applied the systematic study of comparative anatomy of humankind in order to derive a racial classification of our own species based on the morphological differences among geographically separated populations. He published his original findings in anthropometry as a volume entitled *On the Natural Variety of Mankind* (1775) and is therefore credited as the founder of physical anthropology as the science of human biology.

In England, the botanist and successful physician Erasmus Darwin (1731-1802) speculated on the origin of all organisms from one primordial living filament that existed millions of ages ago.[19] The philosophical naturalist held that species gradually evolve and improve over vast periods of time. He emphasized the factors of mutation, adaptation, struggle for existence, reproduction of the strongest, and even the role of sexual selection as well as artificial selection in the transformation of plant and animal forms. These ideas were presented in his major work, the two-volume *Zoonomia: or, the Laws of Organic Life* (1794, 1796), which apparently did not impress his grandson Charles Darwin, even though it had anticipated the scientific theory of biological evolution but in an unsystematic, vitalist, and deist framework. Nevertheless, Charles's own interests in nature were without a doubt closer to those of his paternal grandfather than to those of his own father.

In 1802, Erasmus Darwin's long poem, *The Temple of Nature*, was published posthumously; this same year saw Lamarck, Treviranus, and Oken independently outline their principles of biology and views on evolution. Lamarck was the first naturalist to write a book solely for the purpose of explicating the theory of evolution in the animal world. Treviranus claimed that in the flux of nature even man has not as yet reached the highest term of his existence but will progress to still higher regions and produce a nobler type of being. Similar to Schelling and Hegel, among others, Oken presented and defended the conception of an evolving God whose development is manifested in the progressive complexity and diversity of living things: in this pantheist interpretation of the gradually evolving ascent of all life from some primeval mucus through plants to animals, man is a god in a universe created by a God that metamorphosed itself into the cosmos: as such, mind is merely a mirror in which Time or the Absolute and human consciousness as the goal and end of creative nature can behold themselves.[20] In evolution theory, Oken's process metaphysics is an advance over Hegel's dialectic synthesis.

Although important conceptual advances were being made in the emerging science of biology before the nineteenth century, one does find the following representative views in the literature of that time: fallen leaves on water slowly transform into fishes and those on land slowly transform into birds; rats emerge from decaying wheat and flies are formed from decaying meat; orchids touching the ground give birth to both birds and very small men; birds are derived from flying fishes, lions from sea-lions, and man from the metamorphosis of the husband of a mermaid. Most early biologists accounted for the origin of life as a result of divine special creations or the spontaneous generation of the organic from the inorganic, and they taught the immutability of plant and animal species. Nevertheless, by the middle of the nineteenth century, the Darwinian theory of biological evolution would radically alter forever man's view of organic history in terms of science and reason.

NOTES AND SELECTED REFERENCES

[1]Cf. S. Radhakrishnan, *Principal Upanishads* (London: George Allan and Unwin, 1955, passim). I am grateful to Professor S. Chandrasekhar for bringing this reference to my attention. Also see Shree Purohit Swami and W. B. Yeats, eds., *The Ten Principal Upanishads* (London: Faber and Faber, 1970, esp. pp. 79-83), and Clive Johnson, ed., *Vedanta* (New York: Bantam Books, 1974, pp. 17-38).

[2]Cf. Betty Radice, ed., *Lao tzu or Tao te ching* (New York: Penguin Classics, 1981, trans. by D. C. Lau). Also refer to Wing-Tsit Chan, *A Source Book in Chinese Philosophy* (Princeton: Princeton University Press, 1969, pp. 136-176), and Max Weber, *The Religion of China: Confucianism and Taoism* (New York: The Free Press, 1968, pp. 171-225).

[3]Cf. Wing-Tsit Chan, *A Source Book in Chinese Philosophy*, p. 204, (also see pp. 177-210, 743-745). Also refer to Arthur Waley, *Three Ways of Thought in Ancient China* (Garden City, New York: Anchor Books, 1939, "Chuang Tzu" pp. 3-79).

[4]Cf. John Blofeld, ed., *I Ching: The Book of Change* (New York: E.P. Dutton, 1968).

[5]Cf. Wing-Tsit Chan, *A Source Book in Chinese Philosophy*, pp. 743-745.

[6]Cf. F. M. Cornford, *Principium Sapientiae: The Origins of Greek Philosophical Thought* (New York: Harper Torchbooks, 1965, esp. pp. 159-201); W. K. C. Guthrie, *The Greek Philosophers: From Thales to Aristotle* (New York: Harper Torchbooks, 1960, esp. pp. 22-62); Charles H. Kahn, *Anaximander and the Origins of Greek Cosmology* (New York: Columbia University Press, 1964, esp. pp. 68-71); G. S. Kirk, ed., *Heraclitus: The Cosmic Fragments* (Cambridge: Cambridge University Press, 1962, esp. pp. 14-15, 17, 19, 95n, 244, 335, 369-380 *passim.*); G. S. Kirk and J. E. Raven, *The Presocratic Philosophers: A Critical History with a Selection of Texts* (Cambridge: Cambridge University Press, 1966, esp. pp. 73-215, 319-361); and Philip Wheelwright, ed., *The Presocratics* (New York: Odyssey Press, 1966, esp. pp. 39, 125-126, 134-135, 152-153, 173).

[7]Cf. Benjamin Farrington, *Aristotle: Founder of Scientific Philosophy* (New York: Praeger, 1970, esp. pp. 36-40, 83); Marjorie Grene, *A Portrait of Aristotle* (London: Faber and Faber, 1963, esp. pp. 137, 145-148, 151, 227-234, 244-247); W. K. C. Guthrie, *The Greek Philosophers: From Thales to Aristotle* (New York: Harper Torchbooks, 1960, esp. pp. 140, 144-145); G. E. R. Lloyd, *Aristotle: The Growth and Structure of His Thought* (Cambridge: Cambridge University Press, 1968, esp. pp. 82, 88-90, 303); Richard McKeon, ed., *The Basic Works of Aristotle* (New York: Random House, 1941, esp. pp. 470-531, 633-680); and A. E. Taylor, *Aristotle* (New York: Dover, 1955, esp. pp. 26-27, 56-57, 61, 74-75). Also refer to John Herman Randall, Jr., *Aristotle* (New York: Columbia University Press, 1960, esp. pp. 129, 138, 166, 211, 237).

[8]Cf. Arthur O. Lovejoy, *The Great Chain of Being: The Study of the History of an Idea* (New York: Harper Torchbooks, 1960).

[9]Cf. Lucretius, *On the Nature of the Universe* (New York: Penguin Classics, 1951, trans. by R. E. Latham, esp. pp. 171-256).

[10]Cf. A. H. Armstrong, *Plotinus* (New York: Collier Books, 1953).

[11]Cf. Lloyd L. Morain, "Avicenna: Asian Humanist Forerunner" in *The Humanist*, 41(2):27-34.

[12]Cf. Jean Paul Richter, ed., *The Notebooks of Leonardo da Vinci* (New York: Dover, 2 vols., 1970, esp. volume two, pp. 105-133, 135-221, 283-311). Also refer to Bentley Glass, Owsei Temkin, and William L. Straus, Jr., eds., *Forerunners of Darwin: 1745-1859* (Baltimore: The John Hopkins Press, 1968, pp. 12-17); Pamela Taylor, ed., *The Notebooks of Leonardo da Vinci* (New York: Plume Books, 1971, esp. pp. 121-122, 136-138, 141-157, 166-167, 182-188); and V. P. Zubov, *Leonardo da Vinci* (Cambridge: Harvard University Press, 1968, esp. pp. 210, 221-263).

[13]Cf. Jean Paul Richter, ed., *The Notebooks of Leonardo da Vinci*, volume two, p. 118.

[14]Cf. Antoinette Mann Paterson, *The Infinite Worlds of Giordano Bruno* (Springfield, Illinois: Charles C Thomas, Publisher, 1970, pp. 101-102, 124-125, 127), and Ksenija Atanasijević, *The Metaphysical and Geometrical*

Doctrine of Bruno, translated from the French original by George V. Tomashevich (St. Louis, Missouri: Warren H. Green, 1972).

[15]Cf. Wilfrid Blunt, *The Compleat Naturalist: A Life of Linnaeus* (New York: Viking Press, 1971), and Heinz Goerke, *Linnaeus* (New York: Charles Scribner's Sons, 1973).

[16]Cf. R. H. Lowe-McConnell, *Speciation in Tropical Environments* (London: Academic Press, 1969, pp. 98-105, 116-117, 123-130, 144-147, 192-193, 197-203). These papers are published in *Biological Journal of the Linnean Society,* Vol. 1, Nos. 1 & 2, (double issue), April 1969.

[17]Cf. Bentley Glass, Owesei Temkin, and William L. Straus, Jr., eds., *Forerunners of Darwin: 1745-1859* (Baltimore: The John Hopkins Press, 1968, esp. pp. 51-83, as well as pp. 44n., 104, 107, 112, 116, 118-119, 122, 125-133, 138-141, 144, 163-164, 167, 172, 175, 179, 237, 239, 281, 289, 414n.).

[18]Cf. Otis E. Fellows and Stephen F. Milliken, *Buffon* (New York: Twayne Publishers, 1972). Also refer to Alexander B. Adams, *Eternal Quest: The Story of the Great Naturalists* (New York: G.P. Putnam's Sons, 1969, pp. 84-116); Bentley Glass, Owesi Temkin, and William L. Straus, Jr., eds., *Forerunners of Darwin: 1745-1859* (Baltimore: The Johns Hopkins Press, 1968, esp. pp. 84-113, as well as pp. 23, 44, 51, 60n., 68, 77-82, 116, 122, 124, 126, 128-131, 133, 135, 143-144, 153, 160, 170, 172, 175, 179, 190, 205n., 208, 228-229, 232-237, 241, 244, 247, 250, 254, 276, 279, 281, 324, 364, 382n., 394-398); and Donald Culross Peattie, *Green Laurels: The Lives and Adventures of the Great Naturalists* (New York: Simon and Schuster, 1936, pp. 56-77).

[19]Cf. Hesketh Pearson, *Doctor Darwin* (New York: Walker and Company, 1930, esp. pp. 198-199).

[20]Cf. Arthur O. Lovejoy, *The Great Chain of Being: A Study of the History of an Idea* (New York: Harper Torchbooks, 1960, esp. pp. 320-321).

Chapter 3

THE AGE OF ENLIGHTENMENT

The eighteenth century is often referred to as the Age of Enlightenment: a period of time in western intellectual history somewhat reminiscent of the Presocratic age and the Italian Renaissance (and, *mutatis mutandis*, of the midnineteenth-century intellectual efflorescence of America's New England). In general, the philosophers of the French Enlightenment rejected traditional religious beliefs and metaphysical systems that failed to recognize the freedom, equality, and moral dignity of the human being, as well as the powers of science and reason along with the simple fact that humanity is totally within the process of nature itself.[1] These philosophers embraced the objective method of empiricism and the invaluable use of mathematics, subscribed to the psychobiological basis of human experience and its resultant knowledge, strove for sociopolitical freedom of the general public from the oppression of both Church and State governments, took history and change seriously, and interpreted process as progress (besides writing about sociocultural development, some of these thinkers even anticipated the theory of evolution).[2]

Strongly impressed with the great successes resulting from the use of the scientific method and mathematics in the physical sciences, especially astronomy and physics (e.g. the major accomplishments of Copernicus, Kepler, Galileo, and Newton), and upholding the need for free inquiry within an open society, these enlightened critics of their age envisioned extending the methods of science and reason to a rigorous investigation of both human thought and human behavior. For them, the human intellect is capable of discovering the basic principles or natural laws of human history and, more importantly, those of the mental and social worlds of our own species as well.

These naturalists and humanists criticized the Church and State governments, calling for educational reforms to improve the human condition of the masses. They optimistically thought that further advances in science and technology (particularly psychology and medicine) would lead to the eradication of most, if not all, of the problems facing mankind. Their essays, articles, and treatises on a wide range of subjects in the special sciences and humanities were published in the voluminous *Encyclopédie*, edited by Denis Diderot (thus, the intellectual advances of the few were made available to the growing interests of the many in order to enlighten and improve the overall condition of the multitudes).

For the most part, the Encyclopedists interpreted the material universe as a cosmic machine operating according to determinist laws or principles and, likewise, saw the human sociocultural animal as a part of this mechanist/materialist machine. The philosophers boldly presented this-worldly deistic, agnostic, or atheistic views of reality that excluded any meaningful and purposeful consideration of the alleged supernatural realm. In their quest for truth, happiness, and a world civilization, they gave preference to science, reason, and nature.

Charles-Louis de Secondat, Baron de La Brède et de Montesquieu, (1689-1755) was a philosopher and political theorist. He taught that the freedom of the individual could best be guaranteed by the separation of the powers of the state among three distinct organs that could balance and check one another, i.e. freedom was preserved by setting "power against power" (he held that the establishment of a potent aristocracy would limit the despotic tendency of both the monarch and the common people). Montesquieu's masterpiece is *De l'esprit des lois* (1748), a major contribution to political theory that, as is well known, powerfully influenced a number of modern constitutional structures.

François-Marie Arouet de Voltaire (1694-1778) lived an extraordinarily active life of exiled travel and intellectual productivity. As a humanist, he campaigned against religious oppression and barbarous judicial procedures: he abhorred the evils of super-stition, fanaticism, and cruelty within the Church and State governments of the French Enlightenment. Yet, Voltaire admitted that, "if God did not exist, it would be necessary to invent him," for, "the heavens declare the glory of God" (he remained a deist and, consistent with his humanism, sheltered on his private country estate persecuted Jesuit scholars). However, Voltaire vigorously fought for social justice and humane reforms, claiming that political liberty could best be achieved under the rule of a philosopher-king. In general, his thoughts reflect genuine awe and veneration derived from his personal poetic and mystical experience of cosmic grandeur, although his undisputed humor-ous and satirical masterpiece is *Candide* (1759), a hostile attack on the philosophical optimism of Leibniz and Pope.

Georges-Louis Leclerc, Comte de Buffon, (1707-1788) was a naturalist whose comprehensive 44-volume *Histoire naturelle, général et particulière* (1749-1804) recognized both the vast expanses of geological time and the changes within biological varieties. He called for the strict separation of science from religion, and his natural-historical orientation anticipated the Lamarckian and Darwinian perspectives. However, Buffon him-self was not a rigorous evolutionist (contrast with Pierre Louis Moreau de Maupertuis).

Julien Offray de La Mettrie (1709-1751) was a physician and unorthodox philosopher who boldly advocated atheism, materi-alism, and determinism. He was the first significant biomedically oriented philosopher. As a rigorous naturalist, La Mettrie's empirico-physiological theory of mind was a refutation of the existence of a nonmaterial immortal soul. His major work is *L'homme machine* (1747), which presented his man-machine theory. His view of a "thinking machine" went beyond the inadequacies of Descartes's dualistic system of psychophysical parallelism. Finally, La Mettrie recognized man's close relation-ship to all other animals (especially the apes).[3]

The father of Romanticism, Jean-Jacques Rousseau (1712-1778) presented a philosophical view contrasting (1) the alleged free, innocent, happy, prehistoric "noble" savage with (2) the anxious, unhappy, civilized, contemporary man. As a reaction to the Christian-held doctrine of man's inherent wickedness, he believed in the natural goodness of man (not yet able to realize that, at birth, man is actually neither and potentially both). Rousseau was critical of social corruption, recognizing the close connection between the structure of society and the psychological and moral condition of the individual. He advocated progressive education and religious beliefs as equally necessary for developing the moral man within a democratic society. Rousseau's *Émile* (1762) is representative of his thought, which, unlike that of most of his contemporaries who overstressed the power of reason, tended to emphasize the strength of the emotions.

Denis Diderot (1713-1784) was a versatile and critical thinker with an encyclopedic mind and range of information.[4] As a rational empiricist, he advocated strict adherence to the scientific method. Diderot taught sensationalism, materialistic monism, atheism, and humanism: nature is pervasive matter in constant motion in space (birth, life, and decay are merely changes of form). His thoughts foreshadowed both Darwin and Freud; he arrived at the principles of transformation and natural selection, as well as sensing the dangers of sexual repression in modern society. Diderot is remembered as editor-in-chief of the *Encyclopédie*, a job that lasted over twenty years. His book, *Thoughts on the Interpretation of Nature* (1754), presents his dynamic materialist philosophy.[5] Among his successors and admirers were Feuerbach, Marx, and Engels. Diderot was one of the greatest thinkers of the French Enlightenment.

Claude-Adrien Helvetius (1715-1771) devoted himself primarily to philosophical and literary pursuits. His two major works are *De l'esprit* (1758), which was condemned by the French Parliament and publicly burned in Paris, and *De l'homme* (1772). They support environmental psychology and behaviorism grounded in a naturalist view of man. Helvetius's philosophy stressed sensationalism, self-interest, the equality of intellects, and the moral improvement and happiness of humankind through reformed public education (the latter was a major tenet of the French Enlightenment). Helvetius argued that by devoting one's self to the general welfare of society one actually promoted one's own interest and can hence be credited with having fathered the modern notion of enlightened self-interest.

Indebted to John Locke, Etienne Bonnot de Condillac (1715-1780) was a priest who believed that the secular and rational systematization of human experience was his primary task. His deductive inquiry into epistemology (empirical sensationalism) led him to language analysis and logic. Condillac's important *Traité des sensations* (1754) established him as a rigorous thinker (refer to his marble statue hypothesis). He held to a dualistic ontology between matter and spirit (mind) and believed in the existence of God as the ultimate source of understanding and morality.

Jean-Le-Rond d'Alembert (1717-1783) was a mathematician and admirer of Bacon's experimental and inductive method. He discounted metaphysical truths as inaccessible through reason. D'Alembert was representative of the eighteenth-century viewpoint expressed in the *Encyclopédie*. Like Newton, he viewed the cosmos as a clock that necessarily required a clockmaker. As such, he at first maintained a skeptical deism (under Diderot's influence, d'Alembert later adopted a materialist and atheist interpretation of the universe). His most important work in philosophy is *Éléments de philosophie* (1759).

Of German ancestry, Paul-Henri Thiry, Baron d'Holbach, (1723-1789) was the foremost exponent of atheist materialism and determinism in the French Enlightenment. He wrote about the natural sciences (geology, mineralogy, chemistry, and metallurgy) and against religion. His outspoken thoughts are best expressed in *Système de la nature, ou des lois du monde physique et du monde moral* (1770). D'Holbach also wrote 376 articles for Diderot's *Encyclopédie*. Despite his opposition to all organized religion, he was benevolent, unselfish, altruistic, and tolerant.

Marie-Jean-Antoine-Nicolas Caritat, Marquis de Condorcet, (1743-1794) was a mathematician, social scientist, and moderate revolutionary. He was especially interested in the application of probability theory to the special sciences, particularly the social sciences. His major work is *Sketch for a Historical Picture of the Progress of the Human Mind* (1795), in which he argued for the continuous progress of the human species to a phase of ultimate perfection.[6] Condorcet saw the story of humanity as passing uninterruptedly through nine great stages of universal history: (1) hordes of hunters and fishers united in tribes, (2) pastoral peoples, (3) the progress of agriculture up to the invention of the alphabet, (4) the progress of the human mind in Greece up to the division of the sciences about the time of Alexander the Great, (5) the progress of the sciences from their division to their decline, (6) the decadence of knowledge to its restoration about the time of the Crusades, (7) the early progress of science from its revival in the West to the invention of printing, (8) from the invention of printing to the time when philosophy and the sciences shook off the yoke of authority, and (9) from Descartes to the foundation of the French Republic.

Condorcet also envisioned a future tenth epoch of universal equality and freedom brought about through scientific enlightenment and social reform. In this final stage the physical, intellectual, and moral aspects of humankind would be perfected through the awesome advances in science and technology (particularly medicine). The world would become free of war, error, prejudice, and superstition. Through endless progress, his extraordinary and somewhat wishful optimism foresaw the ongoing perfectibility of mankind to the eventual attainment of virtual (though never literal) immortality by everyone on this planet! In his elaborate *Sketch*, written ironically under the shadow of death while he was hiding from Robespierre and other Jacobins, Condorcet prophesized the following:

> Our hopes for the future condition of the human race can be subsumed under three important heads: the abolition of inequality between nations, the progress of equality within each nation, and the true perfection of mankind.... The time will therefore come when the sun will shine only on free men who know no other master but their reason. ... Would it be absurd then to suppose that this perfection of the human species might be capable of indefinite progress; that the day will come when death will be due only to extraordinary accidents or to the decay of the vital forces, and that ultimately the average span between birth and decay will have no assignable value? Certainly man will not become immortal, but will not the interval between the first breath that he draws and the time when in the natural course of events, without disease or accident, he expires, increase indefinitely?[7]

Condorcet's speculations on things to come make this volume a great monument of liberal thought and daringly hopeful vision.

Under the influence of his friend Turgot, the tolerant economist and reformer, Condorcet as a futurist propounded a sweeping interpretation of human history that offered a key to a comprehensive understanding of mankind's unfolding potentialities for indefinite growth, development, and perfection.[8]

In Germany, the progressive ideas of the French Enlightenment were first championed by Gotthold Ephraim Lessing (1729-1781), who fought for the freedom of thought and the education of the human species as a distinguished dramatist, literary critic, and cultural historian. In science, however, Friedrich Heinrich Alexander, Baron von Humboldt, (1769-1859) was the father of scientific exploration.[9] He grew up on his father's Tegel estate in the Berlin of Prussian Germany, developing an insatiable curiosity for the natural sciences (especially exotic flowering plants). At the age of twenty-seven, he decided to pursue a lifetime career as a scientist and researcher. In fact, he became the greatest naturalist traveler of his time and a universal scholar dedicated to a deep understanding of and holistic appreciation for the physico-biological world. Humboldt's personality and genius welded together to produce a remarkably great man of ideas and action. His many interests included astronomy, geology, geography, and mineralogy (as well as climatology, botany, zoology, and medicine). He even studied etching, drawing, and mathematics. Likewise, there was his ever-present pervasive geophilosophical attitude that emphasized the creative totality of things.

Independently wealthy, Alexander von Humboldt was able to devote all his time to research, travel, and writing. The greatest influence on the young naturalist was the German scientist Johann Friedrich Blumenbach (1752-1840), the founder of physical anthropology. Humboldt's own special concerns were for understanding both terrestrial magnetism and the geographical distribution of plants. Ultimately, he attempted to synthesize a universal and systematic philosophy of nature grounded in his own organismic view of the planet: the earth as a whole is like a living entity, with all of its parts interrelating and functioning as an active and creative single being (this same view was expressed by Leonardo da Vinci)!

By the age of thirty, Humboldt had already spent several years preparing for the scientific voyage that would make him a world-famous figure in the natural sciences: a five-year journey throughout the New World (especially South America) that would mark him as the first significant scientific explorer in the history of the Western Hemisphere. It was Humboldt's desire to combine extensive travel and systematic research in order to contribute to a fuller scientific discovery of the Americas. At the time, this was an entirely new concept in the history of descriptive inquiry. Accompanied by the French botanist and devoted friend Aimé Goujaud Bonpland, Humboldt boldly started his great and unique exploration of the Western World (1799-1804). With his assistant Bonpland, Alexander traced the course of the Orinoco River in Venezuela, thereby demonstrating that the Casiquaire natural canal links the Orinoco with the Rio Negro and thus with the Amazon River. He also traveled across South America from Cuba through Columbia and Ecuador to Peru. Finally, he visited Mexico (followed by a brief stay in the United States). He both observed and collected plant, animal, and mineral specimens (including fossils). Like his friend Goethe, he assumed the existence of universal forms or archetypes within the plant and animal kingdoms and, like the former, was not actually an evolutionist. He enjoyed preparing maps, climbing volcanoes (e.g. Mt. Chimborazo of the Andes in 1802), making astronomical observations, and traveling through dense tropical jungles with their exotic plants and strange animals.

Humboldt's voyage to the Americas had been a tremendous triumph for the now-famous German naturalist. His thirty-five volumes on the trip include the theory of the morphological geography of plants, particularly the existence of climatic zones (his law of the third dimension). In 1829, at the age of sixty, he led a nine-month journey through Russia and Siberia (it was only a moderate success). Unfortunately, a planned extensive scientific voyage through Asia was never realized.

Humboldt's masterpiece is his five-volume work entitled *Cosmos: A Sketch of a Physical Description of the Universe* (1834-1859).[10] In it, he presents a "cosmic natural painting" of the earth and its various parts. Special attention is given to both the planet's magnetic force and the vegetation of the world as a biosphere (demonstrating the close connection between flora and fauna). The illustrious German naturalist's chosen motto in life was, "Man must only strive for the good and the great!" As a cosmic philosopher, he dedicated his own life to the greater understanding of and deeper appreciation for this planet as a dynamic unity. He lived his philosophy. His expedition to the Americas was a major contribution to modern geography, and the publication of *Cosmos* was a serious treatment of scientific inquiry within a holistic view of the world. He died of a stroke on 6 May 1859, the year of the publication of Charles Darwin's *On the Origin of Species*. As a result, it may be safely assumed that the Darwinian theory of evolution inadvertantly eclipsed Humboldt's own monumental work.

In summary, the Age of Enlightenment was an outgrowth and elaboration of the worldview presented primarily by Newton and Locke. It was characterized by an emphasis on reason, science, and history, resulting in a turn from religion to mechanist materialism, sensationalism, and determinism. Although at the time there was little empirical evidence from geology, paleontology, archaeology, biology, and ethnography to document the views of the philosophers, the Enlightenment nevertheless provided the conceptual framework for the modern scientific orientation.

NOTES AND SELECTED REFERENCES

[1]Cf. Carl L. Becker, *The Heavenly City of the Eighteenth-Century Philosophers* (New Haven: Yale University Press, 1959); J. B. Bury, *The Idea of Progress: An Inquiry into its Origin and Growth* (New York: Dover, 1955, esp. chapter eleven on Condorcet pp. 202-216 and chapter nineteen on progress in the light of evolution pp. 334-349); Ernst Cassirer, *The Philosophy of the Enlightenment* (Boston: Beacon Press, 1955, esp. pp. 3-92); Alfred Cobban, *In Search of Humanity: The Role of the Enlightenment in Modern History* (London: Jonathan Cape, 1960, esp. chapter sixteen on Diderot, pp. 139-146); Julian M. Drachman, *Studies in the Literature of Natural Science* (New York: Macmillan, 1930, esp. the references to Alexander von Humboldt); Kingsley Martin, *The Rise of French Liberal Thought: A Study of Political Ideas from Bayle to Condorcet* (New York: New York University Press, 1954, esp. pp. 286-299 for a discussion of Condorcet's theory of progress); Romain Rolland, André Maurois, and Edouard Herriot, *French Thought in the Eighteenth Century* (New York: David McKay, 1953, esp. the section on Denis Diderot by Edouard

Herriot pp. 249-271); and Albert Salomon, *The Tyranny of Progress: Reflections on the Origins of Sociology* (New York: Noonday, 1955).

[3]Cf. Bentley Glass et al., eds., *Forerunners of Darwin: 1745-1859* (Baltimore: The Johns Hopkins Press, 1959, esp. chapters three, four, five, and six).

[3]Cf. Julien Offray de La Mettrie, *Man à Machine* (La Salle, Illinois: Open Court, 1961, esp. pp. 98, 100-101, 103, 113-115, 135, 140, 144-145, 148).

[4]Cf. Alfred Cobban, *In Search of Humanity: The Role of the Enlightenment in Modern History* (London: Jonathan Cape, 1960, esp. chapter sixteen, pp. 139-146), Lester G. Crocker, *Diderot: The Embattled Philosopher* (New York: Free Press, 1966), Romain Rolland et al., eds., *French Thought in the Eighteenth Century* (New York: David McKay, 1953, pp. 249-420), and Aram Vartanian, *Diderot and Descartes: A Study of Scientific Naturalism in the Enlightenment* (Princeton, New Jersey: Princeton University Press, 1953).

[5]Cf. Jonathan Kemp, ed., *Diderot: Interpreter of Nature* (New York: International, 1963, esp. pp. 43-48).

[6]Cf. Antoine-Nicolas de Condorcet, *Sketch for a Historical Picture of the Progress of the Human Mind* (New York: Noonday, 1955, esp. pp. 173-202).

[7]*Ibid.*, pp. 173, 179, 200.

[8]Cf. J. B. Bury, *The Idea of Progress: An Inquiry into its Origin and Growth* (New York: Dover, 1955, esp. pp. 202-216).

[9]Cf. Douglas Bottling, *Humboldt and the Cosmos* (New York: Harper & Row, 1973, esp. pp. 253-262), and Helmut de Terra, *Humboldt: The Life and Times of Alexander von Humboldt, 1769-1859* (New York: Alfred A. Knopf, 1955). Also refer to Julian M. Drachman, *Studies in the Literature of Natural Science* (New York: Macmillan, 1930, pp. 162-167, 182, 237, 339-342, 373, 402).

[10]Cf. Alexander von Humboldt, *Cosmos: A Sketch of a Physical Description of the Universe* (New York: Harper & Brothers, 5 vols., 1865). Also refer to *Alexander von Humboldt: 1769-1969* (Bonn/Bad Godesberg: Inter Nationes, 1969, esp. pp. 7-94) and Desmond Wilcox, *Ten Who Dared* (Boston: Little/Brown, 1977, pp. 118-147).

Chapter 4

LEIBNIZ, KANT, GOETHE, AND HEGEL

T
the German Enlightenment (1700-1780) also gave birth to
new ideas and perspectives. Cardinal Nicholas of Cusa
(1401-1464) had already offered an intriguing system of thoughts
that represented a significant break from the static Christian
worldview grounded in the concepts of Aristotle, Ptolemy,
Augustine, and Aquinas.[1] Cusa, the first transalpine Renaissance
philosopher, spoke of "learned ignorance" and taught the
doctrine of the coincidence of opposites. He boldly presented a
moral cosmology of involution or convergence, which emerged
out of his interests in mathematics and mysticism. It claimed man
to be the existential midpoint between God as the universal
circumference or the macrocosm and Christ as the universal
center or the microcosm in an indeterminately vast world. Cusa's
panentheistic view of process and reality anticipated both the
dynamic cosmology of Whitehead and the cosmic mysticism of
Teilhard de Chardin.

Acknowledging the growing evidence in astronomy for sudden
changes, physical imperfections, and irregular motions among
the celestial bodies (especially as a result of the pioneering
investigations of Kepler and Galileo), some naturalists and
philosophers of the eighteenth century took earth history and
organic development seriously. In general, change first recognized
in the heavens was now extended to include our planet and later
its inhabitants as well (including the human animal).

Just before the German Enlightenment, the rationalist philosopher Gottfried Wilhelm Leibniz (1646-1716) taught an organismic

view of the dynamic universe as a whole: there are worlds within worlds, and an infinite number of imperfect perspectives of the cosmos.[2] He had been influenced by the new discoveries in cellular biology and also the thoughts of Bruno, Descartes, and Spinoza. In 1676, independent of Newton, Leibniz discovered the infinitesimal (differential and integral) calculus, which provided the mathematical foundation for his process metaphyics. Later, in 1695, he coined the term "monad" to refer to a metaphysical unit of psychic force more or less analogous to a mathematical point.

In his writings, e.g. the *Protogaea* (1691, 1749) and elsewhere, after having observed rock strata and the fossils of extinct ammonites, Leibniz speculated on the appearance and history of life on earth as well as the mutability of species.[3] He even hazarded the thought that there are living species intermediate between the apes and man, but cautiously relegated the existence of such "missing links" in the unbroken continuum of life to some other planet beyond our own!

In the *Monadology* (1714), Leibniz finally presented his rich and profound interpretation of the teleology, harmony, and order in the unfolding cosmos.[4] The Leibnizian system incorporates several important principles: monadic plenitude, identity of indiscernibles, dynamic panpsychism, process continuum, pervasive causality (necessity or determinism), ongoing fulguration, preestablished harmony, and sufficient reason. Briefly, this process metaphysics argued for a psychic reality of endless progress composed of an infinite number of independently unique monads eternally developing by degree in an unbroken hierarchy of ever-increasing awareness toward God's transcendent perfection. Leibniz's monadic continuum is a natural chain of ceaseless becoming that advances never by leaps but always in linear gradation by degrees (*natura non fecit saltus*). Like space, time is merely a relational concept abstracted from within the subjective manifold of each monad. There are preconscious (inorganic), conscious (organic), self-conscious (human), and superconscious (angelic) levels of monadic development. Thus spirits are linked with men, men with the animals, animals with plants, and these with the fossils, which in turn merge with those material bodies that our senses and our imagination represent to us as absolutely inanimate: all these beings form a single chain of dynamic continuity. God, the hypercategormatic absolute monad, transcreates this universe and (at each instant) holistically apperceives all things within all space and throughout all time.[5]

Like Bruno, Leibniz taught that this actual universe of simple laws and maximum complexity is the best of all possible worlds: the ongoing but always incomplete cosmos manifests the greatest degree of diversity while, at the same time, having the least degree of evil.

In short, Leibniz gave a vitalist philosophy of activity and development (but not a philosophy of evolution as such). Nevertheless, in his critical and original view of things, he did pave the way for the German Enlightenment. Johann Christian Wolff (1679-1754) systematized and popularized the Leibnizian worldview in his *First Philosophy* (1729) and *General Cosmology* (1731). Similar to Leibniz, Wolff held philosophy to be the science of possibles so far as they can be known in terms of reason.[6]

Immanuel Kant (1724-1804), the rational philosopher and critical idealist of Königsberg, East Prussia, was interested in the natural sciences (e.g. physics, astronomy, and geography), mathematics and formal logic, and the major areas of philosophy (metaphysics, epistemology, ethics, and aesthetics).[7] Although he lived an uneventful life, never traveling beyond the Königsberg countryside, despite his love for physical geography, two things did fill his great mind with awe and admiration: the starry heavens above and the moral law within.[8] Interestingly enough, the key philosopher of the German Enlightenment presented a process cosmology but maintained the immutability of species and clung to a fixed ethics.

Kant's entire philosophy is grounded in a strictly Newtonian/ Euclidean framework. His most notable early work of a pre-critical period is *Universal Natural History and Theory of the Heavens* (1755), a milestone in speculative cosmology that was first published anonymously.[9] It is a grand and remarkably bold conception of the origin and natural history of the eternal and infinite universe considered as a single unified system grounded in laws, atoms, and the void (with special attention given to the genesis and development of galaxies). In fact, the mechanical universe as a whole is said to be composed of a vast number of galaxies.

The sun, planets, satellites, and comets of our solar system arose slowly through a process of certain necessary mechanical laws of mathematical determination (rather than mere blind chance), resulting in the contraction of the universal diffusion of primordial stuff (elemental matter) of a nebulous mass of chaos into ever-greater cosmic order, beauty, and harmony. By analogy, the systematic constitution of our solar system mirrors this Milky Way Galaxy and all other extragalactic star-systems and world-systems (it must be noted that in Kant's time there was as yet no astronomical verification of such extragalactic systems). The disk-shaped universe as a totality is endlessly perfecting itself through the two cosmic forces of attraction and repulsion, although solar systems are continuously perishing in some places while being created in others. As such, there is no final end or ultimate goal to cosmic creativity (similar to Kant's nebular hypothesis are the cosmic views of Swedenborg, Buffon, and Laplace).

Kant did claim that life, including intelligent rational beings, exists elsewhere in the expanding cosmos of increasing perfection. Yet, although recognizing the dynamic nature of the sidereal universe, he did not (interestingly enough) seriously consider even the possibility of biological evolution on earth.[10]

Kant taught that both space and time are pure and fundamental *a priori* forms of intuition comprising a transcendental aesthetic independent of and prior to immediate human experience as sense perceptions of the phenomenal world of knowable appearances (in sharp contrast to the noumenal realm of unknowable reality).[11] In short, in this context, space and time make up the essential subjective and ideal schema of the knowing mind. In this Kantian framework, the world is not independently existent but has its being as phenomena only in relation to human consciousness.

Kant's anthropology deals primarily with psychology and ethics. He was also interested in taxonomy and embryology and wrote that there were four original races of humanity. In sharp contrast to his own process view of the mechanical cosmos and unlike the biological ideas of Maupertuis and Buffon, Kant never developed a hypothesis of organic transformism. His natural philosophy vehemently rejected the emerging theory of biological evolution with its inescapable implications for the origin of man (including human mental, emotional, and behavioral phenomena). He simply recognized only the "unity of type" among similar (seemingly immutable) species, none of which for him ever had any common ancestors.

In 1785, Kant wrote, "The slightness of the degrees of difference between species is (since the number of species is so great) a necessary consequence of their number, but a relationship between them (such that one species should originate from another and all from one original species, or that all should spring from the teeming womb of a universal mother) would lead to ideas so monstrous that reason shrinks from them with a shudder."[12]

In 1790, the German philosopher clearly rejected the hypothesis that every kind of organism is derived from another earlier kind of organism, though the one may differ from the other in species (as if, for example, certain water-animals transformed themselves little by little into marsh-animals and these in turn, after some generations, into land-animals). He saw this hypothesis as a daring adventure of reason that is neither self-contradictory nor absurd. Yet, he unequivocally emphasized that no experience of nature supports the biological transformation of real species into other species (not to mention the generation of the organic out of the inorganic).[13]

However, in 1798, recognizing the similarities between man and the apes, Kant did (like Charles Bonnet) speculate that this epoch may, on the occasion of some great revolution of nature, be followed by a third epoch in which the orangutan or the chimpanzee would perfect the organs that serve for walking, touching, and speaking into the biological structures of an articulate human being, with a central organ for the use of the understanding, and thereby this ape would gradually develop itself through social culture into a civilized human being! Despite this wild speculation, Kant held that like begets like and kind begets kind. In fact, he even maintained that the "first parents" of humanity had themselves no parents but just suddenly began to exist full-grown with no natural cause explaining their abrupt appearance on this planet![14]

Although not a forerunner of Darwin, Kant was a philosopher of humanity. He attempted to demonstrate the limitations of both science and reason in order to substantiate the practical need for moral theology. The major critical works of the first professional philosopher include *Critique of Pure Reason* (1781, 1787), *Critique of Practical Reason* (1788), and *Critique of Judgment* (1790).[15]

Given a community of free beings as noumenal selves, each having an autonomous holy will, Kant's synthetic *a priori* ethical maxim (or ultimate and timeless formal principle of reverent duty) of practical reason is the apodictic proposition known as the categorical imperative grounded in moral theology: I ought never to act except in such a way that I can also will that the maxim of my action must be a universal law.[16] For Kant, morality is an end in itself (in sharp contrast to actions based on sensuous inclinations). He places the moral aim of human life within a subjective and theistic framework forever barred from experience in general and both science and reason in particular.[17] In essence, a personal God is the author of the starry heavens above and the moral law within.

In 1784, Immanuel Kant had speculated that a philosophy of history might suggest that nature in general (and our species in particular) may be pursuing a long-term but hidden plan of struggle whose final purpose is to produce an international society of perpetual peace.[18] Such an interpretation would require a teleological (rather than merely a mechanistic) overview of human history and physical nature. Finally, if he were alive today, it would certainly be interesting to know what Kant would say about the theories of relativity and evolution.

Johann Wolfgang von Goethe (1749-1832), perhaps the last universal man of western civilization, is primarily remembered as Germany's greatest lyric poet and the author of the two-part drama *Faust* (1808, 1832), his literary masterpiece.[19] The work is provocative, of timeless value, and one of the greatest accomplishments in all literature. In its pages, the inexhaustible genius and intense humanism of this "Sage of Weimar" and citizen of the world give an optimistic study of the striving ascent and unending quest of mankind for both pleasure and understanding. In essence, *Faust* is a universal hymn to human action in the loving service of a progressing society.

Essentially a poet, Goethe's own approach to things emphasized a reliance upon feelings, intuitions, and the imagination (rather than experimentation, mathematics, and reason). In science, however, his own interests ranged from magnetism and mineralogy to chemistry and medicine. Yet, Goethe's Neoplatonic worldview resulted in serious errors of judgment that pervade his theoretical interpretations in several special sciences. In fact, he was even prejudiced against instruments (e.g. the microscope, telescope, and eyeglasses).

Although Goethe intuited a general glimpse of the developmental unity of the natural realm in his own scientific writings, which emphasized metamorphosis in the hierarchically arranged upward sweep of all life in terms of morphology, he certainly did not specifically contribute to (or even anticipate) an early formulation of an actually evolutionary viewpoint to account for the origin, history, and diversity of the living world.[20] As such, despite his own biological studies, Goethe was not a direct precursor of Darwin (the distinction between individual development and the evolution of species is, of course, crucial).

A naturalist, Goethe had a passion for comparative science (especially osteology and botany). In geology, as a Neptunist, his theory of earth history ignored the role of fire in causing changes in rock strata throughout time. In biology, to support his metamorphosis theory, he searched unsuccessfully for the ideal plant type (*Urpflanze*), which he assumed to be ancestral to all plant varieties and species in the world. The truth is that such a pure, essential, and concrete Platonic archetype or eternal model does not exist in the flux of reality. In anatomy, he did discover that the intermaxillary bone in apes and other animals is also present in the upper human jawbone (but unknown to Goethe, this fact was independently discovered by Vesalius, Vicq d'Azyr, Oken, and later Owen). This fact did help to demonstrate clearly man's anatomical kinship with the apes. Yet Goethe's theory of 1790 for the origin of the skull, as a variation and continuation of the vertebrae that encloses the brain as the spinal column encloses the spinal cord, is not accepted by modern science.

For Goethe, a creative force pervades and is responsible for the organic activity within nature. He taught a philosophical outlook based upon a theme-with-variations view of things, claiming that fixed primal forms or original types still exist that are the basic patterns or sources for all plants and animals in the living world. As such, he did not suggest the historical origin and organic evolution of species from common ancestral forms. At the end of his life, however, Goethe was deeply concerned over the scientific dispute between the antievolutionist Georges Cuvier and the evolutionist Étienne Geoffroy Saint-Hilaire.

In summary, the philosophical poet of Weimar taught an organismic interpretation of the cosmos that advocated a holistic view of man within nature and emphasized the pervasive polarity-in-unity of reality. Reminiscent of thinkers from Xenophanes and Bruno to Spinoza and Hegel, Goethe believed that God is the entire Universe itself: this dynamic pantheism satisfied his interests in universal creativity and cosmic unity. Lastly, Goethe wrote, "one who has science and art does not need traditional religion."

After the German Enlightenment, the objective idealist Georg Wilhelm Friedrich Hegel (1770-1831) dominated modern philosophy of the recent past.[21] He is, of course, famous for his *a priori* dialectical method: a process logic of endless progress involving a qualitative change through the interaction of opposites (thesis and antithesis) with the negation of the negation so that the temporary result is a synthesis or, in actuality, a new thesis of higher truth and value retaining only the positive aspects of the earlier thesis and antithesis. In this ongoing triadic construction, being and nothing generate becoming, thought and nature are more and more "lifted up" (*aufgehoben*) and synthesized in mind throughout social history, and science and religion are becoming unified in philosophy as a comprehensive and rational view of everything in reality.

Following his own early writings in theology and under the

influences of both Fichte and Schelling, Hegel wrote *The Phenomenology of Mind* (1807).[22] However, his most extensive treatment of the philosophy of nature is found in the two-volume *Science of Logic* (1812-1813, 1816) and especially in the *Encyclopedia of the Philosophical Sciences* (1817, 1827, 1830).[23]

Hegel's triadic view of reality, grounded in the necessary and universal logic of the dialectical method, pervades his whole system of teleological process as endless progress spiraling toward ever greater degrees of freedom, unity, and reason (i.e. the Absolute Idea is the unobtainable but essential goal of both the historical development of mind and the continuous becoming of God within time). This unfolding of the World-Spirit is manifested throughout the social history of states, the ongoing growth of *Weltanschauung* philosophy, and the directed emergence of human thought from consciousness through self-consciousness and free concrete mind to absolute knowledge. Hegel even attempted to show that these three areas of development mirror each other, their common goal being the fulfillment of universal reason. The dialectical process of becoming is at the same time logical, ontological, and chronological: one has the cosmic triad of Idea, Nature, and Spirit.

"Germany's Aristotle" presented a philosophy of nature that dealt with actually existent things (minerals, plants, and animals). The transition from mind or logic to these objects of nature is, of course, the crucial point in this comprehensive system of ideas. The deductive leap from pure abstract thoughts to sensuous material things is, in fact, merely a transition from one set of fluid concepts or categories or universals to another. Briefly, objective things are thoughts and nothing but thoughts. These thoughts are, in turn, ultimately rooted in the Absolute Idea. From this objective idealist perspective of things in the world, the thought of plant is deduced from the prior thought of inorganic matter, just as the thought of animal is deduced from the prior thought of plant. Therefore, the external world of things is the consequent of the antecedent logical sphere of internal abstractions. The Idea of thought and the subsequent Idea of nature are essentially one and the same in a process monism ultimately grounded in the endless creative advance of the Absolute Idea or

World-Spirit as God or History (clearly a pantheist view of reality).

Hegel is aware of the caprice, contingency, irrationality, particularity, and infinite variety (mad productivity) found throughout nature in space. As one logically moves from the inorganic through the organic to human consciousness, nature as absolute externality passes over into the rational mind of man as absolute internality. It must be emphasized, however, that this objective idealist philosophy of nature is a deductive system of thoughts in formal order but not a theory of biological evolution to explain the emergence of life from earlier to later forms during the whole course of earth history.

Like Aristotle, Hegel was a process philosopher who, although concerned with change and development, never realized that evolution is a fact in both space and history as well as a temporal process of unfolding logical ideas in human thought. In his *Encyclopedia* (1817), Hegel wrote,

> Nature is to be regarded as a system of grades, of which the one necessarily arises out of the other, and is the proximate truth of the one from which it results; but not so that the one were naturally generated out of the other. ... It has been an inept conception of earlier and later 'Naturphilosophie' to regard the progression and transition of one natural form and sphere into a higher as an outwardly actual production.... Thinking consideration must deny itself such nebulous (at bottom sensuous) conceptions, as is represented in the hypothesis holding to the so-called origin, for example, of plants and animals from water, and then the origin of the more highly developed animal organizations from the lower.[24]

Just as his process logic gives a series of mental categories of increasing truth, Hegel's conception of nature gives a series of natural forms of increasing value (although, of course, this system at best merely suggests a philosophy of evolution): human beings are more valuable than the lower animals, animals are more valuable than the lower plants, and plants are more valuable than the lower minerals. Yet, it never occurred to Hegel that organic nature could evolve from a worm through an apelike creature to man. Concerning the material world, it is clear that the German thinker never accepted even the possibility of the biological evolution of plant and later animal species in the spatiotemporal sphere of external nature.

Similarly, in *The Philosophy of History* (1837), Hegel stated:

> The changes that take place in nature, how infinitely manifold soever they may be, exhibit only a perpetually self-repeating cycle; in nature there happens 'nothing new under the sun' and the multiform play of its phenomena so far induces a feeling of ennui; only in those changes which take place in the region of Spirit does anything new arise. This peculiarity in the world of mind has indicated in the case of man an altogether different destiny from that of merely natural objects (in which we find always one and the same stable character, to which all change reverts), namely, a real capacity for change, and that for the better; an impulse of perfectibility.[25]

In *The Philosophy of Right* (1821), Hegel claims the state itself to be the holistic activity of rational mind on earth, the unfolding actuality of the ethical Idea, the progressive march of God in the human world, and therefore, a secular deity.[26]

Later, in *The Philosophy of History* (1837), as the spokesman for the Prussian State, Hegel presented a philosophy of history (but neglected prehistory). He wrote "The East knew and to the present day knows only that one is free; the Greek and Roman world, that some are free; the German world knows that all are free. The first political form therefore which we observe in history is despotism, the second democracy and aristocracy, the third monarchy."[27] In short, world history is the development of Spirit in time, just as material nature is the unfolding of the Idea in space.[28]

Hegel did not contribute to the emergence of the scientific theory of biological evolution. Instead, he saw external nature as the objectification of Spirit manifested in the hierarchical categories of the inorganic and organic realms.

In rigorous German thought, one moves from the process cosmologies of Leibniz and Kant through Goethe's interests in nature (e.g. rocks, plants, and bones) to Hegel's dialectical view of the World-Spirit within a rational philosophy emphasizing human history. None of this, however, directly influenced Charles Darwin or the biological sciences. Yet, one may argue that the eternal visions of process as offered in the metaphysics of Leibniz and Hegel (not to forget Kant's dynamic cosmogony) represent a philosophical link between Aristotle's static world-view and Darwin's evolutionary framework.

NOTES AND SELECTED REFERENCES

[1]Cf. Lewis White Beck, *Early German Philosophy: Kant and His Predecessors* (Cambridge, Massachusetts: Belknap Press, 1969, pp. 57-71). Also see material on Leibniz (pp. 196-240) and Kant (pp. 426-501).

[2]Cf. Herbert Wildon Carr, *Leibniz* (New York: Dover, 1960); Arthur O. Lovejoy, *The Great Chain of Being: A Study of the History of an Idea* (New York: Harper Torchbooks, 1960, pp. 144-182); Nicholas Rescher, *The Philosophy of Leibniz* (Englewood Cliffs, New Jersey: Prentice-Hall, Inc., 1967); Anna Teresa Tymieniecka, *Leibniz' Cosmological Synthesis* (Assen: Royàl Van Gorcum, 1964), and R. S. Woolhouse, ed., *Leibniz: Metaphysics and Philosophy of Science* (Oxford: Oxford University Press, 1981, esp. p. 134).

[3]Cf. Bentley Glass, Owsei Temkin, and William L. Straus, Jr., eds., *Forerunners of Darwin: 1745-1859* (Baltimore: The Johns Hopkins Press, 1968, pp. 22-23, 233).

[4]Cf. Gottfried Wilhelm Leibniz, *The Monadology and Other Philosophical Writings* (London: Oxford University Press, 1965, pp. 215-277, trans. by Robert Latta); *Leibniz: Basic Writings* (La Salle, Illinois: Open Court, 1962, pp. 251-272, trans. by George R. Montgomery); *Monadology and Other Philosophical Essays* (New York: Bobbs-Merrill, 1965, pp. 148-163, trans. by Paul Schrecker).

[5]Cf. Gottfried Wilhelm Leibniz, *Theodicy* (New York: Bobbs-Merrill, 1966).

[6]Cf. John V. Burns, *Dynamism in the Cosmology of Christian Wolff: A Study in Pre-critical Rationalism* (New York: Exposition Press, 1966).

[7]Cf. Stephan Körner, *Kant* (New Haven: Yale University Press, 1982); Kurt Rossmann, *Immanuel Kant: A German Philosopher* (Bonn-Bad Godesberg: Inter Nationes, 1974).

[8]Cf. Immanuel Kant, *Critique of Practical Reason* (New York: Bobbs-Merrill, 1956, p. 166, trans. by Lewis White Beck).

[9]Cf. Immanuel Kant, *Universal Natural History and Theory of the Heavens* (Ann Arbor: University of Michigan Press, 1969, trans. by W. Hastie).

[10]Cf. Arthur O. Lovejoy, "Kant and Evolution" in *Forerunners of Darwin: 1745-1859*, Bentley Glass, Owsei Temkin, and William L. Straus, Jr., eds. (Baltimore: The Johns Hopkins Press, 1968, pp. 173-206).

[11]Cf. Sadik J. Al-Azm, *Kant's Theory of Time* (New York: Philosophical Library, 1967); Jonathan Bennett, *Kant's Analytic* (Cambridge: Cambridge University Press, 1966, esp. pp. 61-67); and Justus Hartnack, *Kant's Theory of Knowledge* (New York: Harcourt, Brace & World, 1967, esp. pp. 17-30).

[12]Cf. *Forerunners of Darwin: 1745-1859*, p. 190.

[13]Cf. *Ibid.*, pp. 196-200. Also see Immanuel Kant, *Critique of Judgment* (New York: Hafner, 1961, pp. 266-271, trans. by J. H. Bernard).

[14]Cf. *Ibid.*, pp. 200-206.

¹⁵Cf. Immanuel Kant, *Critique of Pure Reason* (New York: St Martin's Press, 1965, trans. by Norman Kemp Smith), *Critique of Practical Reason* (New York: Bobbs-Merrill, 1956, trans. by Lewis White Beck), and *Critique of Judgment* (New York: Hafner, 1961, trans. by J. H. Bernard).

¹⁶Cf. Immanuel Kant, *Groundwork of the Metaphysic of Morals* (New York: Harper Torchbooks, 1964, trans. by H. J. Paton), *Lectures on Ethics* (New York: Harper Torchbooks, 1963, trans. by Louis Infield), *The Metaphysical Elements of Justice* (New York: Bobbs-Merrill, 1965, trans. by John Ladd), and *The Metaphysical Principles of Virtue* (New York: Bobbs-Merrill, 1964, trans. by James Ellington). Also refer to H. J. Paton, *The Categorical Imperative: A Study of Kant's Moral Philosophy* (New York: Harper Torchbooks, 1967) and *The Moral Law: Kant's Groundwork of the Metaphysic of Morals* (London: Hutchinson University Library, 3rd ed., 1965).

¹⁷Cf. Immanuel Kant, *Prolegomena to Any Future Metaphysics* (New York: Bobbs-Merrill, 1950, trans. by Lewis White Beck) and *Religion Within the Limits of Reason Alone* (New York: Harper Torchbooks, 1960, trans. by Theodore M. Greene and Hoyt H. Hudson).

¹⁸Cf. Immanuel Kant, *On History* (New York: Bobbs-Merrill, 1963, trans. by Lewis White Beck, Robert E. Anchor, and Emil L. Fackenheim) and *Perpetual Peace* (New York: Bobbs-Merrill, 1957, trans. by Lewis White Beck). Also refer to Eduard Gerresheim, ed., *Immanuel Kant (1724-1974): Kant as a Political Thinker* (Bonn-Bad Godesberg: Inter Nationes, 1974).

¹⁹Cf. Joseph-Francois Angelloz, *Goethe* (New York: Orion Press, 1958); Johann Peter Eckermann, ed., *Words of Goethe: Being the Conversations of Johann Wolfgang von Goethe* (New York: Tudor, 1949); Ronald Gray, *Goethe: A Critical Introduction* (Cambridge: Cambridge University Press, 1967, esp. pp. 114-125); and Henry Hatfield, *Goethe: A Critical Introduction* (New York: New Directions, 1963, esp. pp. 13-14).

²⁰Cf. Goethe's poem *Nature* (1780) and his book *The Metamorphosis of Plants* (1790).

²¹Cf. J. N. Findlay, *Hegel: A Re-Examination* (New York: Collier Books, 1962, esp. pp. 269-290); Walter Kaufmann, *Hegel: A Reinterpretation* (Garden City, New York: Anchor Books, 1966, esp. pp. 129-133, 234-243); Karl Löwith, *From Hegel to Nietzsche: The Revolution in Nineteenth-Century Thought* (Garden City, New York: Anchor Books, 1967, esp. pp. 59, 217-218); W. T. Stace, *The Philosophy of Hegel: A Systematic Exposition* (New York: Dover, 1955, pp. 97-98, 131-133, 297-317 (esp. 313-317), 323-324); and refer to Franz Wiedmann, *Hegel: An Illustrated Biography* (New York: Pegasus, 1968). Recent works on Hegel include: Jean Hyppolite, *Genesis and Structure of Hegel's Phenomenology of Spirit* (Evanston: Northwestern University Press, 1974), Alexandre Kojeve, *Introduction to the Reading of Hegel: Lectures on the Phenomenology of Spirit* (Ithaca: Cornell University Press, 1980), Stanley Rosen, *G.W.F. Hegel: An Introduction to the Science of Wisdom* (New Haven: Yale University Press, 1974), and Charles Taylor, *Hegel* (Cambridge: Cambridge University Press, 1978, esp. pp. 91n, 354, 542, 551). Also refer to

Lucio Colletti, *Marxism and Hegel* (London: Verso, 1979) and John Somerville, *The Philosophy of Marxism: An Exposition* (Minneapolis: Marxist Education Press, 1981, esp. pp. 3-39, 46, 54, 61, 113, 150).

[22]Cf. Georg Wilhelm Friedrich Hegel, *On Christianity: Early Theological Writings* (New York: Harper Torchbooks, 1961, trans. by T. M. Knox) and *The Phenomenology of Mind* (New York: Harper Torchbooks, 1967, trans. by J. B. Baillie).

[23]Cf. H. D. Lewis, ed., *Hegel's Science of Logic* (New York: Humanities Press, 1969, trans. by J. N. Findlay, esp. pp. 761-774).

[24]Cf. W. T. Stace, *The Philosophy of Hegel: A Systematic Exposition*, p. 313. Refer to pp. 313-317.

[25]Cf. Georg Wilhelm Friedrich Hegel, *The Philosophy of History* (New York: Dover, 1956, trans. by J. Sibree, p. 54).

[26]Cf. Georg Wilhelm Friedrich Hegel, *The Philosophy of Right* (New York: Oxford University Press, 1967, trans. by T. M. Knox, esp. pp. 155, 279, 285, 287).

[27]Cf. Georg Wilhelm Friedrich Hegel, *The Philosophy of History*, p. 104.

[28]Cf. Georg Wilhelm Friedrich Hegel, *Reason in History: A General Introduction to the Philosophy of History* (New York: Bobbs-Merrill, 1953, trans. by Robert S. Hartman, p. 87).

Chapter 5

LAMARCK AND SCHOPENHAUER

Jean-Baptiste-Pierre-Antoine de Monet, Chevalier de Lamarck, (1744-1829) was a naturalist, evolutionist, and deist. his work represented the first significant attempt at a comprehensive theory of biological evolution. Although he recognized the complementary value of science and philosophy, his evolutionary view is not systematically unified. He divided nature into three separate kingdoms: minerals, plants, and animals. Hence, there is a gap between the inorganic and organic realms, as well as between plants and animals.

Lamarck had prepared for the priesthood, did army service as a lieutenant, studied medicine, and was especially interested in meteorology and chemical speculations. As a naturalist, he held the position of Professor of Zoology and is responsible for the distinction between invertebrate and vertebrate animals by the presence of a vertebral column in the latter. He also was the first to establish the groups of Crustacea, Arachnida, and Annelida. Under the influence of Rousseau, he took an ardent interest in botany and wrote *Flore Française* (1778). Later, his interest turned to zoology, and he coined the term "biology" in 1802. His lifelong specialty remained the invertebrates.

Lamarck's observations and reflections led him to postulate an evolutionary framework. His belief in transformationism or the mutability of species went against the prevailing biological teachings of the time. He opposed Cuvier's theory of geological catastrophism and special creations, as well as the dogma of the permanent fixity of biological forms. However, aware of the

81

shortcomings of taxonomy, he held that classification systems are artificial but necessary constructs imposed upon the fluidity of nature.

Lamarck's major work is *Zoological Philosophy* (1809).[1] It held that the supreme and sublime Author of the independent universe had created, by will and infinite power, matter and natural laws as well as given to the cosmos universal motion and pervasive change.

Lamarck gave an evolutionary interpretation of organic nature. The historical order of things in the organic world revealed a single, continuous, linear succession of natural objects from the simplest and most imperfect to the more complex and perfect. There is a grand scale of an infinitely graded series of branching forms manifesting the increasing diversity of organization and accumulative faculties by degree from "monas" to man. According to Lamarck, nature is generally and gradually perfecting itself. It is interesting to note that he starts with man's assumed perfection as the standard for evolutionary judgment projected backwards (an obvious anthropocentric position). From this perspective, nature represents a degradation in organization as well as a proportionate diminution in the number of faculties as one descends the ladder of evolution from man to unicellular beings.

Lamarck's philosophy of evolution contains the following elements: spontaneous generation, vitalism, teleology, and deism. He held that there is an inherent perfecting power running through nature, and a separation between the organic realms of plants and animals and the inorganic realm of chemical materials. The simplest plants and animals are said to arise from inorganic chemical materials by spontaneous generation. In summary, Lamarck believed that (1) a grand linear series of organisms existed from the simple and imperfect to the complex and perfect without gaps and (2) the particular series of evolutions resulted in the small, collateral, branching networks of species. He denied that there had been the extinction of any organic types, assuming that the gaps in classification are merely the result of a lack of empirical evidence to account fully for the continuous chain of living things in nature.

In order to explain the process of biological evolution, Lamarck formulated two causal laws in his *Zoological Philosophy*: (1) in every animal that has not passed the limit of its development, a more frequent and continuous use of any organ gradually strengthens, develops, and enlarges that organ and gives it a power proportional to the length of time it has been so used, while the permanent disuse of any organ imperceptibly weakens and deteriorates it and progressively diminishes its functional capacity until it finally disappears, and (2) all the acquisitions or losses wrought by nature on individuals through the influence of the predominant use or permanent disuse of any organ are preserved by reproduction to the new individuals that arise, provided that the acquired modifications are common to both sexes or at least to the individuals that produce the young.[2]

In *The Natural History of Invertebrate Animals* (1822), Lamarck restated his views on organic history in four laws of biological evolution. Also, he boldly suggested that man originated and slowly evolved from an orangutanlike hominoid form somewhere in the vastness of Asia.[3] In short, Lamarck's opinions represent the most advanced pre-Darwinian evolutionary position in the natural science that he named. Although primarily speculative, his defense of the theory of the mutability of species against the then almost universally held belief in special creations represented an outstanding contribution to conceptual science. He opposed the. permanent fixity of species, as well as the appearance of sudden major mutations, and adhered to the slow continuous directional evolution of animals from protozoans to the human zoological group.

Arthur Schopenhauer (1788-1860) emerged as a major speculative philosopher who greatly influenced Germany's Friedrich Nietzsche. His major work is *The World as Will and Representation* (Vol. I, 1818; Vol. II, 1844).[4] It presented a metaphysical system that clearly anticipated cosmic evolution, gave artistic preference to music, remained philosophically grounded in the Will-to-Live (*Wille zum Leben*), and supported both irrational process and pervasive pessimism. He was influenced by Plato, Kant, Schelling, and Oriental philosophy (particularly the Upanishadic and the later Buddhist texts). At the University of

Berlin, Schopenhauer deliberately lectured at the same hours as
the rationalist philosopher Hegel. But Schopenhauer's teaching
was a failure against Hegel's authority and popularity, and he
even regarded Hegel as a charlatan. Schopenhauer was a misan-
thrope who lived a solitary and resentful life. However, he did
receive academic recognition just before his death.

Schopenhauer's metaphysics has its starting point with the
essential aspect of man: his will. The essence of man (the
microcosm) is held to be the essence of nature (the macrocosm).
Schopenhauer's metaphysical synthesis asserts that the ultimate
reality of nature is the irrational and negative Absolute as the
cosmic Will-to-Live or the thing-in-itself beyond the Kantian
subjective intuitions of space and time and the twelve categories
of human understanding. The cosmic Will is presented as an
eternal and irrational force manifesting the restless movement of
change. The appearance or representation of the timeless Will is a
gradual developmental process of expansion and diversification.
However, the cosmic Will represents a monistic metaphysical
unity (Schopenhauer held that space and time constitute the
principium individuationis in contrast to the ultimate unity of
the Will beyond space, time, and causality).

For Schopenhauer, the Platonic Ideas as archetypal essences
mediate between the cosmic Will and illusory phenomena. These
Platonic Ideas determine and limit the phenomenal hierarchically-
ordered series of the objectification of the ever-restless Will. There
are two basic perspectives upon reality: the scientific attitude,
which is concerned with the appearances of things or natural
phenomena as representation, and the aesthetic attitude, which is
concerned with the Ideas (or music) that mediate between the
Will-to-Live and its representation. Like Plato, Schopenhauer
held to the eternal fixity of Ideas; unlike Plato, he held that these
Ideas are temporally manifested as a successive hierarchy of forms
within an evolving physical world. The arts (especially music)
represent an attitude of aesthetic awareness, will-less contemplation, and disinterested perception of the Ideas. Music has, as its
subject, the Will itself and is therefore the very language of the
metaphysical realm.[5]

Schopenhauer's philosophy taught a hierarchy of art forms

from matter-oriented architecture to will-oriented music. Of all the arts, music alone has a metaphysical significance. His metaphysical glorification of music was due to the latter's unique ability to penetrate into the essence of man: the unconscious Will. Again, music is a universal language, for it represents the Will itself. The great German pessimist denied that there are *a priori* innate ideas in the human mind. However, he held that both space and time are the only *a priori* forms of all perceptions and that causality is the only category of the human understanding. The Kantian scheme is hereby reduced to only three elements: space, time, and causality. These are merely three functions of the human brain. The origin of the human intellect and its relationship to the universal Will are left unclear. The finite human mind perceives the phenomenal world as a representation of the Will-to-Live (a blind, irrational, struggling, and evil cosmic force).

Since the Will is evil, it necessarily follows that the universe and life are evil, too. Man is a product of, and totally caught up in, this evil world that manifests two basic drives: the unconscious instinct to survive and the conscious urge to sexuality. Schopenhauer held that the good conduct of each man results from the renunciation of life and the denial of the irrational Will: human life is the unsuccessful struggle against finitude, irrationality, and pervasive evil. Concerning ethics, his compassionate moral maxim was "Injure no one; on the contrary, help everyone as much as you can."[6]

Schopenhauer was influenced by Robert Chambers's *Vestiges of the Natural History of Creation* (1844, 6th ed. appeared in 1847), the argument of recapitulation (the parallelism between ontogenetic and phylogenetic development later elaborated by Ernst Haeckel), and the homologous structures of the vertebrates. His mature metaphysical system presented a thorough scheme of cosmic evolution. He held to planetary evolution, distinguishing among the following three major temporal divisions: inorganic (chemical), organic (plants and the nonhuman animals), and humanity. Interestingly enough, he saw the appearance of humanity as the final stage of planetary evolution, since he believed that without mankind the continuation of the world would be purposeless. His process philosophy of nature accepted

Cuvier's theory of geological catastrophism, supported the appearance of major saltatory mutations, offered a vitalistic explanation grounded in the pervasive activity of the unfolding but ever-unsatisfied Will, and claimed that cosmic evolution manifests blind purposiveness.

Concerning man, Schopenhauer claimed that the human phylum had a simian origin in the tropics of the Old World. He wrote that the first men were suddenly born from the chimpanzee in Africa and the orangutan in Asia.[7] He also taught that man is trapped in universal unhappiness because of the irrational change of things and the temporal finitude of his joys in general and his life in particular. Like Thomas Hobbes, he referred to man as *Homo homini lupus*, i.e. man is a wolf to man. As already noted, the human animal manifests a will-to-survive and a will-to-sex. For Schopenhauer, salvation consists in a total mystical rejection of the irrational nature of reality (a liberation from the Will itself into nothingness). Unfortunately, Schopenhauer is remembered less for his interpretation of cosmic evolution than for his ultra-pessimistic doctrine of the Will.[8]

NOTES AND SELECTED REFERENCES

[1]Cf. J. B. Lamarck, *Zoological Philosophy: An Exposition with Regard to the Natural History of Animals* (New York: Hafner, 1963), translated by Hugh Elliot.

[2]*Ibid.*, p. 113.

[3]*Ibid.*, pp. 169-173.

[4]Cf. Arthur Schopenhauer, *The World as Will and Representation* (New York: Dover, 2 vols., 1958), translated by E. F. J. Payne.

[5]*Ibid.*, esp. chapter thirty-nine, pp. 447-457.

[6]Cf. Arthur Schopenhauer, *On the Basis of Morality* (1841) (New York: Bobbs-Merrill, 1965, p. 69).

[7]*The World as Will and Representation*, vol. 2, p. 312.

[8]For an introduction to his philosophy see Patrick Gardiner, *Schopenhauer* (Baltimore, Maryland: Penguin Books, 1963) and V. J. McGill, *Schopenhauer: Pessimist and Pagan* (New York: Brentano's Publishers, 1931).

Chapter 6.

CHAMBERS AND GOSSE

Between the publication of Lamarck's *Zoological Philosophy* (1809) and the appearance of Darwin's *On the Origin of Species* (1859), at least two volumes of note treated the concept of organic history in an interesting as well as amusing fashion. The two authors were Robert Chambers (1802-1871) and Philip Gosse (1810-1888), whose views were diametrically opposed to each other: Chambers was a theistic evolutionist while Gosse was a fundamentalist creationist. The lasting intellectual merit of their opinions resides only in that they clearly demonstrated that the idea of evolution was in the air before the scientific writings of Darwin, Huxley, Haeckel, Kovalevskii, and Gray were seen in print.

In his book entitled *Vestiges of the Natural History of Creation* (published anonymously in 1844, with a sequel called "Explanations" added in 1846), Robert Chambers championed a vitalistic and deistic theory of adaptive organic transformism and cosmic development.[1] The successful Edinburgh publisher and prolific author had acquired an interest in geology and biology, became convinced that plants and animals must have evolved from common ancestors throughout the eons of life on earth, and wrote the volume to express his unorthodox interpretation of a process universe and the progressive advancement of living things on our planet. His work was widely and avidly read by the general public but, at the same time, was immediately rejected outright by such learned critics as Agassiz, Herschel, Sedgwick, Silliman, and Whewell (including the naturalists Huxley, Lyell, and Owen).

Although Darwin himself had correctly identified the unknown writer, the authorship was not certainly known until the twelfth and last edition appeared in 1884.

In *Vestiges*, Chambers writes, "The idea, then, which I form of the progress of organic life upon our earth—and the hypothesis is applicable to all similar theaters of vital being—is, that the simplest and most primitive type, under a law to which that of like-production is subordinate, gave birth to the type next above it, that this again produced the next higher, and so on to the very highest, the stages of advance being in all cases very small— namely, from one species only to another; so that the phenomenon has always been of a simple and modest character."[2] His rejection of divine special creations, acceptance of the principle of cosmic and terrestrial uniformity, and defense of the pervasive law of necessary and progressive development or evolution amounted to an attempt at reconciling traditional fideist theology with the emerging natural sciences (e.g. historical geology, paleontology, and comparative studies in embryology and zoology). Yet he gave no definitive scientific explanation for the mechanisms of bio- logical evolution or clear explication of the assumed design in cosmic unity and throughout organic history, grounded some- how in an essential law below and beyond gravitation and development that he purported to evidence a natural theology. Strangely enough, he dismissed the insightful writings of both Malthus and Lamarck.

Unfortunately, *Vestiges* is filled with errors of fact and interpre- tation; for example, Chambers claims that those crystalline designs of frosted vapor on a windowpane that are reminiscent of fern leaves illustrate the identity of inorganic and organic matter, lime laid on waste ground will spontaneously spring up as clover, plant forms are determined by the laws of electricity, oat plants cropped before maturity can metamorphose into rye plants while in the ground during a single winter season, and a fish might suddenly develop a reptile heart (or a reptile a mammalian one)! He also accepted as true the biological hypothesis of recapitula- tion and even the claims of comparative phrenology. Hence, it is not surprising that the scientific and philosophical communities at that time did not take his most general idea of transformism seriously. In short, despite his impressive style and good intentions, the author presented neither a rigorous nor a convincing process cosmology or theory of organic development on earth. Chambers died in 1871, the year that saw the publication of Darwin's *The Descent of Man*.

Of interest is the fact that not until the sixth and last edition of the *Origin* (1872) did the always honest and considerate Darwin write in his own "An Historical Sketch" the following sentence about Chambers's *Vestiges*: "In my opinion it has done excellent service in this country in calling attention to the subject, in removing prejudice, and in thus preparing the ground for the reception of analogous views."[3] Of course, by this date, the Darwinian theory of biological evolution was being accepted by a growing number of important scientists and natural philosophers as well as a few process theologians.

In light of modern science and critical reason, Robert Chambers's *Vestiges* is an intriguing but deficient, inconsistent, and inconsequential literary fossil from a gifted but amateur naturalist. Chambers's sincere enthusiasm for unorthodox speculation in organic history, more or less orthodox beliefs in theology, and perhaps even subtle but revered materialism prompted him to write such a synthesis of man, life, and the universe within the framework of ongoing cosmic development and biological evolution.

A fellow Scotsman and naturalist/journalist, Hugh Miller, Chambers's most damaging critic (not to forget Adam Sedgwick), had opposed the evolutionary movement in various works, including *The Old Red Sandstone* (1841), *The Footprints of the Creator* (1847), *The Testimony of the Rocks* (1857), and *Popular Geology* (1859).

Philip Gosse, a leading member of that most fundamentalist of all sects in Victorian England, the Plymouth Brethren, wrote a book entitled *Omphalos: An Attempt to Untie the Geological Knot* (1857).[4] It appeared in print just three years before the publication of Charles Darwin's major work on the theory of evolution. Gosse firmly believed in a literal interpretation of the story of divine creation as given by revelation in *Genesis* of the Old Testament in the Holy Scriptures and, at the same time, just as firmly accepted the impressive scientific evidence for planetary evolution as documented by the growing empirical facts in the geological and paleontological records of the middle of the last century. He hoped to resolve this glaring contradiction between creationist religion and evolutionist science in his unique interpretation of the earth, which would seemingly (at least for him) do justice to both conflicting conceptual worldviews.

Although Gosse accepted the biblical age of our planet as being merely a few thousand years old, he nevertheless also maintained that the earth had been divinely created as an already ongoing terrestrial world that contained a pattern of fossils in rock layers at the very moment of the act of creation to suggest both an extensive physical past and creative organic development before its actual origin in space and time. This interpretation allowed him simultaneously both to admit and dismiss all the geological, paleontological, and biological data for the enormous antiquity of the earth itself and the vast evolutionary history of life upon it.

As if this view were not a sufficiently strange form of natural theology, Gosse even went as far as to propose that all those fossils of so-called early life found in the alleged ancient rocks of our planet were purposefully placed in these geological strata by a personal God in a plausibly misleading sequence in order to deliberately test the faith of their discoverers or perhaps even deceive the faithful (more or less just as Adam's navel had been created by the same clever God to simulate a natural birth). For Gosse, there is less than meets the eye! In brief, he argued that naturalists are simply studying the divinely imprinted appearances of a seemingly precreational globe. From his ingenious compromise, but clearly erroneous perspective (a view contrary to the ideas of Hutton and Lyell as well as Lamarck and Darwin), Gosse claimed that there had been no gradual change of earth's surface over the supposed vast eons of planetary time and no slow evolution of plant and animal species throughout the equally assumed ages of organic history. Needless to say, the bizarre argument of *Omphalos* was not taken seriously and had no major effect on the scientists, theologians, or general public of that time. However, its position does surface now and then in the fundamentalist creationist literature of today.[5] Clearly, the whole desperate scheme remains as the residue of a pathetic attempt to synthesize the beliefs of an outmoded religious cosmogony with the advancements in the new empirical sciences.

As a respected specialist in marine biology, however, Philip Gosse is favorably remembered as the inventor of the aquarium and the author of numerous works on natural history (particularly those publications dealing with sea anemones and corals).[6]

NOTES AND SELECTED REFERENCES

[1]Cf. Robert Chambers, *Vestiges of the Natural History of Creation* (New York: Wiley and Putnam, 4th ed., 1846), including the sequel "Explanations" by the author.

[2]*Ibid.*, p. 170 (the chapter is entitled "Hypothesis of the Development of the Vegetable and Animal Kingdoms"). For further references to Robert Chambers, see the numerous citations in Charles Couston Gillispie, *Genesis and Geology: A Study in the Relationships of Scientific Thought, Natural Theology, and Social Opinion in Great Britain, 1790-1850* (New York: Harper Torchbooks, 1959, especially chapter six, pp. 149-183) as well as those in Henry Fairfield Osborn, *From the Greeks to Darwin: The Development of the Evolution Idea Through Twenty-Four Centuries* (New York: Charles Scribner's Sons, 2nd ed., 1929, pp. 16, 305-306, 312-316, 352).

[3]Charles Darwin, *The Origin of Species* (New York: Pelican Classics, 1968, p. 58), edited with an introduction by J. W. Burrow.

[4]Cf. Philip Henry Gosse, *Creation (Omphalos): An Attempt to Untie the Geological Knot* (London: John Van Voorst, 1857).

[5]Cf. Robert Price, "The Return of the Navel: The 'Omphalos' Argument in Contemporary Creationism" in *Creation/Evolution* (issue two), 1 (2): 26:33. Ably edited by Frederick Edwords, *Creation/Evolution* is the only scholarly journal devoted exclusively to the current controversy between fundamentalist creationism and scientific evolutionism, and all relevant issues and publications.

[6]Cf. Philip Henry Gosse, *The Aquarium: An Unveiling of the Wonders of the Deep Sea* (London: John Van Voorst, 1854).

Chapter 7

FEUERBACH AND WALLACE

F euerbach and Wallace were both interested in anthropology and accepted the theory of evolution. Yet, their final
views of humankind's place in natural history were diametrically
opposed: Feuerbach turned from theology to materialism and
advocated a philosophical anthropology grounded in science and
reason, while Wallace turned from naturalism to spiritualism and
(on that basis) argued for the uniqueness of the human species
within the cosmic scheme of things.

Ludwig Andreas Feuerbach (1804-1872) was a unique moralist,
philosopher, and theologian.[1] He was born in Landshut, Bavaria,
and studied theology at Heidelberg and later philosophy under
Hegel in Berlin. In 1825, a dissatisfaction with Hegelianism
turned Feuerbach's interest from theology to anthropology. In
1828, he studied the natural sciences at Erlangen. His unorthodox
lectures and writings became a symbol of liberal thought, and
they called for a new philosophy based on anthropology,
psychology, and physiology. This new philosophy would provide
the foundation for a materialist naturalism and a humanist ethics
as the secularized outgrowth of a humanized theology. For the
German philosopher, the essence of religion originates in man's
self-consciousness of the infinite, and God is merely the outward
projection of man's finite inward nature (views later held by
Sigmund Freud).

Feuerbach wrote that to think one must first eat: thereby clearly
pointing out that in material nature ontology precedes epistemology (evolving man as the knower and doer is a newcomer

totally within the dynamic reality of the physical universe). In brief, our cosmos precedes and is independent of human experience.

Rejecting the Hegelian synthesis of pantheist idealism, Feuerbach's philosophy of the future advocates a materialist metaphysics, sensationalist epistemology, and humanist ethics. Within this view, man achieves his essence only in community with others: the social unity of man with man is nothing but the unity of I and Thou grounded especially in the emotion of love (refer to Martin Buber's I-It and I-Thou forms of dialogue).

In *The Essence of Christianity* (1841), Feuerbach adopted a materialist position.[2] He saw theology as a mystical dream, a perverted anthropology, an imaginary psychology, and an esoteric pathology. He considered it to be responsible for the dichotomy between God (who is infinite, perfect, eternal, omnipotent, and holy) and man (who is finite, imperfect, temporal, weak, and sinful). As a result, he advocated that this false theology should be replaced by a true anthropology.

Feuerbach taught that the theological beliefs in a personal God, the personal immortality of the human soul, the freedom of the human will, Heaven, the Incarnation and Resurrection, miracles, prayer, and revelation are all merely the realizations of human wishes of the heart: the essence of faith is the idea that what man wishes actually is. Since theological beliefs are the products of man's subjective nature, religion is ultimately grounded in human feelings and emotions rather than metaphysical correlates. There is no direct human experience of the supernatural God through sense perception, and there is not any experience that would indicate His existence. In short, theology leads to the ontological objectification of the wishful ideas generated by the deep psychosocial needs born in the imagination of the heart.

For Feuerbach, the objective physical world in process as a differentiated whole of heterogeneous matter is the essential starting point for an empirically sound and rationally valid interpretation of humanity in reality. As a straightforward naturalist and explicit materialist, he clearly taught the great geological age of our planet, vast environmental changes of the globe throughout earth history, evolutionary origin of organic life from inorganic matter, extinctions of various plants and animals in the remote past, a strictly scientific account of the emergence of new flora and fauna (including the human species), and the importance of accepting the cosmic perspective and its far-reaching philosophical consequences. The insightful German thinker flatly and rigorously rejected all teleological and theological explanations of man, life, and nature.[3]

In *Principles of the Philosophy of the Future* (1843), Feuerbach advocated a philosophical anthropology that was fundamentally antitheological.[4] Not beyond controversy, he claimed that the idea of a personal God is a deceit perpetuated by the privileged class to counteract the struggle of the underprivileged to liberate and emancipate themselves. Similarly, theology is anthropomorphic, for God's attributes are merely the projection of man's finite qualities. As such, man is clearly the beginning, middle, and end. of religious discourse, man alone created religious language, and he alone uses it and is the object to which it refers. In other words, the denial of the metaphysical correlate of faith is grounded in Feuerbach's principle of wishful thinking, i.e. religious objects and events are merely the projection and objectification of psychology or the creative imagination: religion is anthropocentric, anthropomorphic, and grounded in the human heart (not in reason).

Unfortunately, Feuerbach's interpretation of man and nature gave a passive view of human individuals while retaining a relational concept of God as the interrelationships among persons in a loving collective. It lacked an objective, critical concern for historical and social activity and failed to emphasize that the religious feeling of individuals is a sociohistorical product. Nevertheless, Feuerbach does represent a chronological and philosophical link between Hegel and Marx.

As the founder of scientific philosophical anthropology, Ludwig Feuerbach inadvertently prepared the German intellectual atmosphere for the advent of the theory of evolution as taught by the natural scientist Ernst Haeckel and echoed in the poetic philosophy of Friedrich Neitzsche.

Alfred Russel Wallace (1823-1913), an outstanding naturalist in his own right, did research in the Amazon (1848-1852) and the

Malay Archipelago (1854-1862): "Wallace's Line" (1863) refers to a zoogeographical boundary, separating Indian and Australian floral and faunal regions, that passes through the Malay Archipelago.[5]

There are astonishing parallels between Darwin and Wallace: each was an English naturalist who had explored South America with interests in both geology and biology, visited an archipelago (Galapagos and Malay, respectively), studied orchids and collected beetles, revered life while doing original research (e.g. the barnacle and butterfly, respectively), read Humboldt and Lyell as well as Malthus, enjoyed reflecting in solitude, accepted the theory of biological evolution as a true framework, and independently founded the explanatory principle of natural selection to account for the origin and divergence of new species throughout organic history.[6] Although both held natural selection to be the primary mechanism of biological evolution, Darwin grounded his theory in mechanist materialism while Wallace eventually turned to spiritualism and argued for the essential uniqueness of the human being. He ultimately rejected Darwin's rigorous naturalism, holding that the appearance of man on earth required a supernatural interpretation to give an adequate account of man's essence.

Wallace had first presented a theory of evolution in two published works, *The Sarawak Law* (1855) and the *Ternate Essay* (1858).[7] Darwin had merely written an abstract in 1842 and a manuscript in 1844: *On the Origin of Species* (1859) appeared only after the Darwin/Wallace positions had been read at the Linnean Society meeting in London on the evening of 1 July 1858.

Wallace maintained that there were distinct but successive stages in cosmic evolution, i.e. inorganic, organic, conscious, and spiritual levels (matter is ultimately will-force). He also claimed that Darwinism is necessary but not sufficient to account for biological evolution in general and the emergence of the human species in particular. For Wallace, rigorous naturalism is replaced by a spiritualist interpretation of the evolving universe and the emergence of humankind.

To be more specific, Wallace as philosopher did not believe that man is totally the product of material evolution. In fact, he clung to a more or less religious interpretation of human existence. He emphasized the biological direction toward greater complexity (essentially, his writings advocated a spiritualist monism). He firmly argued that man's mental abilities are superior to his needs for mere survival, and therefore a spiritualist explanation is necessary to account for this discrepancy. Elaborating his argument, he further claimed that man's mathematical, musical, artistic, and metaphysical faculties (as well as his speech, wit, humor, morality, naked and sensitive skin, and specialized and perfected brain, hands, and feet) could not have progressively developed in continuity by degree from animals as a result of natural selection or the ability of the fittest to survive and reproduce (differential adaptation). He held that man differs significantly in kind, not merely in degree, from the great apes (contrast with Huxley, Haeckel, and Darwin). Finally, Wallace believed in the personal immortality of the human soul. For him, man's superior mental abilities allegedly demonstrate the existence of a Supreme Mind or higher intelligences guiding human evolution to its end.

In *Man's Place in the Universe* (1903), considering exobiology, Wallace as spiritualist claimed dogmatically that unique humankind on earth must be absolutely alone in the great diversity of a limited universe.[8] In glaring contrast, the contemporary naturalist and evolutionist Ernst Haeckel totally rejected such an anthropic perspective as being merely the unfortunate philosophical residue of an outmoded medieval theology.

Wallace explicitly held to a religious orientation. He defended Darwinism but maintained that it is not sufficient to interpret cosmic evolution. In *The World of Life* (1910), he clearly replaced naturalism with vitalism, teleology, and spiritualism.[9] Beyond all the phenomena, causes, and laws of nature there is Mind and Purpose: the ultimate purpose being the further development of mankind for an enduring spiritual existence. This universe is the best of all possible worlds calculated to bring about this predestined end (compare with Bruno and Leibniz). Despite Huxley's agnosticism and Darwin's religious doubts, Wallace advocated theism: the infinite Deity designed the whole cosmos and is the acting power within all life. The evolving world of matter is altogether subordinate to an unseen universe of Spirit. Finally, the universe is the Will of higher intelligences or of one Supreme Intelligence.

In brief, Alfred Russel Wallace's views are personal impressions grounded in religious beliefs rather than rigorous philosophical arguments grounded in empirical evidence and logical procedure. His lasting contributions are the cofounding of the principle of natural selection and his sincere defense of biological evolution (particularly as presented in his early scientific writings).[10]

NOTES AND SELECTED REFERENCES

[1]Cf. Frederick Engels, *Ludwig Feuerbach and the Outcome of Classical German Philosophy* (New York: International Publishers, 1941).

[2]Cf. Ludwig Feuerbach, *The Essence of Christianity* (New York: Harper Torchbooks, 1957) translated from the German by George Eliot. Also refer to Ludwig Feuerbach, *Lectures on the Essence of Religion* (New York: Harper Torchbooks, 1967) translated by Ralph Manheim.

[3]Cf. Marx W. Wartofsky, *Feuerbach* (Cambridge: Cambridge University Press, 1977, esp. pp. 397-401).

[4]Cf. Ludwig Feuerbach, *Principles of the Philosophy of the Future* (New York: Bobbs-Merrill, 1966) translated by Manfred H. Vogel.

[5]Cf. Alfred Russel Wallace, *The Malay Archipelago: The Land of the Orangutan and the Bird of Paradise, A Narrative of Travel with Studies of Man and Nature* (New York: Dover, 1962). This volume first appeared in 1869, and the work was dedicated to Charles Darwin.

[6]Cf. Alfred Russel Wallace, *Contributions to the Theory of Natural Selection: A Series of Essays* (New York: Macmillan, 1870). Also refer to Barbara G. Beddall, ed., *Wallace and Bates in the Tropics: An Introduction to the Theory of Natural Selection* (London: Macmillan, 1969).

[7]Cf. Arnold C. Brackman, *A Delicate Arrangement: The Strange Case of Charles Darwin and Alfred Russel Wallace* (New York: Times Books, 1980).

[8]Cf. Stephen Jay Gould, "Mind and Supermind" in *Natural History* (May, 1983), 92(5): 34, 36, 38. Wallace's volume was devoted to a study of the results of scientific research in relation to the unity or plurality of worlds.

[9]Cf. Alfred Russel Wallace, *The World of Life: A Manifestation of Creative Power, Directive Mind and Ultimate Purpose* (New York: Moffat/Yard, 1910).

[10]Cf. Alfred Russel Wallace, *My Life: A Record of Events and Opinions* (New York: Dodd/Mead, 2 vols., 1905, esp. vol. 2., chapter twenty-five on Darwin, pp. 1-22). Also refer to Wilma George, *Biologist Philosopher: A Study of the Life and Writings of Alfred Russel Wallace* (New York: Abelard-Schuman, 1964, esp. pp. 64-74 on natural selection and man, pp. 251-262 on Darwinism, and pp. 274-278 concerning life on other worlds).

Chapter 8

DARWIN

I n the evolution of the idea of evolution during the last century, Charles Robert Darwin (1809-1882) is undoubtedly the key figure among the intellectual giants of the natural sciences.[1] He boldly asked a crucial question: How do new plant and animal species appear within organic history? His answer, supported primarily by massive empirical evidence and unique perceptual experiences as well as intuitive insights and rational reflections, brought about a scientific revolution that gravely challenged the traditional Judeo-Christian static worldview and influenced forever our interpretation of earth history. His remarkable glimpse of process reality, which focused on rocks and fossils as well as floral and faunal populations, now encompasses everything from the origin of sidereal galaxies to the development of human ethics. It is to his lasting credit that modern biology rests upon a comprehensive and intelligible view of life within an evolutionary framework.

Charles Darwin was born at The Mount in Shrewsbury, England, on 12 February 1809; he was the fifth of six children. Like Aristotle, whom he admired, Charles was raised in a wealthy family with a line of physicians. Dr. Robert Waring Darwin, his atheist father, had a domineering character that molded the psychological development of the inquisitive youth. Interestingly enough, he seems never to have been greatly influenced by his siblings, although he shared a teenage interest in chemistry with his older brother Erasmus (who later accompanied Charles at the University of Edinburgh to study medicine).

Fortunately, Charles was born into both prosperity and social prestige (in sharp contrast to the biologists Wallace, Huxley, and Mendel). His paternal grandfather and father were both success-ful medical doctors with lucrative practices, which allowed for educational advantages and experiential opportunities that would have otherwise been closed to the young enthusiast of unspoiled nature.

Despite the pervasive conservativism of Victorian England, his upbringing was not religious (his tolerant father kept his own atheism silently to himself). He was a very ordinary boy, presumably somewhat below the common standard of intellect; there was no indication of his latent genius. As he developed, Charles enjoyed the beauties of the quaint Shropshire countryside. Not dedicated to book learning, he was a mediocre student. Instead, he was to acquire a lifelong dedication to natural history.

Charles had an insatiable curiosity about the objects of nature. At the age of ten, he exhibited an interest in insects (especially beetles as well as moths and butterflies). His early concern for entomology would soon extend to include historical geology and marine biology. He became an obsessive collector, which proved to be a significant asset in later years when the descriptive analysis of empirical evidence played a crucial role in his supporting his own scientific theory of biological evolution (descent with modification).

Erasmus Darwin (1731-1802) was a versatile genius who, besides his medical career, found time to write poetry and speculate on nature.[2] In his book *Zoonomia* (1796), the naturalist anticipated the scientific theory of biological evolution: he wrote that irritability or sensibility pervades the organic world, and all life had its origin from a single primordial filament that existed millions of years ago on the surface of our earth. As such, Charles's curiosity and intellectual propensity clearly owed more to his paternal grandfather than to his living father.

The natural philosopher and invertebrate specialist Lamarck was the first thinker to write a volume solely for the purpose of presenting a speculative theory of biological evolution as it applies to the organic history of the animal kingdom.[3] Entitled *Zoological Philosophy* (1809), it appeared in the year of Darwin's birth and exactly fifty years before his publication of *On the Origin of Species* (1859). Lamarck's controversial arguments for organic evolution were metaphysical rather than empirical and, as a result, went unaccepted and unappreciated by the naturalists of his time. Instead, the biologists supported divine intervention, special creations, forms of vitalism, or spontaneous generation.

Charles progressed in his passion for entomology and more and more turned his interest to natural history. He collected rocks, minerals, fossils, shells, plants and insects (particularly beetles) and spent time catching rats and shooting birds for sport, e.g. pheasants and partridges (taking the time to study their habits and eggs). He relished hunting for foxes with dogs and taking solitary walks in North Wales, learned ornithology and taxidermy, and even took note of glacial striations and erratic boulders. In all this, he developed an intense devotion to the accuracy of naturalist study and scientific thought grounded in empirical facts, rigorous logic, and hypothetico-deductive inferences. In 1825, having finished school with no definite plans for a career, Darwin was sent by his father to the University of Edinburgh, Scotland, to study medicine. Not interested in committing himself to this prestigious profession, he often cut classes to attend lectures in natural history or participate in field trips. Several factors were decisive in destroying what slight medical ambition Charles had: he could not stand the sight of blood or bear the agony of having to watch a child patient undergoing the amputation of a limb without an anesthetic. To avoid such a ghastly circumstance, he fled from the operating room convinced he could never practice medicine. After less than two years at the university, he returned to The Mount; having no heart for medicine, his mind gravitated even more to natural history (especially geology).

Disappointed, Dr. Robert Waring Darwin now recommended that his son attend Christ's College, Cambridge, to prepare for the ministry. For four years (1828-1831), Charles studied theology as well as mathematics and the classics (for the most part, what today are the special sciences were, at that time, considered to be areas in natural philosophy relegated to private pursuits). Perhaps fortunately, his academic training excluded those subjects in which he would later excel and distinguish himself: geology, paleontology, taxonomy, botany, zoology, and comparative anatomy/physiology, as well as entomology and community ecology. He never had laboratory training in either invertebrate or vertebrate morphology, although he achieved on his own an uncanny skill for analyzing and describing living things from apes to worms. He even acquired a taste for art, music, and poetry but always remained poor in languages, mathematics, and metaphysics.

While at Cambridge, Charles made important contacts with two prominent professors of science: the geologist Adam Sedgwick and the botanist John Stevens Henslow (both were religionists, the latter becoming a lifelong intimate friend). Numerous field trips with the two naturalists reinforced his love of nature. During these years of formal study, he was also influenced by five books: *The Elements* (c. 300 BC) by the mathematician Euclid, *Paradise Lost* (1667) by the poet John Milton, *Natural Theology: or, Evidences of the Existence and Attributes of the Deity, Collected from the Appearances of Nature* (1802) by the theologian William Paley, *Preliminary Discourse on the Study of Natural Philosophy* (1830) by the astronomer Sir John Herschel, and especially *Personal Narrative of Travels to the Equinoctial Regions of the New Continent* (1805-1834) by the naturalist Alexander von Humboldt; the great German was the father of scientific exploration as well as one of the pioneers in ecology, geography, and meteorology. Apparently Charles did not take the writings of Buffon, Diderot, Cuvier, Lamarck, and Saint-Hilaire seriously (he seems to have even ignored the ideas of his grandfather Erasmus Darwin and later those of Robert Chambers).

Finally, Charles received a Bachelor of Arts degree from Cambridge; he had been an undistinguished student at school, the university, and college. As a rural parson, when not reading the biblical story of creation as found in *Genesis* of the Old Testament in the Holy Scriptures or preaching the Christian beliefs of the Anglican Church from the pulpit to his devout parishioners, he imagined there would be leisure time to spend in the English countryside as an enthusiastic naturalist. The would-be minister could collect rocks and fossils, study beetles, and absorb natural history (he was already having religious doubts about points of traditional theology). However, the self-made scientist was never to become a clergyman; one wonders why he never pursued a legal career, which was the chosen occupation of some important nineteenth-century naturalists (e.g. Lyell and Morgan).

Darwin was shy, gentle, tolerant, and inquisitive and an admirable young man of patience and zeal, as well as emerging genius and wisdom. He had an insatiable curiosity and penetrating intellect. Yet, at the age of twenty-two, the graduate of Cambridge still had not decided on a career. He had no formal degree in geology or biology and, in fact, never taught those subjects at a college or university. As such, however, he was not indoctrinated into any of the scientific hypotheses of the time that dogmatically precluded the idea of evolution. For Darwin, experience was his best teacher and nature was the classroom. Also, from our perspective and in the light of our benefit of hindsight, many of the very instruments he used in his scientific investigations were faulty. To his immortal credit, however, he always remained open and hospitable to the theoretical implications and physical consequences of his own experiences and experiments as well as those of other naturalists.

During his life, Darwin was influenced by a wide variety of accidental incidents and ironic twists of fate. In retrospect, it was as if an unlikely series of improbable events was to inevitably guide him to a position aboard H.M.S. *Beagle* and, subsequently, to develop his theory of evolution and publish his major work on the subject.

Darwin, the disillusioned medical student and halfhearted theologian, was not to remain for long back at The Mount. On August 29, in a letter from Henslow dated 24 August 1831, he received an invitation to become both an unpaid naturalist and traveling companion aboard H.M.S. *Beagle* (actually, the young gentleman had been nominated for this dual position by George Peacock, a mathematician and astronomer at Cambridge; four individuals had already turned down the unique appointment). It would not be the last letter to suddenly change the course of his life. Since he still had no definite plans for a career, Darwin was elated at the prospect of taking a two- or three-year trip around the world, as it would provide enough opportunities to satisfy the most ardent scientist. Unfortunately, however, Dr. Robert Waring Darwin was reluctant to support this astonishing undertaking. Yet, should the now disappointed naturalist find someone whom his father respected to support the daring venture, he would be allowed to go.

When visiting Josiah Wedgewood II ("Uncle Jos") of the famous pottery family at Maer Hall, Darwin found a sympathetic friend who thought the global trip would be beneficial to the emerging scientist, although, in truth, perhaps Josiah Wedgewood had the fate of his two young daughters more at heart than Darwin's naturalist ambitions. Both returned to The Mount and persuaded its successful physician that this unusual invitation should be accepted and monies provided. In finally granting permission, the father was hoping that the voyage would mature his indecisive son (after all, Dr. Darwin would be financing the trip). The journey would prove to be a very worthwhile endeavor.

Having overcome his first hurdle, Charles encountered a second one: Captain Robert FitzRoy of H.M.S. *Beagle* (a military ship renovated for scientific surveys) was an aristocrat, conservative member of the Tory political party, outspoken fundamentalist creationist, and young eccentric with manic-depressive and suicidal tendencies. At first, FitzRoy R.N. was reluctant to allow Darwin to fill the position of companion/naturalist; he claimed that one could determine the mental qualities and character of a man by his outward appearance (especially facial features), and he disliked Darwin's nose. Nevertheless, after several days, the devotee of the fashionable alleged science of physiognomy consented to have the geologist aboard as his shipmate (as such, Charles overcame the second hurdle).

Actually, Captain FitzRoy's personal desire was to have a naturalist amass empirical evidence to scientifically support a literal interpretation of the story of creation as presented in *Genesis* of the Old Testament in the Holy Scriptures. Ironically enough, this task fell to the would-be minister but confirmed naturalist Charles Darwin, who viewed the upcoming worldwide voyage as an incredibly fortunate opportunity to further pursue his love of geology and various related sciences. The approaching journey of discovery through space and time would not only determine his own thoughts and writings but also represent a major turning point in the scientific ideas of western history. It would alter forever the view of the age of our planet, the origin and development of life on earth, and the antiquity of humankind in the world: rocks, fossils, and artifacts were becoming the subject matter of serious inquiry. The dogmatic adherence to the eternal fixity of species would be soon superceded through the growing awareness of the mutability of plant and animal forms throughout organic history.

Before the voyage, Charles Darwin was primarily an amateur geologist who accepted the then taught doctrine of the immutability of species and believed in the biblical account of the origin of things (although he never took religion seriously). After sailing around the world, he became a specialist in biology who developed an evolution theory of organic history and espoused agnosticism (if not silent atheism). This considerable shift in his intellectual outlook from amorphous opinions to reasoned convictions was due to three major influences: his careful reading of Sir Charles Lyell's three-volume tome *Principles of Geology* (1830-1833), which argued for the theory of uniformitarianism (the gradual but continuous alteration of the stratigraphic structures of the crust of our planet as a result of the same pervasive natural forces operating throughout earth history), the scientific circumnavigation of H.M.S. *Beagle* (especially its five-week visit to the Galapagos Islands during the fall of 1835), and the chance but beneficial reading of Thomas Robert Malthus's classic work *An Essay on the Principle of Population* (1798).

After several unsuccessful attempts to leave Plymouth harbor due to severe storms, as if nature did not wish to easily surrender the evolutionary evidence to the young scientist, the *Beagle* finally set sail on 27 December 1831 from England to the Canary Islands.[4] At this time, the energetic Darwin had no idea just how remarkably influential this unique adventure would be on his own view of life. He had looks, brains, and a pleasant personality as well as social status, financial security, and considerable good fortune. His powerful intellect and enlarged curiosity, unencumbered by theology or metaphysics, was free to keenly observe and meticulously describe the details of nature. With an open mind during the entire expedition, he would analyze and synthesize a vast range of facts and concepts conducive to bringing about a major revolution in science and natural philosophy within the evolutionary perspective.

In the four decades immediately preceding the voyage of the *Beagle*, the impact of scientific theories and empirical discoveries in both geology and paleontology seriously challenged the religious beliefs and social values of the Judeo-Christian tradition. An essential contribution to the shift in Charles Darwin's own conceptual view of earth history was his reading of Sir Charles Lyell's *Principles of Geology* (volume one had been given to the young naturalist by his friend Henslow, and the remaining two volumes were sent to him from England). Actually, this materialist theory of geological evolution had been first conceived by James Hutton (1726-1797) and published in his book *Theory of the Earth* (1795); it first appeared as an article entitled "Investigation of the Laws Observable in the Composition, Dissolution, and Restoration of Land Upon the Globe" in the transactions of the Royal Society of Edinburgh and, seven years later, as the two-volume work that marked the beginning of dynamic concepts in historical geology.

Hutton had appealed directly to nature for an understanding of geological history in terms of rational thoughts and natural forces.[5] With a sweeping imagination, the Scottish theorist and physician offered the earliest comprehensive treatise providing a geological synthesis: the Huttonian worldview taught the endless, gradual, and imperceptible cycles of rising and falling continents as well as the antiquity of life on earth. This Vulcanist espoused the ongoing uniformity of geological dynamics, although they vary in intensity from age to age. His lofty vision maintained the uninterrupted continuity of past and present natural causes and effects throughout everlasting time and cyclic change, especially the force of internal heat from a central subsurface source (not until 1912 would the German geophysicist Alfred Wegener father the scientific theory of plate tectonics).

Hutton's thoughts represented a process view in natural philosophy supporting a world-machine interpretation of planetary development: our earth is a terrestrial system subject to physical laws and historical alterations and ultimately exists for the benefit of humankind. The fossils in rock strata are the remains of life forms in previous ages deposited in sediments over vast periods of time. Summarizing, subterranean heat has slowly but repeatedly transformed the surface of our planet (the principle of decay) resulting in a continuous and unending series of new worlds (the principle of renovation) during the eternality of our earth (the principle of endurance). In contemplating earth history, the great geological theorist found no vestige of a beginning and anticipated no prospect of an end.

Hutton did not openly deny the teachings of the Sacred Scriptures or, on the other hand, advocate atheism. Clearly, he did not support a literal belief in the biblical account of divine creation; for him, there was simply no empirical evidence that demonstrated the supernatural origin of our planet earth. His materialistic interpretation of world history in the emerging science of geology dealt a fatal blow to the fundamentalist creationist belief in the rigid acceptance of the Mosaic cosmogony as described in *Genesis* of the Holy Bible. As such, he instilled an attitude conducive to the further advancement of both descriptive and historical geology. Two other significant books in this new area of study were William Smith's *Stratigraphical System of Organized Fossils* (1817) and Leopold von Buch's *Description of the Canary Islands* (1825).

In his two works, *Illustrations of the Huttonian Theory of the Earth* (1802) and *Outlines of Natural Philosophy* (1812), John Playfair, the Vulcanist and mathematician at the University of Edinburgh, rigorously popularized the Huttonian worldview, which boldly opposed religious dogmatism: the traditional six days of theistic creation as believed by the majority were challenged by a naturalist theory supporting millions of years of earth history as documented by rock strata and fossil evidence, which convinced only a minority (Hutton's geologic uniformitarianism with its endless time and mechanistic operations left no room for spiritual intervention).

In Scotland, with ironic appropriateness, Sir Charles Lyell (1797-1875) was born in the year Hutton died.[6] Although he also had an interest in zoology, Lyell became the distinguished founder of modern historical geology. In his *Principles of Geology* (which rejected outright mystery, cataclysms, supernatural violence, and theological fantasies), he provided empirical evidence as best he could at the time to substantiate the theory of uniformitarianism: the slow and continuous evolution of geological structures on the surface of our earth over vast periods of time due to the action of constant and existing physical causes of change within the planet itself. Eventually to Darwin, the Hutton-Lyell viewpoint suggested the mutability of species throughout organic history. Likewise, a growing awareness of changes in the starry heavens above resulted in some naturalists considering the probability of the flux of life on the earth below (fossils in rock strata were now recognized to be traces of prehistoric life). The deeper scientists dug into the crust of our earth and the farther they searched into the depths of outer space, the more apparent it became that cosmic history spans across millions if not even billions of years.[7]

Darwin became a thoroughgoing uniformitarian or actualist who overcame the orthodox geological creed of global catastrophism and, later, even the biological beliefs in essentialism and progressive creationism: Cuvier, Sedgwick, Henslow, Blyth, Owen, Buckland, and Agassiz were among those who resisted the emerging evolutionary framework. Ironically, although he had a profound influence on Darwin, Lyell never became a thoroughgoing evolutionist; he did not take the insights of Lamarck, Chambers, or even Darwin seriously (his reluctance was grounded in a mental residue of the man-centered geotheological drama popular among some naturalists of the last century, who saw no evidence for a materialist explanation for the origin of new plant and animal species).

The ensuing debate in France between Georges Cuvier the theistic catastrophist and defender of special creations and Geoffrey Saint-Hilaire the deistic protoevolutionist foreshadowed the bitter controversy that would later emerge between rigidly closed fundamentalists and scientifically open evolutionists. The Lyellian position provided Darwin with a vast temporal framework of ongoing change and development that suggested, if only unconsciously, the mutability of species (even Immanuel Kant, whom Darwin read without benefit, suggested that the earth could be millions of years old).

Unlike FitzRoy, Darwin acknowledged the awesome age of our planet. He realized that the geology of the earth is of immense duration: this planet is not a solid orb of static elements but, instead, is a world of ceaseless change both violent and gradual. This radically new conception of time and events implied that floral and faunal organisms could also be gradually and continuously transformed through countless epochs due to natural causes discoverable by the human intellect. A dynamic explanation for biological history was clearly forthcoming.

After a two-month delay, the *Beagle* had sailed from Plymouth Sound with about seventy crew members along with Darwin and FitzRoy. This military ship had been overhauled for use as a scientific survey vessel. The numerous practical purposes of the unexpected five-year global voyage (1831-1836) included: charting the extensive coastlines of South America, including Patagonia, Tierra del Fuego, and more especially the Straits of Magellan; taking ocean soundings and getting more accurate longitudinal measurements via worldwide chronological reckonings; and both surveying and mapping sea currents as well as the rivers of numerous volcanic islands and continents. These undertakings were sponsored by the British Admiralty in order to improve navigational maps for obvious economic and mercantile reasons.

Captain Robert FitzRoy (1805-1865), age twenty-five and only three years older than his companion/naturalist, looked forward to achieving a splendid survey of the southern hemisphere (the previous captain of the *Beagle* had committed suicide, an ironic coincidence). Only twenty-two and largely self-taught, Darwin could not envision how this lengthy trip aboard a floating microcosm would radically alter both his view of things and the development of natural science.

Through various circumstances, Charles Darwin had managed to attain this unique opportunity to expand his inquisitive intellect. In a brief time, he was transformed from an English gentleman in proper attire (suit with double-breasted waistcoat and tails, high-collared shirt and cravat, top hat and an elegant topcoat) walking the deck of the ship with a marine collecting net or telescope in his hand to an explorer riding horseback across the extensive pampas of South America sharing an evening campfire with rugged gauchos or, later, an excited naturalist astride the back of a giant tortoise on the Galapagos Islands in the Pacific Ocean. Deep under a projecting forehead with its bushy eyebrows, Darwin's clear and alert gray-blue eyes steadily surveyed all of nature with incredibly disciplined observational powers. He gave steadfast, patient, and practical attention to details in observing, analyzing, and describing the objects of the world. The inexperienced naturalist had an uncanny ability to determine which geological features and biological characteristics are of scientific value in terms of crucial inferences for understanding and appreciating earth history. One marvels at his stamina, humanism, and scientific perceptiveness. The crew referred to him as "the dear old philosopher" (he himself greatly venerated Aristotle),[8] and, despite his constant seasickness, he exuded determination and energy.

The amateur scientist came prepared with the following: a compass, magnifying glass, geological hammer, dissecting instruments, binoculars, telescope, notebooks, jars for preserving specimens, and even a case of pistols. He quickly became aware of the stupendous dimensions of our planet in terms of space, time, and change. His unending discoveries were made against a panorama of endless ocean, immense time, and awesome diversity.

Darwin seems to have been interested in everything that moves in water, crawls on land, and flies in the air. His interests in geology encompassed icebergs and glaciers, atolls, coral reefs, fringing reefs, barrier reefs, rocks and minerals, fossils, and even volcanic ash. He collected and examined invertebrates, marine life (especially luminous fish), insects, amphibians, reptiles, birds, mammals (including vampire bats), and various plants as well as exotic seeds. With great zeal the industrious and curious young naturalist found time to explore the jungles of Brazil, cross the windswept grasslands of Patagonia, climb the impressive Andes of Chile, and traverse volcanic islands with their unusual flora and fauna. He was not one to pass up the opportunity to widen his grasp of nature, and his great efforts were not doomed to oblivion.

Darwin became aware of the common ancestoral origin of differentiated but similar varieties inhabiting ocean islands and a nearby continent, as demonstrated in the isolated populations of the Cape Verde Islands off the coast of Africa as well as, later, the independently varying plants and animals among the Galapagos Islands off the coast of South America. The geographical distribution of representative types (biogeography) would, in retrospect, argue for the adaptive radiation of life from a common source in organic history.

Darwin loved the biological exuberance of the South American tropics; in the jungle, his inquisitive mind was a chaos of excitement and delight. He was enthralled with the novelty and complexity in the teeming life of the Brazilian rain-forest (particularly the hummingbirds and beetles), the richest area of organic forms on earth, and became increasingly aware of the important roles of camouflage and mimicry in the living world. Likewise, he developed a lasting and deep respect for life. He abhorred slavery, vivisection, and cruelty; he opposed violence, barbarity, ideological controversy, and the infliction of physical suffering of any sort (as such, he found naval discipline distasteful). In fact, he even lost his earlier passion for hunting as a sport.

Free from the intellectual confines of Victorian England, Darwin was able to reflect on the empirical evidence he was amassing and its implications; he was continuously sending unique data back to Henslow at Cambridge. In sharp contrast, as a fundamentalist creationist, FitzRoy had his own personal religious motive for bringing along a gentleman naturalist: Darwin was to collect empirical evidence to scientifically substantiate as true a strict and literal interpretation of the story of creation as offered in the book of *Genesis* in the Holy Bible. Unable to predict the ironic consequences, the captain had given this task to the would-be minister from Cambridge.

In South America, Darwin became aware of the startling resemblance between the fossil remains of certain extinct animals preserved in rock strata and those living species now surviving on the surface of the earth. To the scientist, the evidence suggested the historical continuity and biological unity of all life on earth. With enlarged curiosity, Darwin unearthed the giant fossil remains of prehistoric mammals from the cliffs of Punta Alta in Argentina. His two visits to this paleontological site resulted in his discovering numerous specimens: *Glyptodont, Macrauchenia, Mastodon, Megalonyx, Megatherium, Mylodon, Scelidotherium,* and *Toxodon.* For Darwin, these huge fossils represented an impressive and undeniable record of previous existence; they shed light on the appearance and disappearance of life on earth. There was a "secret" held in these remains of organic creatures from the remote past, and the ship's naturalist rejected their being interpreted as representing antediluvian quadrupeds that had been suddenly drowned as a result of the "Noachian Deluge" less than 6,000 years ago (a position religiously defended by the intellectually myopic FitzRoy).

Darwin must have wondered about the origin, survival, adaptation, distribution, history, and extinction of these enormous animals of distant eons. Gradually, the geobiologist realized that our planet is a grand museum containing an incomplete record of the history of life on earth. Even the sea-worn pebbles strewn across the pampas of Argentina suggested the geological activity of prehistoric times. But the arid pampas were now the land of rodents, ostriches, guanacos, and condors where once mighty dinosaurs had freely roamed. FitzRoy was displeased and dismayed at Darwin's interpretation of rocks and fossils, which challenged the biblical account of earth history. As the months passed, the captain's growing religious fundamentalism became firmer. Giving preference to science, however, the naturalist's immersion into earth history seriously questioned the divine creation and worldwide flood of biblical belief. Inevitably, the viewpoints of Darwin and FitzRoy became diametrically opposed, and a personal conflict ensued. One may even argue that the intensifying war between science and religion was personified by Darwin and FitzRoy, respectively.

FitzRoy believed the divine creation of our material planet to have occurred in the year 4004 BC (a figure that had been determined by Archbishop Ussher, an Anglican prelate in seventeenth-century Ireland). In sharp contrast, however, Darwin now thought that the world could be millions of years old. As a result of this new perspective, he asked, is there throughout organic history a biological series of changes between the past forms of life and those plants and animals now inhabiting our earth? The growing evidence was clearly suggesting a succession of floral and faunal types from epoch to epoch without any sudden discontinuities or supernatural interventions.

After a brief visit to the Falkland Islands, in January 1833 the *Beagle* set sail for the bleak world of Tierra del Fuego at Cape Horn of the southern tip of South America. Darwin considered those naked and cannibalistic natives who barely eked out an existence in this desolate environment of ice, snow, harsh winds, and freezing waters to be the most savage and primitive people on earth. The naturalist claimed that the paleolithic and nonliterate Fuegians represented human degradation; he argued that they demonstrated how wide the distance is between savage natives and civilized men and wondered why humans would live under such inhospitable conditions (on at least this point, Darwin was less an anthropologist than Alfred Russel Wallace, who also held to the psychic unity of mankind but took a deeper interest in those sociocultural differences among distinct human populations).

An overzealous evangelical fundamentalist, FitzRoy had hoped to convert the Fuegian bands to Christianity and thereby start a religious community of civilized natives among the so-called backward peoples of Tierra del Fuego. The captain was returning to this rugged world three Fuegians whom he had already taken to England in 1829 to be civilized in the beliefs, practices, and values of modern Christianity. These three passengers were York Minster, Jemmy Button, and Fuegia Basket. Unfortunately, in a short period of time after they were returned to their homeland, these three hostages reverted to their primitive ways. As a result, the captain's religious plan had failed miserably, and he was understandably depressed. Unlike FitzRoy, Darwin had a more humane tolerance for the wide range of sociocultural thoughts and behavior patterns he experienced while traveling around the world.

During the stay at Cape Horn, special attention was given to surveying the Straits of Magellan; Darwin was awed by the impressive icebergs and massive glaciers of this foreboding area of the world. After overcoming a vicious storm while rounding the tip of South America, the scientific vessel next headed to Chile (Darwin was still a victim of unending seasickness).

Reminiscent of Leonardo da Vinci in the Alps, Charles Darwin was easily persuaded to ride muleback into the Andes. While crossing this magnificent mountain range 13,000 feet above sea level, the naturalist developed a strange assemblage of ideas: geological gradualism and the web of life, along with his growing awareness of the enormous periods of time in earth history and the struggle for existence in the living world, clearly suggested to the young scientist a dynamic worldview that radically departed from the biblical story of a recent divine creation. In brief, to the immensity of space revealed by astronomy was now being added the immensity of time revealed by geopaleontology. For Darwin, this grasp of rock and life was intensified when he discovered petrified conifer trees *in situ* and fossil seashells in the peaks of the Andes. Without doubt, natural forces had slowly and continuously elevated this range of mountains from beneath an ocean of prehistoric times. Such an astonishing geological change on the surface of our malleable planet gave the needed empirical evidence to support Lyell's scientific theory of uniformitarianism and perhaps even implied to the acute naturalist (if only unconsciously) the transformation of living things over vast periods of time.

Returning from the Andes, Darwin became seriously ill due to the bite of a Benchuca bug, a large black insect of the pampas that carries Chagas' disease. It may have been this particular sickness that undermined the naturalist's health throughout the remainder of his life.

At Valdivia and later at the towns of Concepción and Talcuhano as well as the surrounding areas, Darwin saw the devastating effects as a result of a series of violent earthquakes. He also became more aware of the destructive forces of hurricanes, volcanic eruptions, and tidal waves. It was now very obvious to the geobiologist that our world is not solid and secure but always subject to physical changes, whether slight or major.

After a brief stay in Peru, the *Beagle* set sail for the Galapagos Islands. One wonders what ideas and visions must have passed through Darwin's imagination during those many hours he paced the deck of the small survey ship as it crossed the expansive ocean beneath a night sky of bright stars in the seemingly fixed heavens; with a touch of irony, the Southern Cross dominated the horizon.

The remote and primeval Galapagos Islands, also known as the Encantadas or the Enchanted Isles, appear to be removed from both the ravages of time and destructive encroachment of human civilization.[9] They take their name from the Spanish word *galapagos* for giant tortoises, those huge reptiles that once thrived throughout this extraordinary archipelago and that reminded Darwin of antediluvian monsters or creatures from another planet. The archipelago, isolated from the South American mainland, consists of fifteen major islands along with numerous islets, uplifts, and reefs located on the equator about 600 miles west of Ecuador in the Pacific Ocean. Sharks, swift currents, and mild winds isolate life among these islands. Darwin's eerie "cradle of life" appeared as an inhospitable world of lava flows and desolate vistas: a cluster of igneous islands looking like a group of enormous chunks of lunar landscape haphazardly floating upon the swift Humboldt Current. When surrounded by mists and low clouds, these deceptively foreboding islands present a vivid scene reminiscent of our reconstruction of prehistoric times. In fact, this archipelago actually represents a delightful anachronism: present-day life forms eking out an existence in a pristine environment. Fortunately, the inhabitants enjoy a subtropical climate that is cool, due to fresh trade winds and several capricious ocean currents (especially the Humboldt Current).

The static first appearance of the Galapagos Islands is an illusion. An extremely strange but special place on the earth, this slowly changing archipelago testifies to the pervasive influences of evolutionary forces. This area of our planet is a geological hot spot of frequent and intensive volcanic eruptions reflecting the restless and convulsive entrails of the earth: one sees collapsed cliffs, craters, and calderas everywhere. About fifteen years from now, wind erosion should finally cause the famous and official geological sentinel of these islands (Pinnacle Rock of Bartolome Bay, which rises 200 feet above sea level) to topple back into the ocean, which waits to once again reclaim its own. In fact, this entire archipelago is almost imperceptibly drifting eastward on the Nazca Plate due to tectonic movement. The eastern uplift islands are oldest and may sink back into the sea, while the western volcanic islands may suddenly explode. Yet, further geological building is likely to result in the eventual formation of new islands. The ages of these islands range from 600,000 to at least three million years (the discovery of fossil invertebrates and fishes in uplifted marine limestone layers between lava flows allows geologists to date the ancient strata of these rocks). Although creatures differed distinctly from island to island throughout this geologically recent archipelago, they bore a marked resemblance to species found on the vastly older South American mainland.

At the end of summer in 1835, the H.M.S. *Beagle* reached the Galapagos Islands; it was exactly 300 years after their first discovery by man! From September 15 to October 20, as the ship sailed from island to island, Darwin explored and familiarized himself with this puzzling world of unusual geological structures as well as unique plants, peculiar birds, and fascinating animals. He especially marvelled at the giant tortoises, land and marine iguanas, and those intriguing varieties of ground finches that held the key to the understanding and appreciation of that "mystery of mysteries": the origin of new species on our earth. No doubt, Darwin's philosophical curiosity and scientific exactitude were put to the test. Actually, his visit to this unique archipelago proved to be the most significant event of the entire trip: his conceptions of space and time were abruptly altered, and his puzzles and doubts about divine creation were once again reawakened.

From the summit of a volcano on Bartolome Island, one may experience the breathtaking panoramic view of a moonlike landscape of craters and spatter cones against the blue ocean. Populations of surface colonizer plants and lava·cacti need only soil, moisture, and oxygen to survive in this rugged and demanding volcanic environment. To the north, Tower Island is a veritable ornithologist's paradise. This isolated island supports a rich bird population that includes the mockingbird, great and magnificent frigatebirds, yellow warbler, red-footed ground dove, black lava heron, yellow crowned night heron, and the world's only nocturnal swallow-tailed gull (which is now found throughout this archipelago). On the other islands, Darwin found the penguin, pelican, waved albatross, and vermillion flycatcher as well as the flightless cormorant, red-billed tropicbird, and even the flamingo. The Galapagos microcosm also has its own hawk and two species of owl. Yet, the most beloved birds are the masked, red-footed and blue-footed boobies. The comical appearance along with the humorous sounds and behavior patterns of the blue-footed booby have endeared this particular bird to every visitor of these islands. Today, like some other animals, the boobies often seem to be deliberately posing for a camera.

Throughout this archipelago, the traditionally recognized fourteen species of Galapagos finches of Darwin's time actually represent a far more complex situation of about fifty-two distinct overlapping types with genetic, physical, and behavioral similarities and differences attributed to adaptive radiation, natural selection, and extinction. These enigmatic "Darwin's finches" (as they are now called) remain a puzzle to evolutionists, particularly taxonomists and geneticists, because certain populations prematurely defined as species are in fact capable of interbreeding. However, other groups of finches are reproductively indifferent and still others even display altruistic behavior; the iguanas and tortoises do exhibit clear biological variations to the point of speciation. Darwin had haphazardly collected specimens of finches among the islands, not being careful to label each according to its own particular niche; with an element of irony, it was FitzRoy who had made detailed notes of these birds and their environments.

Until recently, all the wild birds and animals had few if any predators and as a result at first were unafraid even of man. As a sad demonstration of their capacity to learn and thereby acquire a new instinct, one may now observe that some finches and iguanas already do exhibit a fear of man (as do other animals among the once fearless wildlife of these islands).

The struggle of life on the Galapagos Islands demonstrates

both the creative and destructive forces always at work in nature. The fragile ecosystems and precarious niches of this changing archipelago have been drastically altered by recently introduced plants and animals that are harmful to the endemic botanical and zoological specimens. Such harmful feral animals include Norway rats, mice, pigs, goats, dogs, cats, horses, cattle, and burros. There are already an estimated 200,000 wild goats on James Island. Of course the most dangerous animal is man, and around 8,000 people now inhabit the Galapagos Islands.

About 600,000 years ago violent eruptions formed the impressive Alcedo Volcano of Isabela Island, and Darwin was the first to describe the resultant double crater (strangely, Captain FitzRoy's stone engraving marking the historic visit of H.M.S. *Beagle* to this island has never been found). A population of giant tortoises now lives among steaming fumaroles in this volcano's wet and grassy caldera valley one mile above sea level. In fact, it is possible that a tortoise still survives that was living when the young geologist explored this captivating area in the last century! With driving curiosity, the "unfinished naturalist" probably climbed to the top of a high rugged outcrop to obtain an awesome view of this island. Here, on these lava flows at the base of this shield volcano, he came close to experiencing the beginning of our earth.

The world's only seagoing iguana is found at the Galapagos Archipelago and may even inhabit the same island with the land iguana. The black-red marine iguanas are found in seething colonies camouflaged against lava rocks while basking in the warm equatorial sun. These "imps of darkness" represent an evolutionary shift from land to water adaptation, and they feed on seaweed and marine algae to a depth of thirty-six feet. They remain underwater for up to one hour. Retaining a land reptile metabolism, the marine iguanas eject salt water from their systems. The larger brown-yellow land iguanas feed primarily upon the buds and spiny pancake pads that fall from the giant prickly-pear *Opuntia* cactus (Darwin was baffled and no doubt disappointed by the scarcity of insects in the archipelago).

The clear Galapagos waters are inhabited by fur seals, sea lions, and green turtles. There are even rays, sharks, dolphins, and

whales further out at sea. Coastlines are usually spotted with red
Sally Lightfoot crabs. At Tagus Cove (Isabela Island) one finds
soft-colored sponges, beautiful cup corals, and brown sea
anemones.

Under the Southern Cross, life has been evolving on these
islands for millions of years. One may even speak of the evolution
of the Galapagos Islands themselves. Yet, if incisive measures are
not immediately taken, this unique archipelago as the great
naturalist once knew it will soon no longer exist. The loss would
be irreplaceable. To paraphrase the words of Darwin, at the
Galapagos Islands, both in space and time, one seems to have been
brought somewhat nearer to those first appearances of new beings
on this earth.

In retrospect, the Galapagos Islands were the catalyst that
finally convinced Darwin that plant and animal species are in fact
mutable. After the worldwide voyage and as a consequence of his
serious investigation of this archipelago and subsequent critical
reflections upon its significance, he was transformed into a
rigorous scientist who rejected both the two mutually contra-
dicting and incompatible biblical myths of divine creation as
recorded in *Genesis* of the Old Testament and the religious belief
in a series of hierarchically arranged and ascending acts of special
creation (along with the then popular biological explanations of
spontaneous generation and vitalism). He became convinced that
descent with modification does take place within the biosphere:
in organic evolution, survival is the exception and extinction is
the rule.

The Charles Darwin Research Station was established in 1964
at Academy Bay on Santa Cruz Island. Despite its status as a
legally protected national park and wildlife preserve since 1959,
the ongoing conservation efforts of this scientific station have not
prevented the archipelago from being exploited for personal gain
and profit. In fact, international scientists are more destructive
than tourists in upsetting the normal natural conditions of the
delicately balanced ecosystems and fragile niches throughout
these renowned islands. In touching and banding rare birds as
well as netting and branding land iguanas, scientists have altered
the total behavior of these life forms and even contributed to many

needless deaths.

Fortunately, the Charles Darwin Research Station is successfully taking steps to restore the populations of giant tortoises. These huge reptiles seem like antediluvian animals reminiscent of the age of dinosaurs. One may weigh up to 1,000 pounds and live to be 200 years old. In the past, they were killed by the thousands for their meat and the fresh water from their stomachs. Today, there may be only about 15,000 wild tortoises living on these islands. At the research station, tortoise eggs of this endangered species are incubated and hatched, and the young raised until they are from five to ten years old and then released only when they are capable of protecting themselves from predators in their appropriate natural habitats. This scientific station also helps eradicate the introduced pests, conserve other unique species and the differing natural environments of these islands, and maintain *Beagle III* for essential patrolling and communication throughout the Galapagos Islands.

Leaving the fascinating archipelago, the *Beagle* was now enroute to Tahiti, New Zealand, Australia, Tasmania, and several islands in the Indian Ocean (including the Cocos-Keeling Island). Darwin observed the vividly tatooed Maoris of New Zealand and later, in Australia, was fascinated by the duck-billed platypus as well as the kangaroo, other marsupials, and the aborigines of this geographically isolated part of the world. During this time, he also developed a theory to explain the sequential origin of atolls or lagoon islands of coral reefs (a view that differed from Lyell's account). At Cape Town, Darwin met the astronomers Sir John Herschel and Sir Thomas Maclear. Then, after a brief return to Brazil, the scientific survey headed back to England; nearly five years had elapsed during this great and exhilarating voyage of discovery.

By the time Charles Darwin arrived back at Plymouth Sound on 2 October 1836, ending a global journey that covered about 40,000 miles, his invaluable collections and scientific letters had already earned him the reputation as one of the greatest naturalists of the day. In fact, the voyage of H.M.S. *Beagle* proved to be the most important event in Darwin's life, as it determined his entire career in the natural sciences. He had glimpsed the gradual emergence, creative force, staggering diversity, historical continuity, and essential unity of all life on planet earth: nature is an ongoing theater of birth, change, struggle, and death.

Only in retrospect, after returning to England, did Darwin appreciate the significance of his five weeks of discoveries at the Galapagos Islands. The marvelous display of creative force throughout this archipelago now represented to the naturalist a living laboratory as an isolated microcosm in nature that demonstrated the organic results of evolutionary forces. The adaptive complexity of life could be seen in the populations of giant tortoises whose shells varied in size and shape from island to island, the groups of land and sea iguanas that varied in coloration from environment to environment, and especially among the fourteen species of finches (appropriately called Darwin's finches) that varied in the size and shape of their beaks: this curious adaptive and successful gradation among the beaks of the finches is directly related to their different diets and feeding habits within a range of specific ecological niches throughout the archipelago.

Darwin was the first naturalist to seriously study rocks, fossils, plants, insects, birds, and animals from a worldwide perspective (not even Alexander von Humboldt or Alfred Russel Wallace came close to his extensive voyage and intensive work). Actually, Darwin had seen more of the world in terms of science and natural philosophy than any other human being up to his time. As a result, he could no longer ignore the evolution and adaptive radiation of life on earth. Interestingly enough, the naturalist never made another major trip during his lifetime. Today, geological structures and biological species throughout the world bear Darwin's name and thereby attest to his global trip of scientific discovery and its lasting significance: e.g. the rare ostrich *Rhea darwini* of the pampas and Mount Darwin at Tierra del Fuego.

Darwin probably did not become an evolutionist until March 1837, about six months after the historic voyage of H.M.S. *Beagle* had ended. No doubt, he had assembled the most unique collection of scientific specimens in the world. Yet, only after reflecting on all the data and experiences of his five-year global journey did he finally recognize and accept the evolutionary perspective that had slowly emerged in his creative genius. He had come to totally reject both the Mosaic story of divine creation and the Aristotelian account of earth history as the monotonous continuance of the same without creation, creativity, or even extinction. Likewise, the fantastic claims of spontaneous generation must have now seemed truly ludicrous: maggots from apples, tadpoles from leaves, rats from decaying wheat, mice from old clothes, and crocodiles from mud! Fortunately, the scientist had time to ponder the larger consequences of his dynamic overview of organic development.

Free from all preconceptions of eternal nature as a static hierarchy of fixed types in which both continuous changes and accidental extinctions were inconceivable, Darwin was able to recognize the evidence for and patterns in biological evolution. He became preoccupied with the overwhelming diversity of life on earth as a creative unity of mutable plant and animal species. As a true scientist, he was rapidly developing a comprehensive and intelligible vision of the organic history of our entire planet (in religion, however, disbelief crept over him at a slow rate but was eventually to become complete as silent atheism).

In July of 1837, Darwin envisioned organic history as an irregularly branched tree of life and, as a result, had developed the principle of divergence before Alfred Russel Wallace. Also at this time, Darwin started to write his first of several early notebooks on the transmutation of species (1837-1839). He filled it with facts, observations, and opinions: the theory of evolution was clear in his mind, although the working out of its mechanisms and implications still awaited completion.

On 28 September 1838, Darwin merely for amusement happened by chance to read casually but critically Thomas Robert Malthus's *An Essay on the Principle of Population* (1798).[10] The Malthusian treatise on population growth claimed as follows: animal numbers (including man) have a potential to increase at a geometrical rate, whereas the food resources from plants have a potential to increase (at best) only at an arithmetical rate; therefore, since geometric progression outstrips arithmetic growth (and despite the checks on endless human population growth such as disease, famine, and war), the English economist and social philosopher argued that there will be inevitably a pervasive struggle for existence in the living world. And Malthus had emphasized the sober facts of the human condition. Suddenly, upon reading this first of two studies, Darwin intuited the explanatory principle of natural selection as the primary driving force of biological evolution (with random chance variation and severe blind competition playing their roles). The discrepancy between the rate of increase in means of subsistence and animals guaranteed a ruthless struggle for existence in which only a favored few organisms are destined to survive and thereby perpetuate their kind within the limited economy of nature. To Darwin, any beneficial because it was useful slight variation or major modification enhancing the adaptation of an organism to its own niche would provide it a survival advantage that would (more likely than not) allow for its producing offspring to insure future generations. Malthus's schema was the theoretical catalyst that gave Darwin the idea of natural selection which his contemporary Herbert Spencer referred to as the "survival of the fittest" to adapt and reproduce (among others, both Erasmus Darwin and Comte de Buffon had also glimpsed the struggle for existence). In the human realm, the most important argument against Malthus is that he had completely underestimated the gain in food productivity made possible by the industrial revolution (not to mention the coming of birth control procedures, to which he objected, having favored voluntary moral restraint). Be that as it may, Malthus did give Darwin his major explanatory principle or causal mechanism of natural selection to account for the origin of species. For the brilliant biologist, the task was now to synthesize facts and concepts in such a way as to clearly demonstrate the truth of organic evolution.

Darwin's scientific theory of biological evolution was primarily a synthesis of Lyell and Malthus: Lyellian historical geology provided the vast temporal framework of changes and natural causes, while the Malthusian principle of population offered the major causal explanation of natural selection that crystallized Darwin's thoughts on descent with modification, i.e. the transformation of species understood in terms of science and reason.

Briefly, geological change implied biological evolution. The pervasive competition in life and ruthless struggle for existence ensure that only the strongest and most vigorous organisms in populations adapt, survive, and reproduce. Throughout time, there is the survival of the fittest within changing environments as a result of some individuals having by chance favorable because beneficial variations in physical characteristics and/or behavioral patterns: since habitats are always subject to change, most organisms must also change to adapt and thereby survive to reproduce or face extinction. With the explanatory principle of natural selection, the Darwinian evolutionary viewpoint was now at least a plausible working hypothesis to be taken seriously by scientists and natural philosophers.

In 1838, Charles Darwin was elected to the secretaryship of the Geological Society of London (a position he held until 1841) and became a ten-year member of the Philosophical Club. During his later years, however, he was never tied to any professional duties.

On 29 January 1839, Charles Darwin married his first cousin Emma Wedgewood; Charles was a loving and devoted husband, but Emma did not share his scientific preoccupation and religious doubts. In August of the same year, Darwin's first book, entitled *The Voyage of the Beagle,* appeared in print.[11] Written in a literary style, it is the most readable of his naturalist volumes (the second edition was dedicated to Sir Charles Lyell). This journal of researches was published independently of FitzRoy's manuscript.

Eventually, in 1842, the Darwins left London to settle at Down House in the Village of Downe in the Kent countryside about twenty miles southeast of the city.[12] This move proved to be more suitable to Darwin's health and work. Here, the young couple found refuge from the endless demands of a sometimes thoughtless world. They lived a remote and secluded life of peace and security, raising a large family that provided comfort and happiness.

A timid and gentle genius as semi-invalid and quasirecluse, Charles Darwin remained for the rest of his life more or less isolated from the scientific community. He limited his personal associations to only a few intimate friends (e.g. Lyell, Hooker, Huxley, and at least on one occasion a visit from FitzRoy). There was ample time to consider the evolutionary implications of his panoramic glimpses into natural history during his five-year global voyage of scientific discovery: the teeming life of the primeval jungle in Brazil, those gigantic fossil mammals from a cliff in Argentina, the vast grassy plain of Patagonia, those primitive natives of Tierra del Fuego, the geology of the Andes, the zoology of the Galapagos, the unique fauna of Australia, and the coral reefs of the Cocos-Keeling Island. The naturalist was now content to work on his own estate, carrying out experiments in his own gardens, greenhouses, and the surrounding fields. He worked slowly but steadily, devoting only four hours each day to writing manuscripts or completing research on a wide range of various subjects in the natural sciences; his three books on geology were published during this time.[13]

Darwin settled into a comfortable routine. Each morning he took a solitary stroll down his private path, which became known as the Sandwalk, and afterwards devoted part of the day to sciencing and family life. Free from menial tasks and mundane responsibilities, not to mention the advantage of being independently wealthy, the naturalist accomplished an enormous amount of work (even though his chronic illness allowed him to do serious writing or research only a few hours each day). For the most part, he seems to have enjoyed his intellectual pursuits in quiet isolation. At this time, the world at large was unaware of the theory emerging in this naturalist thinker and its later impact that would devastate the beliefs of the Christian worldview.

Darwin was now a professional scientist searching for the truth. The "finished naturalist" had time to methodically examine in minute detail his painstaking research, which yielded an impressive body of empirical evidence. He worked cautiously and slowly, paying scrupulous attention to his investigations and recording everything in copious notes. There was also time to become involved in worldwide correspondence with those naturalists who shared his controversial ideas and vision.

Darwin devoted years to empirical research that covered a broad array of subjects. His geological writings contributed to understanding the origin of coral formations, volcanic islands, and lagoons within a dynamic framework. As a botanist he wrote about climbing, twining, flowering, and insectiverous plants; he was particularly interested in the reproductive methods of orchids and experimented with buttercups, snapdragons, and sweetpeas.[14] He also concentrated on the activities and value of the earthworm, oddly enough his favorite animal. Darwin took careful notice of the appearance of varieties due to artificial selection in cultivated plants and domesticated animals (e.g. dogs, sheep, horses, cattle, and especially pigeons as well as cabbages and cauliflowers). He gave a comparative study of the expression of the emotions in higher animals (including man) and, lastly, even investigated the formation of vegetable mold. Attention was paid to goldfish, silk moths, ducks, canaries, donkeys, rabbits, bumblebees, and even the rhinoceros and armadillo! His extensive research contributed to ecology, ethology, psychology, and sexual biology.

Darwin spent eight years of meticulous research analyzing and describing variations among barnacles (subclass Cirripedia). This tedious study demonstrated the variability of a species in nature and resulted in a four-volume monograph on these invertebrates: two volumes on living barnacles appeared in 1851, and two volumes on fossil barnacles appeared in 1854. This intensive and definitive work was based upon the dissection and comparative morphology of over 10,000 specimens of his "beloved barnacles" (as he called them)! In retrospect, after completing this arduous task, Darwin admitted that perhaps he had devoted too much time and effort on such a specialized area of scientific inquiry. One may suspect that this study of barnacles was deliberately self-imposed in order to keep the energetic naturalist away from having to grapple with the theory of evolution. Nevertheless, the project made him an astute taxonomist.

Concerning artifical selection, Darwin took seriously the experiments of gardeners and breeders. He became more and more aware of the fact that the appearances of varieties due to artificial selection (human choices) may perpetuate some characteristics and eliminate others as a result of the deliberate crossing of particular cultivated plants and the selection of partners in the mating of domesticated animals. It dawned on the naturalist that man himself may be an active instrument in the further evolution of life on earth (including the future development of his own species). There was, for him, an essential connection between such practical experiments on the subspecies level and the scientific theory of organic evolution, i.e. the mutability of plant and animal forms on the species level. In fact, Darwin referred to a variety as an incipient species. Yet, his own experiments in sexual reproduction lacked both the use of mathematics and rigorous control required for appreciating the results in terms of the basic principles of heredity. Consequently, Darwin asked, if man can create new varieties of plants and animals through artificial selection, could not nature herself over vast periods of time evolve new species of flora and fauna as a result of natural selection acting on random but beneficial because adaptively useful chance variations?

Before 1842, Darwin did share the larger consequences of his insights into a gradualist theory of descent with modification and its explanatory principle of natural selection with the botanist Sir Joseph Dalton Hooker of the Royal Botanic Gardens at Kew in London: the infinitely slow evolution of life on earth challenged the conceptual foundation of biology, which at that time was grounded in the idea of the immutability of species.

In 1842, Darwin wrote a thirty-five-page abstract that briefly sketched his theory of evolution. In 1844, the naturalist extended this preliminary outline into a 230-page manuscript. Neither essay was published. Still plagued by illness and anxiety, Darwin now took an intriguing precautionary measure: in his will of 5 July 1844, the naturalist left instructions and the financial means so that his scientific theory of biological evolution would be published by his wife Emma should he unexpectedly die before the manuscript was seen into print. One may even argue that Darwin actually had no serious intention of publishing a multivolume work on the theory of evolution during his lifetime.

Also in 1844, a book entitled *Vestiges of the Natural History of Creation* written by an unknown author appeared in print. It was the most significant work on the theory of evolution since Lamarck's *Zoological Philosophy* (1809), but explained organic history in terms of two impulses: one natural for adaptation and the other divine for evolution itself. Charles Darwin did not take this book with its vitalist views seriously, which he correctly guessed to have been written by the Scottish encyclopedist Robert Chambers (1802-1871).

For twenty years, Darwin reflected upon the creative development of life and the questions it raised. During this time, one may speak of the evolution of his views on biological history. It seems strange that he would remain so indifferent to those philosophical and theological issues raised concerning the ultimate nature of the material universe,[15] and his rational speculations remained within the realm of earth history. Nevertheless, as a biologist, he asked, could evolution account for the human eye and hand, insect sterility, and instinctive behavior (among other things such as the giraffe, mistletoe, woodpecker, peacock's tail, the electric organs of fishes, flatfishes, and the whalebone whale)? And what about the Irish elk (an excellent example of allometric principles at work)?

Preoccupied with biological research and his poor health, Darwin lost his aesthetic appreciation for poetry and music as well as politics and religion (he found anything artificial repugnant) but did continue to read widely in science and literature.[16] Yet, his chronic illness remained a puzzlement.[17] He suffered from headaches, eye trouble, nausea, dizziness, attacks of stomach discomfort and nervous indigestion, violent vomiting and shivering, palpitations of the heart, boils and eczema, and understandably periods of sleeplessness and depression. These inflictions caused physical weakness along with mental fatigue; he could only devote four hours each day to his scientific endeavors. Darwin's illness may have been due to long seasickness or more likely from the bite of a poisonous Benchuca bug, *Triatoma infestans* (the large black insect of the pampas in Chile that carries the microorganism that causes Chagas' disease, usually a fatal affliction), or perhaps was merely psychosomatic in origin (possibly due to guilt feelings resulting over the conflict between evolutionary science and traditional religion). Darwin resorted to several methods, some bizarre, to treat his enigmatic illness: chalk, ice packs, snuff taking, ozonized water, damp sheets, frequent showers, the taking of some mineral acids, saline solutions of the water-cure at a hydropathy spa, and even both brass and zinc wires moistened with vinegar and wrapped around his neck and waist!

On 23 April 1851, the beloved young daughter Anne Elizabeth died of a fever. After her death, Charles totally lost his faith in Christianity as a divine revelation. Emma did not share his agnosticism, if not even silent atheism.

Finally, in May of 1856, Darwin actually began to write his projected multivolume work on organic history to be entitled *Natural Selection*: no doubt, the theory of evolution was "in the air," and the naturalist was being urged by close friends (notably Lyell, Hooker, and his older brother Erasmus) to publish the growing manuscript.[18] Yet fearing controversy, persecution, and rejection, the ever-cautious scientist continued to delay offering his manuscript to a publisher. He seems to have deliberately avoided the whole issue.

In general, Darwin may have withheld the publication of his theory of evolution because of the scandal surrounding the appearance of the anonymously published book *Vestiges of the Natural History of Creation* (1844), authored by Robert Chambers, as well as the metaphysical nonsense found in Lamarck's *Zoological Philosophy* (1809). He also anticipated the negative reverberations that his disturbing ideas and their disquieting implications would have in the scientific, philosophical, and theological circles at that time. Nevertheless, his reluctance to publish will always remain a puzzlement.

On 18 June 1858, Darwin received an envelope postmarked from the small island of Ternate in Malaysia; it contained a letter and manuscript from Alfred Russel Wallace. The thunderboldt had struck. After twenty years, Darwin abruptly learned that his theory of evolution and explanatory principle of natural selection had been independently discovered and summarized by another naturalist on the other side of the world![19] His intimate scientific friends had even anticipated the possibility of such an unlikely event. Furthermore, the explanatory principle of natural selection had already been clearly glimpsed by James Hutton, William Wells, and Patrick Matthew.

Wallace first desired Darwin to comment on his four-thousand-word unpublished essay of fifteen pages entitled "On the Tendency of Varieties to Depart Indefinitely From the Original Type," which would then be sent to Lyell, who was now recognized as the dean of natural science in Victorian England. Darwin was astonished to find that Wallace's ideas and views were strikingly similar to his own. The whole situation was an incredible coincidence.

There were remarkable parallels between Darwin and Wallace: each was an English naturalist who had explored South America with interests in both geology and biology; visited a unique archipelago (Galapagos and Malay, respectively); studied orchids and collected beetles; revered life while doing original research on the barnacle and the butterfly, respectively; read Paley, Humboldt, Lyell, and Malthus; enjoyed reflecting on nature in solitude; accepted the scientific theory of biological evolution as a true framework of organic history; and independently arrived at the explanatory principle of natural selection as the key mechanism to account for the origin, survival, divergence, and extinction of species throughout the emergence of life on earth.

Darwin had intuited the concept of natural selection as early as 1838, but never published his views on the subject (neither the spring 1842 abstract nor the expanded summer 1844 version had appeared in print). On 5 September 1857, he had referred to his theory of evolution in a letter sent to the famous botanist Asa Gray in America. Darwin had not established his priority in the scientific literature, whereas Wallace had already presented an introduction to his theory of evolution in two published works: *The Sarawak Law* (1855) and the *Ternate Essay* (1858).[20]

Both Hooker and Lyell sought to establish Darwin's priority. At the Linnean Society meeting in London on the soporific evening of 1 July 1858, they delivered the two papers by Wallace and Darwin under the dry joint title "On the Tendency of Species to Form Varieties; and On the Perpetuation of Varieties and Species by Natural Means of Selection" (on 20 August, they appeared in the *Journal of the Proceedings of the Linnean Society*). Neither Darwin nor Wallace was present. It was agreed that priority be given to Darwin as the scientific father of the biological theory of organic evolution primarily by means of natural selection. Although both naturalists had independently recognized the significance of natural selection as the key mechanism in biological evolution, it must be emphasized that Darwin had a far greater body of documented empirical evidence and a much wider range of both experiences and experiments than Wallace had ever accumulated or could ever claim. A meritorious and respectable scientist in his own right and an honest and grateful man, Wallace loyally conceded the priority to his fellow naturalist Charles Darwin at Down House.

It has recently been suggested that Hooker and Lyell had plotted together against Wallace in order to establish for their friend Darwin a clear victory as regards to the question of priority concerning the theory of evolution by means of natural selection.[21] Be that as it may, the value and significance of this whole affair went unappreciated by the scientific community. Clearly, the Linnean Society meeting had been a nonevent. Likewise, Wallace seems to have never resented the priority having been given to Darwin for discovering the mechanism of natural selection to explain biological evolution (unfortunately, the former was always an unlucky man).

As incredibly similar as these two naturalists were in 1858, in later years their interpretations of evolution increasingly diverged from each other to the point of eventual diametrical opposition: Darwin grounded his theory in mechanistic materialism, while Wallace turned to teleological spiritualism (the latter argued that the essential uniqueness of the human being could not be accounted for merely through the slow accumulation of slight beneficial variations over long periods of time due to natural selection). The shy and gentle Darwin never defended his theory of evolution in public or in print and always viewed life, including the human animal, from a naturalist attitude alone.

On 24 November 1859, Charles Darwin's major work *On the Origin of Species* appeared in print; all 1,250 copies of the first edition were sold on the day of their publication.[22] It should be noted that the great naturalist never finished a detailed multi-volume treatise on organic evolution. This "premature" book focused on that mystery of mysteries, the appearance of new forms of life on earth. Its general thesis was that species are not ("It is like confessing a murder") immutable types in organic nature: structures and functions are not fixed within the flux of biological history. Either a population adapts and survives and reproduces in its changing environment, or extinction is the inevitable outcome. Briefly, the evolutionary imperative is change or perish!

The *Origin* shocked the world, generating a storm of controversy.[23] It is actually one long impressive argument for the theory of evolution, originally intended to be merely a preliminary sketch of this scientific framework. In fact, it is the theoretical core of all of Darwin's following scientific works. What the geologist Lyell had done for inorganic nature (expanding and surpassing Hutton), the biologist Darwin now did for the living world (modifying and improving on Lamarck and many others). Even an essay in Hebrew appeared in which its author attempted to show that the theory of evolution is, after all, contained in the Old Testament! Of course, Darwinism is incompatible with any literal interpretation put upon the Holy Bible. Yet, even today, outrage and furor over the theory of evolution continue in some important and influential religious circles. In sharp contrast to the creation myth in *Genesis*, Darwin's *Origin* is an evidential masterpiece.

The publication of *On the Origin of Species* was a historical event; it has been judged by many to be the most important single scientific book of the last century. Its comprehensive and intelligible explanation for the origin of new species through organic evolution (although incomplete) rests upon a logical arrangement of the documented facts and relationships in the natural sciences. The theory was supported by an impressive body of empirical evidence: geopaleontology, biogeography (especially isolation), taxonomy, comparative embryology and morphology (e.g. homologies), vestigial features or rudimentary organs (e.g. "Darwin's point" of the human ear, hair on the body, nictitating membrane, wisdom teeth, various muscles, male mammilary glands or mammae, veriform appendix, and the immovable os coccyx vertebrae)[24] and experiments in artificial (steadily accumulative) selection. Darwin paid special attention to the appearance of varieties among domesticated dogs, fowl, sheep, cattle, horses, and especially pigeons as well as various cultivated plants. He was also aware of correlated growth and variation in animals. The volume was a remarkable accomplishment in science and logic, free from theology and metaphysics.

Darwin argued that varieties are incipient species. He reasoned

that if the accumulation of favorable variations under domestication as the result of artificial selection could produce new
varieties, then chance variations under natural conditions could
eventually produce new species over long periods of time: in
nature, a series of slight differences over countless generations in
one species may add up to a major change resulting in a new
species in its own right. As such, the adaptive radiation of life
pervades the complex web of the organic world. The severe
competition among organisms and the mutability of species in
the natural world along with the resultant pervasive struggle for
existence give rise to natural selection, the principle of preservation or the survival of the fittest referred to today as differential
adaptation, survival, and reproduction (fecundity is the essential
factor).

Darwin envisioned the ever-branching and ever-growing tree or
coral of life emerging from a trunk of common origin (recognizing
that both the birth and death of species and higher taxa
throughout organic history are replete with creativity and extinction).

It was clear to the naturalist that there is a historical relationship between the vast array of forms from one or several first types
of life on the primitive earth to the existing monkeys, apes, and
man himself. He argued that all plants and all animals throughout planetary space and time are related in groups subordinate to
groups as a result of the branching tree of organic history. The
human mind reels at the unimaginable expanse of time represented by this creative evolution of life on earth.

Darwin went farther than any other evolutionist before him.
Unlike Lamarck and Chambers and Wallace, and all the other
evolutionists before the publication of the *Origin*, he had amassed
an overwhelming body of empirical evidence to factually support
his theory of evolution and even conceived of the materialist
explanatory principle of natural selection to explain the ongoing
process of descent with modification as early as 1839 (although his
major causal mechanism to account for organic history was not a
totally adequate interpretation of biological evolution). Despite
the irresistible evidence and convincing logic, the theory of
evolution is a conceptual framework above and beyond perceptual

experience and, as a result, generates questions underlying its assumptions (especially concerning the place of humankind within natural history). Anyone critically reading the *Origin* easily sees that the evolutionary framework could be extended to account for the emergence and history of the human zoological group as well.

Darwin's overwhelming achievement in the *Origin* was his drawing valid empirical statements of general application from his meticulous analysis of a vast amount of facts and a wide range of personal experiences. Its importance is beyond dispute. Even in light of scientific advancements since its publication, the *Origin* still presents an astoundingly accurate general picture of the evolutionary process as applied to life on earth. Modifications due primarily to modern genetics, new discoveries in paleontology, and more sophisticated dating techniques have not discredited its general validity. With the towering genius of Darwin, evolution ceased to be merely a rational speculation on nature or a philosophical overview on the scheme of things. Instead, it became an all-embracing scientific theory of organic history with explanatory, predictive, and exploratory powers. Not since Galileo had a scientist so altered man's view of concrete nature and his place within it.

Within the sweeping structure of his theory of evolution, Charles Darwin had not included the human animal. Conspicuously absent was any application of his theory and its principle to account for the emergence and development of man. In his *Origin*, he made very few references to apes or monkeys and only one cryptic sentence concerning our own species. The first edition merely contained what must be the scientific understatement of all time: "Light will be thrown on the origin of man and his history." Although Darwin had been strangely reluctant to discuss human evolution in print, it was obvious to any intelligent reader that the same theory and principle that had been applied to all of organic history could easily be extended to include the human zoological group as well. Nevertheless, it was the implication that evolution held for our species in natural history that actually caused the brutal controversy that quickly surrounded this scientific viewpoint. Darwin was attacked by rigid religionists as well as many scientists and philosophers. There were even deliberate attempts to maliciously discredit Darwinian evolution through absurd misrepresentations of its facts and ideas, particularly as they apply to the glaring similarities between man and the great apes (especially the chimpanzee and gorilla).

Interestingly enough, Darwin had paid little attention to his scientific and philosophical predecessors.[25] Not until April of 1861 did he include "An Historical Sketch" to the revised third edition of his *Origin* volume, which went through six editions in all. The author was always unkind to Lamarck and gave no credit to Edward Blyth, who had actually used the principle of natural selection to account only for the appearance of varieties within seemingly fixed species (Blyth's influence on Darwin remains a moot point).

In 1862, treating our planet as a cooling body, the mathematician and physicist William Thomson (later Baron Kelvin) calculated the age of our earth to be probably about 40 million years. If true, this time framework was far too short to allow for the slow and gradual evolution of all life by means of natural selection. Darwin himself thought our planet to be at least 600 million years old (modern science has determined its age at about 5 billion years). To account for the creative advance of organic history in a relatively short geological time span, the great naturalist reluctantly developed and incorporated a Lamarckian explanation of inheritance; this was his provisional hypothesis of pangenesis, with its "blending" inheritance.

Although condemned from pulpit and press as a dangerous discovery in conflict with entrenched natural philosophy and traditional sacred theology, Darwin's *Origin* was without a doubt a popular success. The envious anatomist Richard Owen and the spiteful religionist Samuel Wilberforce (Soapy Sam) were plotting to trounce the Darwinian theory of organic evolution both in public and in print.

Clearly, the crucial turning point in favor of biological evolution came as a direct result of the Huxley-Wilberforce public debate on 30 June 1860. This epic confrontation occurred at the annual meeting of the British Association for the Advancement of

Science held in Oxford's University Museum Library (nearly 1,000 people were present). Professor John Stevens Henslow presided over the four-hour debate (Charles Darwin remained at Down House). The vertebrate paleontologist and natural philosopher Thomas Henry Huxley defended the theory of evolution against the uninformed and irrelevant arguments of Samuel Wilberforce, bishop of Oxford. Huxley's polite but pointed rebuttal to the religionist exposed the theoretically biased and scientifically faulty opinions of the fundamentalist creationist. Huxley was always delighted to have such opportunities to defend Darwin's theory of organic history. His learned and eloquent defense of organic evolution at Oxford based on facts and logic gave the theory its first major victory over dogmatism and duplicity.

After being a member of the British Parliament and then the Governor of New Zealand (both unsuccessful positions), a distressed Rear Admiral Robert FitzRoy had attended this public debate to support the fundamentalist sector. However, FitzRoy failed in his attempt to denounce Darwinism by appealing to the authority of the Holy Bible (he certainly regretted having welcomed the young Darwin aboard the government ship H.M.S. *Beagle* as its unpaid naturalist and his traveling companion). It was evident that the gentle genius of Down House would triumph after all.

Darwin was greatly abused, being held to have both ridiculed traditional religion with heretical ideas and dishonored natural science with fanciful assumptions.[26] His *Origin* had startled almost everyone who read it, and the theory of evolution was bitterly attacked by theologians as well as philosophers and scientists (e.g. Owen in England and Agassiz in America). Nevertheless, Darwin's intellectual reputation was secure. As "Darwin's bulldog" in England, naturalist Thomas Huxley remained the evolutionist's ardent supporter (along with support from Carpenter, Hooker, Kingsley, Lubbock, and Wallace). In Germany, scientist and philosopher Ernst Haeckel did not hesitate to defend the Darwinian worldview; the monist-pantheist was referred to as "Germany's Darwin," while Haeckel, in turn, referred to Darwin as the "Newton of Biology." The American botanist and taxonomist Asa Gray was an early supporter of Darwin, whose ideas were also popularized and defended by the paleontologist Vladimir Kovalevskii in Russia and, soon afterwards, by Broca and Renan in Europe as well as Peirce and Wright in America. It must be noted, however, that Herbert Spencer had adhered to the idea of evolution since 1852.

However, Sir Charles Lyell's *Geological Evidences of the Antiquity of Man* (1863) was a disappointment to Darwin. Lyell vacillated between accepting and rejecting the theory of organic evolution, never acknowledged the growing fossil evidence for the evolution of humankind, and even encouraged Darwin to refer to God as the Creator of life (a concession made by Darwin in the last paragraph of the later editions of his *Origin* in order to still the storm of controversy).

For Huxley, the question of questions concerned the origin and natural history of the human zoological group. In his book *Evidence as to Man's Place in Nature* (1863), he was quick to extend the theory of evolution to account for the origin of the human animal. In fact, Huxley found the idea of evolution so obvious that he remarked how extremely stupid it was that he had not thought of the explanatory principle of natural selection before Darwin and Wallace. As an agnostic (he coined the term) in science, Huxley differentiated between nature "red in tooth and claw" and the realm of human ethics (a sharp distinction that evolutionist Petr Kropotkin rejected).

On 30 November 1864, Charles Darwin was awarded the Copley Medal of the Royal Society of London, England's highest honor for outstanding achievment in science; Darwin was a Fellow member.

Six months later on 30 April 1865, Vice Admiral Robert FitzRoy, now a depressed believer, committed suicide by slashing his throat. This tragic act could have been due, at least in part, to the guilt he felt in his having played a direct role in providing the young Darwin with the unique opportunity to develop a theory of evolution while the ship's naturalist during the voyage of the H.M.S. *Beagle*. Darwin probably felt somewhat responsible for this unfortunate turn of events.

Foreshadowed in the scientific studies of plant sexuality and experiments in plant hybridization by the German botanist Joseph Gottlieb Koelreuter (1722-1806) but superseding them, the Czech monk-priest Gregor Johann Mendel (1822-1884) worked obscurely in a garden of the provincial Augustinian monastery in Brno, Moravia. There, as a result of his original research involving the artificial fertilization of common garden pea plants (genus *Pisum*), he discovered four basic principles of heredity that established the general foundation of the science of genetics. Mendel sought to understand the connection between inheritance and organic history. Through his rigorously controlled experiments and the use of arithmetic symbols, he observed the transmission of specific pairs of characteristics for seven different physical traits in individual plants from generation to generation. The resultant four contributions were a particulate theory grounded in the existence of discrete and indivisible hereditary units in gametes that are clearly distinct from the somatic cells of the organism, the distinction between dominate and recessive characteristics for the same physical trait, and the two laws of segregation and independent assortment.

After eight years of research, Mendel presented his results in two lectures to the Natural History Society of Brno in 1865, and then published them in his monograph *Experiments in Plant Hybridization* (1866). His pioneering work in mathematical genetics surpassed all Lamarckian theories of heredity. It has been suggested that Mendel had deliberately manipulated the results of his pea experiments in order to obtain (or at least approach) the whole number ratios desired in the physical expression of characteristics in the offspring. Be that as it may, his discoveries were of great significance. He was appointed the abbot of the monastery in the spring of 1868 and later, in 1871, he abandoned altogether his hybrid experiments because of poor health and a hectic schedule. Mendel's discoveries went unrecognized by the scientific community, unappreciated by the other monks, and even he failed to understand the awesome consequences of his unique research.

Mendel's manuscript and its value went unappreciated until 1900, when three botanists in Europe independently discovered his work: Hugo De Vries of Holland, Karl Erich Correns of

Germany, and Erich Tschermak von Seysenegg of Austria, as well as, shortly afterwards, the zoologists William Bateson in Britain and Lucien Cuénot in France. Important discoveries were made later by T.H. Morgan, C.D. Darlington, and C.H. Waddington, not to forget the controversial zoologist Richard B. Goldschmidt's "hopeful monster" hypothesis in his major book entitled *The Material Basis of Evolution* (1940).

One thousand miles separated Darwin from Mendel; although the monk-priest knew of the great naturalist and his theory of evolution, the latter was unaware of the former and his contributions to understanding heredity. Mendel had even sent a copy of his monograph to Darwin, who never opened the package to read the enclosed publication. Had Darwin done so, biology may have had the synthetic theory of organic evolution several decades earlier. Actually, with a touch of irony, the Augustinian abbot was never an evolutionist.

Darwin continued to speculate on the mechanisms of sexual heredity. He paid special attention to the selective breeding of chosen domestic pigeons to deliberately produce numerous novelties that satisfied the fancy of farmers. His volume *The Variation of Animals and Plants Under Domestication* (1868) contained his hypothesis of pangenesis with its conception of hereditary particles referred to as gemmules and its acceptance of the explanatory mechanism of the inheritance of acquired characteristics as a result of use and disuse (habit). The great naturalist had reluctantly incorporated a Lamarckian viewpoint into his own theory of evolution: since gemmules produced from organs constitute the hereditary material, the modification of any organ through use or disuse would result in a corresponding modification of the gemmules produced by that organ. Also, Darwin never abandoned the idea that the physical environment is capable of inducing in organisms favorable variations necessary for their adaptation and subsequent survival and reproduction in changing habitats. In general, he held to the blending sexual inheritance of unlimited fortuitous variations.

Although Darwin taught the slow accumulation of continuous but imperceptible modifications (while Huxley argued for the sudden appearance of saltations), it disturbed him that he could not explain the emergence and transmission of fortuitous variations as the raw material of natural selection: their occurrence always remained an inexplicable puzzle to him. Unlike Darwin's theory of pangenesis, the particulate views of both Mendel (1866) and Weismann (1892) clearly separated body cells from sex cells with their hereditary material. Only a few had ever rejected outright the supposed inheritance of acquired characteristics (e.g. Lucretius, Kant, and Bonnet).

The anthropologist and psychologist Sir Francis Galton (1822-1911), Darwin's cousin, was greatly influenced by the publication of *On the Origin of Species*; the volume was a turning point in his scientific career. After studying biology and medicine, and then devoting time to geography and meteorology, Galton focused on the importance of heredity in determining the psychophysical variations, abilities, and characteristics in man. In contrast to euthenics, he formulated the main principles of eugenics in his book *Hereditary Genius* (1869), which stressed the role of nature over the influences of nurture (Darwin's son Leonard was an advocate of eugenics).[27]

The Russian philosopher and naturalist humanist Petr Kropotkin (1842-1921) devoted his scientific curiosity to an inquiry into the geography, botany, zoology, and anthropology of Eastern Siberia and Northern Manchuria. Although greatly influenced by the Darwinian theory of organic evolution, his own studies investigated the roles that mutual aid and mutual support (or group cooperation) play in the living world as important progressive factors throughout biological history (both nonhuman and human). Kropotkin stressed the significance of these factors in his major volume *Mutual Aid: A Factor of Evolution* (1902), pointing out that this interpretation of life did not contradict but merely supplemented the explanatory principle of natural selection.[28]

As a devoted but critical evolutionist, Kropotkin rejected Huxley's extreme separation of the human realm from the rest of the organic world in terms of behavior. The Russian thinker explored human conduct within the framework of biological evolution and the natural environment. He maintained that the ethical and social progress of our own species is the result of instincts and feelings transformed into morality and love. In his posthumously published book *Ethics: Origin and Development* (1922), the philosophical anarchist advocated both human solidarity and human sociability.[29] Like Darwin, Kropotkin saw man in all his aspects as totally within the evolving matter of concrete reality.

On 24 February 1871, Darwin's *The Descent of Man* appeared in print; Huxley and Haeckel had already written about human evolution.[30] In this, his second major work on evolution, the ever-cautious "finished naturalist" had finally extended his own theory of evolution to account for the origin and biological development of mankind (with special attention given to the mechanism of sexual selection in regards to those physical characteristics and behavior patterns of reproductive significance influencing male combat and female choice). Darwin's explanations for evolution included the pressure of natural selection, artificial selection, sexual selection, and his provisional hypothesis of pangenesis.

The question of questions is the origin and nature of the human zoological group. Evolution is a theory encompassing the organic world from life's minute beginnings up, at last, to man himself. Early in the nineteenth century, the prehistorian Boucher de Pertes had discovered hominid fossils and paleolithic artifacts deep in sedimentary layers, which clearly demonstrated that our species once lived thousands of years ago as a tool-making animal that shared its environment with now extinct mammals.

Darwin's major thesis is that man and the two great apes (chimpanzee and gorilla) have descended through organic evolution from a common but distant ancestral group of hominoids in prehistoric Africa. He held that the terrestrial ancestors of our species slowly changed from quadrupeds into erect bipeds; this was accompanied by language, tool use, tool making, and both the quantitative and qualitative increase in the central nervous system and brain. In body, mind, and behavior, the human animal unmistakably bears the indelible stamp of its lowly origin among the earlier primates. Likewise, Darwin argued that all the higher animals exhibit mental traits usually attributed only to man: reason, emotion, attention, memory, imitation, curiosity, association, imagination, and vocal communication. Like Huxley and Haeckel before him, the great naturalist claimed that there is no structure or function in the human brain that is peculiar to it alone, i.e. that is not already found in the apes to some degree.

For Darwin, the superiority and essence of the human animal resides in its moral faculties; it is this moral dimension alone that distinguishes but does not separate man from the apes (unfortunately, the great biologist also clung to elements of racism and sexism). Also like Huxley and Haeckel, Darwin emphasized the striking and undeniable similarities between humans and the pongids, claiming that the human animal is closer to the great apes than the apes are to the monkeys. Nevertheless, it is necessary to point out that Darwin never suggested that man had evolved from an ape: instead, both man and the living apes have descended from a generalized hominoid ancestor in the remote past (Haeckel had thought Asia to be the birthplace of the alleged "missing link" between apes and man).[31] The latest comparative studies in primate biology and ethology demonstrate that, in terms of genetic makeup and behavior patterns, the human animal is closer to the three great apes (orangutan, chimpanzee, and gorilla) than even Huxley, Haeckel, or Darwin had thought in the last century. From the evolutionary perspective, the human animal can no longer be considered the ultimate center of a divine creation or the final goal of cosmic creativity.

Science affects human conduct and human values: as such, following biological Darwinism came social Darwinism. Darwin himself always restricted his explanatory principle of natural selection to the nonhuman biological realm, while Spencer extended it to include human society and its progressive development. Unlike Spencer and Haeckel, Darwin did not extend his theory of evolution to embrace the whole cosmos, and he did not delve into the crucial ethical questions and moral problems generated by his controversial framework as Huxley, Kropotkin, Waddington, Flew, and Farber have done (among others). In fact, Darwin never seriously thought about the origin of life on earth (much less the origin of matter in the universe). He simply concluded that the mystery surrounding both the beginning and the end of all things is beyond the range of the human intellect.

Darwin had never been a devout believer. Yet, the gentle and tolerant scientist was never mocking or cynical in his attitude toward religion. However, one may make a strong case that something deeply bothered him, which was to remain unresolved. In general, the naturalist was silent about his own inner life. His wife Emma feared that Charles was moving farther and farther away from any belief in God and Christianity.

For Darwin, religion was not a matter of spiritual transcendence but rather a social phenomenon of psychological and ethical or moral significance; religious beliefs and practices (including the concept of God) have evolved as aspects of sociocultural development. He saw the complexity of religious devotion consisting of love, fear, submission, dependence, reverence, hope, and awe. Actually, scientific evolutionism established the foundation for naturalism and humanism in this century.

As a solitary scientist, Darwin continued to investigate new areas of biological research. His next book was *The Expression of the Emotions in Man and Animals* (1872).[32] The sixth and last edition of Darwin's *Origin* appeared in the same year. Among his last works were *The Effects of Cross and Self Fertilization in the Vegetable Kingdom* (1876), a preliminary notice in Ernst Krause's *Erasmus Darwin* (1879), and *The Formation of Vegetable Mould, Through the Action of Worms, With Observations on Their Habits* (1881). His last papers were on chlorophyll bodies and roots.

Although the scientific community was finally accepting Darwinism, several important naturalists had rejected the theory of evolution in part or whole: e.g. Cuvier, Lyell, Sedgwick, Henslow, Owen, Whewell, Mivart, Virchow, and Agassiz.[33] Nevertheless, their attacks were merely dust in the wind that would soon settle before the power of facts and reason. Of course, evolutionists themselves may disagree over mechanisms and interpretations, e.g. the debate between selectionists and mutationists early in this century, the emphasis placed primarily on the influence of heredity or the environment, the controversy over chance or design in nature, and the ongoing philosophical differences among materialists, vitalists, and spiritualists (not to forget the opposing views between the gradualists and punctuationists in the current literature).

Darwin was a devoted husband to Emma, a loving father to their seven children (several distinguished themselves in science), a loyal friend to several naturalists, and an appreciative correspondent with those scientists elsewhere who championed his theory of evolution (notably Ernst Haeckel in Germany and Asa Gray in America). Over the years at Down House, his circle of intimate friends included Henslow, Lyell, Hooker, Wallace, Huxley, Lubbock, and Romanes. Urged by Emma to record his life in general for the family and grandchildren, Charles had written an autobiography during the year 1876.[34] Yet, the real Darwin still eludes us. He seems never to have resolved those metaphysical issues surrounding his scientific theory of biological evolution. His innermost thoughts belonged only to the Sandwalk ("my thinking path"). With the gradual acceptance of his controversial ideas by important members of the scientific community, his poor health began to improve (suggesting a psychosomatic component to his chronic illness).

Darwin had been one of the summer vice-presidents of the British Association for the Advancement of Science. He was awarded both the Royal and Wollaston Medals of British science, received honorary doctorates from five universities (including Cambridge and Oxford), and was given honorary or corresponding membership in fifty-seven foreign societies (including the French Academy of Sciences in 1878). International recognition also included his being awarded the highly coveted Prussian *Ordre pour le mérite* and elected to the Imperial Academy of Science in Saint Petersburg (now Leningrad).

Karl Marx venerated Darwin, and was so impressed with *On the Origin of Species* that he expressed the desire in a letter to dedicate the English first volume of his *Das Kapital* (1882) to the esteemed evolutionist. Darwin courteously refused the sincere request, not wanting his family associated with Marx's communistic and atheistic movement. In 1905, Alexander Eric Kohts founded the Darwin Museum in Moscow: early in this century, Russian biology suffered stagnation under a Lamarckian theory of evolutionary genetics dogmatically upheld by Trofim Denisovich Lysenko (1898-1976) for vested sociopolitical interests.

In his intuitive grasp of both the awesome immensity of geological time and the endless creativity of biological evolution, Charles Darwin stood alone. Yet, despite all his scientific accomplishments, he was never knighted by Queen Victoria or officially honored by the British government; orthodox England thought the theory of evolution (which argued for man and the apes sharing a common ancestry) to be too repugnant a discovery to merit a Sir Charles Darwin. After living the last forty years of his life in the serenity and seclusion of Down House, Charles Darwin at the age of seventy-three died on 19 April 1882 of a heart attack. Public sentiment decreed his final resting spot. So on 26 April, the great naturalist was buried at Westminster Abbey near the grave of Sir Isaac Newton.[35]

In his empirical explanations for the theory of biological evolution, Charles Darwin was hindered by the limits of scientific knowledge and technology during his lifetime. The shortcomings of Darwinian evolution in the last century included greatly underestimating the age of the earth; not accounting for the origin of life on our planet; the incompleteness of the geological and paleontological records; the absence of transitional fossil forms in rock strata; the lack of intermediate living types in the biosphere; a slow, gradual, and uniform view of the process of evolution; the origin of new species is not directly observable; the incorporation of Lamarck's two laws of animal evolution; no understanding of the nature of hereditary transmission; no understanding of the origin of heritable variation; emphasis on the individual organism; and the mistaken notion that the external conditions of the physical environment directly stimulate the appearance of biological variations necessary for adaptation and survival. Darwin's work in evolution did not deal with entropy, philosophy, theology, or the cosmic perspective as such. Nevertheless, many of his scientific shortcomings or errors have been corrected, and his evolutionary framework now encompasses the entire history of the physical universe.

In this century, neither Darwinism nor Neodarwinism (the synthetic theory of organic evolution) is synonymous with evolutionism: there is a crucial distinction between the fact of evolution and those explanations and interpretations to account for it, e.g. various evolutionary rates, modes, levels, distances, and perspectives.[36] Modern biology grounds organic evolution primarily in random genetic variability as a result of chance major or minor mutations along with genetic recombinations and the necessity of ongoing natural selection (refer to the writings of Dobzhansky, Fisher, Haldane, Huxley, Kettlewell, Mayr, Rensch, Simpson, and Wright). Evolutionism has also generated the following areas of scientific inquiry: astrochemistry, astrophysics, process cosmology, plate tectonics, ecology, ethology, sociobiology, systematics, biochemistry, genetic engineering, and the emerging science of exobiology (which suggests that life as well as intelligent beings exist elsewhere in our evolving physical universe). Today, no competent person has any doubts about the truth of the mutability of species and human evolution. From the writings of Tylor and Morgan in the last century to the books of Leslie A. White and V. Gordon Childe in this one (among others including Kovalevsky, Hobhouse, Ellwood, Keller, MacIver and Harris), both anthropologists and sociologists have extended the evolutionary framework to incorporate cultural development (not to forget the pioneering work of Sigmund Freud on the natural evolution of the human psyche from earlier animals).[37]

The present controversy between fundamentalist creationism and scientific evolutionism has an interesting history. In the last century, special creationism upheld the religious argument from divine design in a natural theology grounded in the benevolent intelligence of a personal God responsible for the assumed order, beauty, harmony, and efficiency throughout the inorganic and organic cosmos. It supported the following beliefs: the instantaneous creation of the physical universe (including all life) from nothingness, the age of the world is less than ten thousand years, fixed plant and animal kinds or types were created separately, man is unrelated to the apes, and periodic global catastrophies (e.g. the Noachian Deluge) account for the existence of rock strata with their sequence of fossils and artifacts. There are attempts to extend natural theology into evolutionary science and descriptive science into process theology, e.g. the writings of the Jesuit priest and geopaleontologist Pierre Teilhard de Chardin (1881-1955).[38]

In the last century, the ongoing creation/evolution controversy was foreshadowed in the disagreement between Cuvier and Saint-Hilaire as well as the personal conflict between Darwin and FitzRoy (and later the difference of opinion between Charles and Emma). Philip Gosse, struggling with the evident contradiction between evolutionary science and fundamentalist religion, offered a bizzare compromise that satisfied neither the religionists nor the evolutionists. The public debate between Huxley and Wilberforce as well as the arguments offered against evolution by Agassiz and for it by Gray provided a rich background for both the fundamentalists and naturalists involved in the John Scopes "Monkey Trial" in 1925 (the controversy had become a legal issue on the state level).[39] Today, this same controversy is of national if not international concern.

Special creationism is not science, natural philosophy, or even modern theology.[40] In fact, it is essentially a conservative sociopolitical undertaking rather than a religious movement. Myopic fundamentalism, its most severe form, is a clear and deliberate threat to science as well as tolerance, education, and free inquiry. Evolutionists must defend the right to think: truth is a necessary direction in overcoming hate, bigotry, ignorance, and fanaticism.

Charles Darwin is exemplary of the creative and courageous genius who is dedicated to both science and reason in the pursuit of truth wherever it may lead and whatever its consequences. In all his writings, there was a strict adherence to the facts and logic. The celebrated naturalist had something very decisive to say about space, time, life, change, and death. He contributed critical insights into the workings of the world around us, thereby revolutionizing the scientific study of natural history. Few men have so profoundly influenced human thought.

Today, Darwin's scientific theory of biological evolution (with needed modifications) remains the only plausible explanation to account for the history of life on earth. The truth of evolution is incontrovertible to any enlightened mind, and Darwin's contributions to it remain indispensable.

The name of Charles Darwin will be remembered through the ages; his theory and its influence will prevail. Darwin caused an upheaval in the biological sciences, ushered in an intellectual revolution in natural philosophy, and undermined traditional theology and values. As such, he was one of the great forerunners of modernity (along with Marx, Nietzsche, Freud, and Einstein). The need for a continuous adaptation to new challenges and opportunities undoubtedly applies to the evolution of the theory of evolution, which must keep developing ever further beyond Darwin but certainly never without him.

Perhaps Darwin's own words at the end of *On the Origin of Species* best represent his vision: "There is grandeur in this view of life, with its several powers, having been originally breathed into a few forms or into one; and that, whilst this planet has gone cycling on according to the first law of gravity, from so simple a beginning endless forms most beautiful and most wonderful have been, and are being, evolved."

Modest and aloof, not to forget that he was rich with ample leisure time for reflection and research, Charles Darwin had ushered in one of the greatest scientific revolutions in western intellectual history. His life and work are there to inspire all those who really support science, free inquiry, and the dignity of humankind. Today, the inescapable fact of evolution as both the supreme integrative principle of the biological sciences and general condition of this planet in our material universe is beyond dispute.

NOTES AND SELECTED REFERENCES

[1]Cf. Philip Appleman, ed., *Darwin* (New York: W.W. Norton, 2nd ed., 1979), H. James Birx, "Charles Darwin: A Centennial Tribute" in *Creation/Evolution* 3(2):1-10 and "The Theory of Evolution" in *Collections* 60(1):22-33, Peter Brent, *Charles Darwin: A Man of Enlarged Curiosity* (New York: Harper & Row, 1981), John Chancellor, *Charles Darwin* (New York: Taplinger, 1976), C.D. Darlington, *Darwin's Place in History* (Oxford: Basil Blackwell, 1960), Sir Gavin de Beer, *Charles Darwin* (London: Oxford University Press, 1958) and *Charles Darwin: A Scientific Biography* (Garden City, New York: Anchor Books, 1965), Loren C. Eiseley, "Charles Darwin" in *Scientific American* 194(2):62-70, 72, and *Darwin's Century: Evolution and the Men Who Discovered It* (Garden City, New York: Anchor Books, 1961), Benjamin Farrington, *What Darwin Really Said* (New York: Schocken Books, 1982), F.D. Fletcher, *Darwin: An Illustrated Life of Charles Darwin, 1809-1882* (Bucks, U.K: Shire Publications Ltd., 1975, Lifelines 34), Roy A Gallant, *Charles Darwin: The Making of a Scientist* (Garden City, New York: Doubleday, 1972), Wilma George, *Darwin* (U.K.: Fontana, 1982), John C. Greene, *Darwin and the*

Modern World View (Baton Rouge: Louisiana State University Press, 1981), Gertrude Himmelfarb, *Darwin and the Darwinian Revolution* (New York: W.W. Norton, 1968), Jonathan Howard, *Darwin* (New York: Hill and Wang, 1982), Julian Huxley and H.B.D. Kettlewell, *Charles Darwin and His World* (New York: Viking Press, 1965), Stanley Edgar Hyman, *The Tangled Bank* (New York: Atheneum, 1962, pp. 9-78), William Irvine, *Apes, Angels, & Victorians: Darwin, Huxley, & Evolution* (New York: McGraw-Hill, 1972), Walter Karp, *Charles Darwin and the Origin of Species* (New York: Harper & Row, 1968), Jonathan Miller, *Darwin for Beginners* (New York: Pantheon Books, 1982), Christopher Ralling, ed., *The Voyage of Charles Darwin* (New York: Mayflower Books, 1979), Paul B. Sears, *Charles Darwin: The Naturalist as a Cultural Force* (New York: Charles Scribner's Sons, 1950), L. Robert Stevens, *Charles Darwin* (Boston: Twayne, 1978), and Irving Stone, *The Origin: A Biographical Novel of Charles Darwin* (Garden City, New York: 1980).

[2]Cf. Donald M. Hassler, *Erasmus Darwin* (New York: Twayne, 1973, esp. pp. 69-71) and Desmond King-Hele, *Erasmus Darwin* (London: Macmillan, 1963).

[3]Cf. Richard W. Burkhardt, Jr., *The Spirit of System: Lamarck and Evolutionary Biology* (Cambridge: Harvard University Press, 1977), and Stephen Jay Gould, "Shades of Lamarck" in *Natural History* 88(8):22, 24, 26, 28. Also refer to Arthur Koestler, *The Case of the Midwife Toad* (New York: Vintage Books, 1973).

[4]Cf. Bern Dibner, *Darwin of the Beagle* (Norwalk, Connecticut: Burndy Library, 1960) and Alan Moorehead, *Darwin and the Beagle* (New York: Harper & Row, 1969). Also refer to Victor Wolfgang von Hagen, *South America Called Them* (New York: Alfred A. Knopf, 1945, Humboldt pp. 86-168 and Darwin pp. 169-229) and Robert S. Hopkins, *Darwin's South America* (New York: John Day, 1969).

[5]Cf. Charles Coulston Gillispie, *Genesis and Geology: A Study in the Relations of Scientific Thought, Natural Theology, and Social Opinion in Great Britain, 1790-1850* (New York: Harper Torchbooks, 1959, passim).

[6]Cf. Loren C. Eiseley, "Charles Lyell" in *Scientific American* 210(2):98-106, 168, and Leonard G. Wilson, ed., *Sir Charles Lyell's Scientific Journals on the Species Question* (New Haven: Yale University Press, 1970, passim).

[7]Cf. Richard L. Schoenwald, ed., *Nineteenth-Century Thought: The Discovery of Change* (Englewood Cliffs, New Jersey: Prentice-Hall, 1965, esp. pp. 89-128).

[8]Cf. Benjamin Farrington, *Greek Science: Its Meaning For Us* (Baltimore: Penguin Books, rev. ed., 1961, Aristotle pp. 112-133, 156-159).

[9]Cf. H. James Birx, "The Galapagos Islands" in *Collections* 62(1):12-16, Robert C. Eckhardt, "Introduced Plants and Animals in the Galapagos Islands" in *BioScience* 22(10):585-590, *Galapagos: The Flow of Wildness* (New York: Sierra Club & Ballantine Books, 1970, 2 vols.), Roger Lewin and Sally Anne Thompson, eds., *Darwin's Forgotten World* (Los Angeles:

Reed Books, 1978), Tui De Roy Moore, *Galapagos: Islands Lost in Time* (New York: Viking Press, 1980), Alan Moorehead, *Darwin and the Beagle* (New York: Harper & Row, 1969, esp. pp. 186-209), Ian Thornton, *Darwin's Islands; A Natural History of the Galapagos* (Garden City, New York: Doubleday, 1971), and Alan White and Bruce Epler, *Galapagos Guide* (Quito, Ecuador: Libri Mundi, Liberia Internacional, 3rd ed., 1978). Concerning Darwin's finches in particular, refer to: David Lack, *Darwin's Finches: An Essay on the General Biological Theory of Evolution* (New York: Harper Torchbooks, 1961), Suh Y. Yang and James L. Patton, "Genetic Variability and Differentiation in the Galapagos Finches" in *The Auk* 98(2):230-242, and Frank J. Sulloway, "Darwin and His Finches: The Evolution of a Legend" in *Journal of the History of Biology* 15(1):1-53. For another example of microevolution among oceanic islands, see the special feature "Hawaii: Showcase of Evolution" by Hampton L. Carson, et al., in *Natural History* 91(12):16-18, 20-22, 24, 26, 28, 30, 32, 34-44, 48-72. Of literary interest see Herman Melville, "The Encantadas, or Enchanted Isles" in *Herman Melville: Four Short Stories* (New York: Bantam Books, 1959, pp. 43-103).

[10]Cf. Thomas Robert Malthus, *An Essay on the Principle of Population* (New York: W. W. Norton, 1976, ed. by Philip Appleman). Refer to the brilliant introduction by Philip Appleman, pp. xi-xxvii.

[11]Cf. Charles Darwin, *The Voyage of the Beagle* (Garden City, New York: Anchor Books, 1962, ed. by Leonard Engel). First appeared as *Journal of Researches into the Geology and Natural History of the Various Countries Visited by H.M.S. Beagle* (1839) and in the second edition as *Journal of Researches into the Natural History and Geology of the Countries Visited During the Voyage of H.M.S. Beagle* (1845). Note the preference given to natural history over geology in the title of the second edition of this volume. Also refer to R. Alan Richardson, "Biogeography and the Genesis of Darwin's Ideas on Transmutation" in *Journal of the History of Biology* 14(1):1-41.

[12]Cf. Sir Hedley Atkins KBE, *Down, the Home of the Darwins: The Story of a House and the People who Lived There* (Chichester, England: Phillimore, 1976), and Elizabeth Lambert, "Historic Houses: Charles Darwin's Home in Kent" in *Architectural Digest* (February, 1983), pp. 134-140, 150.

[13]Darwin's three works in geology are: *The Structure and Distribution of Coral Reefs* (1842, 2nd ed. 1874), *Geological Observations on the Volcanic Islands Visited During the Voyage of H.M.S. Beagle* (1844), and *Geological Observations on South America* (1846). As a petrologist, two of Darwin's major contributions to geology were his own deformation theory for the origin of metamorphic rocks which he first formulated at Tierra del Fuego as well as his distinguishing between cleavage and sedimentary bedding. During his visit to Chile, he discovered the connections among earthquakes, volcanic activity, and both geological elevation and subsidence as local phenomena. After the voyage of the *Beagle*, he suggested the use of echo-sounding methods for measuring the depths of the oceans. Darwin's two errors in geology were his account of icebergs carrying erratic boulders from the Alps to the Jura

mountains (they were actually transported over great distances by glaciers), and his also assuming that the so-called three sets of parallel beaches on the slopes of the valleys of Glen Roy and Glen Gluoy were due to the action of marine waves (they were actually formed by a glacier-caused inland lake during the last ice age). Learning of these mistakes, he never again invoked the principle of exclusion in scientific matters.

[14]Darwin's works in botany include: *On the Various Contrivances by Which British and Foreign Orchids are Fertilized by Insects, and on the Good Effects of Intercrossing* (1862, 2nd ed. 1877), *The Variation of Animals and Plants Under Domestication* (1868, 2nd ed. 1875), *Insectivorous Plants* (1875), *Climbing Plants* (1875), *The Effects of Cross and Self Fertilization in the Vegetable Kingdom* (1876), *The Different Forms of Flowers on Plants of the Same Species* (1877), and *The Power of Movement in Plants* (with the assistance of his son Francis Darwin, 1880). Also refer to Mea Allan, *Darwin and His Flowers: The Key to Natural Selection* (New York: Taplinger, 1977).

[15]Cf. Charles Darwin, *Metaphysics, Materialism, and the Evolution of Mind: Early Writings of Charles Darwin* (Chicago: The University of Chicago Press, Phoenix Edition, 1980), and Neal C. Gillespie, *Charles Darwin and the Problem of Creation* (Chicago: The University of Chicago Press, Phoenix Edition, 1979).

[16]Cf. Philip Appleman, "Darwin and Literature" in *Free Inquiry* 2(3):25-28.

[17]Cf. Ralph Colp, Jr., M.D., *To Be an Invalid: The Illness of Charles Darwin* (Chicago: The University of Chicago Press, 1977).

[18]Cf. R. C. Stauffer, ed., *Charles Darwin's Natural Selection: Being the Second Part of His Big Species Book Written From 1856 to 1858* (Cambridge: Cambridge University Press, 1975). Also refer to Dov Ospovat, *The Development of Darwin's Theory: Natural History, Natural Theology, and Natural Selection, 1838-1859* (Cambridge: Cambridge University Press, 1981). Between 1837 and 1859, one may speak of the evolution of Charles Darwin's own views on biological evolution.

[19]Cf. Bert James Loewenberg, *Darwin, Wallace, and the Theory of Natural Selection* (Cambridge: Arlington Books, 1959). This work contains the Linnean Society Papers and a list of books pertinent to the text. Also refer to H. Lewis McKinney, *Wallace and Natural Selection* (New Haven: Yale University Press, 1972).

[20]Cf. Arnold C. Brackman, *A Delicate Arrangement: The Strange Case of Charles Darwin and Alfred Russel Wallace* (New York: Times Books, 1980, passim). Reviewed by David Kohn, "On the Origin of the Principle of Diversity" in *Science* 213(4512):1105-1108. Also refer to Stephen Jay Gould, "On Original Ideas" in *Natural History* 92(1):26, 28-30, 32-33.

[21]*Ibid.*

[22]Cf. Charles Darwin, *The Origin of Species* (New York: Penguin Books, 1968). This copy of the 1859 first edition is edited by J. W. Burrow. Also refer to *The Origin of Species by Means of Natural Selection: or, The Preservation of Favored Races in the Struggle for Life* and *The Descent of Man and Selection in*

Relation to Sex (New York: Modern Library, 1936) as well as *The Origin of Species* (New York: Mentor Books, 1958). To empirically support his theory of evolution, Charles Darwin relied upon the following: fossils ("the succession of types"), zoogeography (similar "representative species" of different local environments descended from a common ancestor), biological isolation (especially on oceanic islands), and the principle of divergence (particularly the adaptive radiation of life at the Galapagos Islands) as well as the inexorable pressure of natural selection resulting in differential reproduction (survival or extinction). For a current view of this classic volume, see Richard E. Leakey, ed., *The Illustrated Origin of Species by Charles Darwin* (New York: Hill and Wang, 1979).

[23]Cf. Julian S. Huxley, et al., *A Book That Shook the World: Anniversary Essays on Charles Darwin's Origin of Species* (Pittsburgh: University of Pittsburgh Press, 1958).

[24]These vestigial features or rudimentary organs are discussed in Charles Darwin's *The Descent of Man* (1871) part one, chapter one (2nd ed. 1874).

[25]Cf. Bentley Glass, Owsei Temkin, and William L. Straus, Jr., eds., *Forerunners of Darwin: 1745-1859* (Baltimore: The Johns Hopkins Press, 1968). Refer to "An Historical Sketch" (of the progress of opinion on the origin of species previously to the publication of the first edition of this work) in *The Origin of Species* (New York: Penguin Books, 1968, pp. 53-63) or *The Origin of Species* (New York: Mentor Books, 1958, pp. 17-25): Darwin wrote, "I may add, that of the thirty-four authors named in this Historical Sketch, who believe in the modification of species, or at least disbelieve in separate acts of creation, twenty-seven have written on special brances of natural history or geology" (p. 61 or p. 23, respectively).

[26]Cf. David L. Hull, *Darwin and His Critics: The Reception of Darwin's Theory of Evolution by the Scientific Community* (Cambridge: Harvard University Press, 1973) and Peter J. Vorzimmer, *Charles Darwin: The Years of Controversy, The Origin of Species and its Critics 1859-1882* (Philadelphia: Temple University Press, 1970). Also refer to Michael Ruse, *The Darwinian Revolution: Science Red in Tooth and Claw* (Chicago: The University of Chicago Press, 1979).

[27]Cf. Robert I. Watson, Sr., "Galton: Developmentalism, Quantitativism, and Individual Differences" in *The Great Psychologists* (New York: J. B. Lippincott, 1978, chapter 14, pp. 319-345).

[28]Cf. Petr Kropotkin, *Mutual Aid: A Factor of Evolution* (Boston: Extending Horizons Books, 1914). Also refer to Ashley Montagu, *Darwin: Competition and Cooperation* (New York: Henry Schman, 1952).

[29]Cf. Petr Kropotkin, *Ethics: Origin and Development* (New York: Tudor, 1947).

[30]Cf. Howard E. Gruber, *Darwin on Man: A Psychological Study of Scientific Creativity* (New York: E. P. Dutton, 1974, esp. pp. 177-200).

[31]Cf. Alfred Kelly, *The Descent of Darwin: The Popularization of Darwinism in Germany, 1860-1914* (Chapel Hill: The University of North Carolina Press, 1981).

[32]Cf. Charles Darwin, *The Expression of Emotions in Man and Animals* (Chicago: The University of Chicago Press, Phoenix Books, 1963, with a preface by Konrad Lorenz).

[33]Cf. *Gists From Agassiz: or, Passages on the Intelligence Working in Nature* (Hawthorne, California: Omni Publications, 1973). This work also contains Agassiz's essay on "Evolution and Permanence of Type" (1874), pp. 97-115.

[34]Cf. Nora Barlow, ed., *The Autobiography of Charles Darwin, 1809-1882* (London: Collins, 1958) and Sir Francis Darwin, ed., *Charles Darwin's Autobiography* (New York: Collier Books, 1961). Also refer to Francis Darwin, ed., *The Life and Letters of Charles Darwin* (New York: Basic Books, 2 vols., 1959, esp. the autobiography in volume one, chapter one, pp. 25-86).

[35]Recent tributes to Charles Darwin include: Roger Bingham, "On the Life of Mr. Darwin" in *Science 82* 3(3):34-39; Boyce Rensberger, "Evolution Since Darwin" in *Science 82* 3(3):40-45; H. James Birx, "Charles Darwin: A Centennial Tribute" in *Creation/Evolution* 3(2):1-10; Stephen Jay Gould, "In Praise of Charles Darwin" in *Discover* 3(2):20-25; Stephen Jay Gould, "The Importance of Trifles" in *Natural History* 91(4):16, 18, 20-23; Francis Hitching, "Was Darwin Wrong?" in *Life* 5(4):48-52; William Van der Kloot, et al., "Evolution: 100 Years After Darwin" in *BioScience* 32(6):479, 495-533; and Paul Kurtz, et al., "Science, The Bible, and Darwin" (special issue) in *Free Inquiry* 2(3):3, 5-23, 25-49, 51-61, 63-70.

[36]Cf. Francisco J. Ayala and James W. Valentine, *Evolving: The Theory and Processes of Organic Evolution* (Menlo Park, California: Bengamin/ Cummings, 1979); Theodosius Dobzhansky, *Genetics of the Evolutionary Process* (New York: Columbia University Press, 1970); Douglas J. Futuyma, *Evolutionary Biology* (Sunderland, Massachusetts: Sinauer Associates, 1979); J. B. S. Haldane, *The Causes of Evolution* (Ithaca, New York: Cornell University Press, Cornell Paperbacks, 1966); Mar K. Hecht and William C. Steere, eds., *Essays in Evolution and Genetics in Honor of Theodosius Dobzhansky: A Supplement to Evolutionary Biology* (New York: Appleton-Century-Crofts, 1970); Julian Huxley, *Evolution: The Modern Synthesis* (New York: John Wiley & Sons, Science Editions, 1964); Ernst Mayr, *Evolution and the Diversity of Life: Selected Essays* (Cambridge: The Belknap Press of Harvard University Press, 1976); Ernst Mayr and William B. Provine, eds., *The Evolutionary Synthesis: Perspectives in the Unification of Biology* (Cambridge: Harvard University Press, 1980); Roger Milkman, ed., *Perspectives on Evolution* (Sunderland, Massachusetts: Sinauer Associates, 1982); Bernard Rensch, *Evolution Above the Species Level* (New York: Columbia University Press, 1959); George Gaylord Simpson, *Tempo and Mode in Evolution* (New York: Hafner, 1965); Steven M. Stanley, *Macroevolution: Pattern and Process* (San Francisco: W. H. Freeman, 1979) and Steven M. Stanley, *The New Evolutionary Timetable: Fossils, Genes, and the Origin of Species* (New York: Basic Books, 1981). Also refer to Ernst Mayr, et al., "Evolution" (special issue) in *Scientific American* 239(3): passim.

[37]Cf. Sir Edward Burnett Tylor, *Researches into the Early History of Mankind*

and the Development of Civilization (Chicago: Phoenix Books, 1964, first published in 1865, 3rd ed. rev. 1878), *Primitive Culture* (New York: Harper Torchbooks, 2 vols., 1958, first published in 1871), and *Anthropology* (Ann Arbor: The University of Michigan Press, 1960, first published in 1881); Lewis Henry Morgan, *Ancient Society: or, Researches in the Lines of Human Progress from Savagery through Barbarism to Civilization* (New York: Meridian Books, 1963, first published in 1877); Leslie A. White, *The Science of Culture: A Study of Man and Civilization* (New York: Grove Press, 1949) and *The Evolution of Culture: The Development of Civilization to the Fall of Rome* (New York: McGraw-Hill, 1959); V. Gordon Childe, *Social Evolution* (New York: Meridian Books, 1963, first published in 1951).

[38]Cf. Pierre Teilhard de Chardin, *The Phenomenon of Man* (New York: Harper Torchbooks, 2nd ed., 1965).

[39]Cf. Ray Ginger, *Six Days or Forever? Tennessee v. John Thomas Scopes* (New York: Signet Books, 1960) and Mary Lee Settle, *The Scopes Trial: The State of Tennessee v. John Thomas Scopes* (New York: Franklin Watts, 1972).

[40]Cf. H. James Birx, "The Creation/Evolution Controversy" in *Free Inquiry* 1(1):24-26, Niles Eldredge, *The Monkey Business: A Scientist Looks at Creationism* (New York: Pocket Books, 1982), Douglas J. Futuyma, *Science on Trial: The Case For Evolution* (New York: Pantheon Books, 1983), Laurie R. Godfrey, ed., *Scientists Confront Creationism* (New York: W. W. Norton, 1983), Philip Kitcher, *Abusing Science: The Case Against Creationism* (Cambridge: The MIT Press, 1982), Dorothy Nelkin, *The Creation Controversy: Science or Scripture in the Schools* (New York: W. W. Norton, 1983), Norman D. Newell, *Creation and Evolution: Myth or Reality?* (New York: Columbia University Press, 1982), Robert Root-Berstein and Donald L. McEachron, "Teaching Theories: The Evolution-Creation Controversy" in *The American Biology Teacher* 44(7):413-420, 405, Michael Ruse, *Darwinism Defended: A Guide to the Evolution Controversies* (Reading, Massachusetts: Addison-Wesley, 1982), and J. Peter Zetterberg, ed., *Evolution versus Creationism: The Public Education Controversy* (Phoenix, Arizona: Oryx Press, 1982). Also refer to articles in the *Creation/Evolution* journal, ed. by Frederick Edwords (P.O. Box 146, Amherst Branch, Buffalo, New York 14226). Of general interest are: Ian G. Barbour, *Issues in Science and Religion* (Englewood Cliffs, New Jersey: Prentice-Hall, 1966, esp. pp. 365-418), U. J. Jensen and R. Harre, eds., *The Philosophy of Evolution* (New York: St. Martin's Press, 1981), Mary Long, "Visions of a New Faith" in *Science Digest* 89(10): 36-43, E. C. Olson, *The Evolution of Life* (New York: Mentor Books, 1966, esp. pp. 265-278), Karl E. Peters, "Religion and an Evolutionary Theory of Knowledge" in *Zygon: Journal of Religion and Science* 17(4):385-415, W. Widick Schroeder, "Evolution, Human Values and Religious Experience: A Process Perspective" in *Zygon: Journal of Religion and Science* 17(3):267-291), and Eugenia Shanklin, "Darwin vs. Religion" in *Science Digest* 90(4):64-69, 116.

Chapter 9

HUXLEY AND HAECKEL

T homas Henry Huxley (1825-1895), a versatile scientist and process philosopher, developed his interests in biology while a surgeon aboard the H.M.S. *Rattlesnake* in Australian waters (1846-1850).[1] He later devoted his attention to anatomy and paleontology as well as defended, expanded, and disseminated Darwin's scientific theory of biological evolution.[2] As a naturalist, Huxley championed abduction or scientific inquiry against blind faith: the scientific method requires observation, classification, hypotheses, deductions, and verification. He held the ingenious theory of biological evolution to be logically valid and empirically sound. He was not reluctant to give serious consideration to the philosophical implications of the doctrine of transmutation. Influenced by David Hume, Huxley adhered to epiphenomenalism, phenomenalism, and agnosticism (he coined the term *agnostic* in 1869 to describe his own epistemological position). Although advocating cosmic evolution, he clearly separated evolutionary nature "red in tooth and claw" from man's ethical character, which he realized condemns "ape and tiger methods" in the realm of human interactions. Kropotkin criticized Huxley for caricaturing the image of nature by overemphasizing the allegedly pervasive element of cutthroat competition to the exclusion of the no less important concomitant element of mutual aid as well as interindividual and interspecific cooperation. Likewise, in keeping with this objection, Kropotkin further criticized Huxley for making too sharp a distinction between nonhuman nature and the human sociocultural world.[3]

At the Oxford meeting of the British Association for the Advancement of Science in 1860, Huxley's valiant defense of Darwinian evolution against the attacks of Bishop Samuel Wilberforce resulted in a victory for the progress in biology over dogmatic theology. He also encouraged educational reforms favoring empirical inquiry, the sciences, and reason.

For the first half of the nineteenth century, inquiries into man's proper place in nature lacked scientific evidence. Geology, paleontology, archaeology, and biology were still underdeveloped sciences in contrast to astronomy and physics.[4] Rigorously defending Darwinism, Huxley was the first to write a book for the sole purpose of scientifically substantiating the application of evolutionary principles to the phyletic development of humankind. In *Evidence as to Man's Place in Nature* (1863), Huxley referred to established knowledge from comparative anatomy and embryology, primatology, and human paleontology.[5] Where Darwin had held the "mystery of mysteries" to be the origin of species, Huxley held the "question of questions" to be the origin of man.

Huxley established his "pithecometra thesis," claiming that man is biologically closer to the great apes than the apes are to the monkeys. Hence, man is neither biologically separated from the animal world in general nor absolutely distinct from the apes in particular. Huxley repeated his pithecometra theory in a short essay also entitled "On the Origin of Species" (1863).[6] These two works substantially anticipated Darwin's *The Descent of Man* (1871).

As a member of the Metaphysical Society, Huxley was a critical rationalist. A bold evolutionary scientist and agnostic, he insisted on the immensity and unity of nature. His writings, lectures, and teachings brought him wide exposure and recognition. Although residual doubt pervaded his thoughts, he chose science rather than religion, matter rather than spirit, and saw evolutionary nature as the product of chance rather than design.

Huxley had an enormous influence on science and education. Besides defending Darwinism, he made major contributions to invertebrate zoology (especially the mollusks, coelenterates, and tunicates) and taxonomy. However, his theological and philosophical views did not parallel Ernst Haeckel's bold atheism and pantheism (scientific monism). Huxley had been primarily concerned with the ethical implications of biological evolution. In "Evolution and Ethics" (The Romanes Lecture, 1893), Huxley correctly and optimistically wrote, "Let us understand, once for all, that the ethical progress of society depends, not on imitating the cosmic process, still less in running away from it, but in combating it."[7]

In this connection, Darwin's own cousin, Sir Francis Galton (1822-1911), in 1883 coined the term "eugenics" to refer to the contemplated improvement of species, including deliberate modifications in human heredity (contrast eugenics and euthenics). It becomes clear that the acceptance of the scientific theory of biological evolution quickly raises serious religious and philosophical questions not easily answerable, particularly in the realm of ethics and morals.

In Germany, Ernst Heinrich Haeckel (1834-1919) devoted his life to defending Darwin's theory of biological evolution.[8] He was not reluctant to consider rigorously the philosophical and theological implications of cosmic evolution in general and human evolution in particular. As a natural philosopher, he contributed to a sound scientific understanding of the nature of mankind and its proper place in the universe (despite the unavoidable limitations of his Victorian epoch).[9]

In Haeckel's major philosophical work, *The Riddle of the Universe at the Close of the Nineteenth Century* (1900), he presented a naturalist (materialist) interpretation of evolution within a cosmic perspective.[10] The dogmatic dualism of Kant and Spencer was rejected, showing his familiarity with the chemistry, biology, anthropology, psychology, and the history of philosophy of his time. He was particularly influenced by Spinoza, Goethe, and Darwin. A major zoologist and evolutionist, he is primarily remembered for his early defense of Darwinism and for establishing his "Biogenetic Law." Committed to the view that both science and speculation are necessary for a sound interpretation of the cosmos, he developed his own scientific cosmology. His monistic philosophy rejected the following three theistic dogmas: a personal God, the personal immortality of the human soul, and the freedom of the individual will. Also he rejected the concept of empty space, but supported strict determinism.

Haeckel's philosophy of nature relied upon experience and inference (thought or speculation), both sources of knowledge being held indispensable. It rejected anthropism, i.e. that powerful and worldwide cluster of erroneous opinions that opposes the human organism to the whole of the rest of nature and represents it to be the preordained end of an organic creation, an entity essentially distinct from it, a godlike being. As a result, Haeckel rejected (1) the anthropocentric dogma, which maintains that man is the preordained center and aim of all terrestrial life (Haeckel advocated a cosmic perspective); (2) the anthropomorphic dogma, which maintains that God, the creator, sustainer, and ruler of the world, has human attributes and that man is therefore godlike instead of the other way around (Haeckel rejected all forms of theism but was sympathetic to scientific pantheism); and (3) the anthropolatric dogma, which maintains an ontological dualism and the personal immortality of the human soul (Haeckel's own view was monistic and, as such, materialistic).[11]

Briefly, the antropistic view of the world that sprang from these three dogmas is in irreconcilable opposition to Haeckel's monistic system. Indeed, it is rejected by his new cosmological perspective. Likewise, Haeckel held that his evolutionary monism was capable of resolving the following problems: (1) the nature of matter and force, (2) the origin of motion, (3) the origin of life, (4) the order in nature, (5) the origin of simple sensation and consciousness, (6) rational thought and speech, and (7) freedom of the will.[12] Consequently, Haeckel maintained that the great Law of Substance remained the only simple, comprehensive, mysterious, and enigmatic riddle of the universe. He espoused the view that this law is the fundamental cosmic fact of reality establishing the eternal persistence of matter and force (this was the law that guided his monistic philosophy through the complexity of the universe to a solution of world problems).

What emerges as crucial in Haeckel's philosophy is his severe criticism of the fallacy of anthropism with its anthropocentric, anthropomorphic, and anthropolatric dogmas. Clearly, he denied the basic assumptions of theism, for he held that his monistic, pantheistic, and evolutionary cosmology would eliminate these errors. In place of these dogmas, Haeckel advocated the Law of Substance, with its naturalist implications supported by scientific knowledge. Man is seen as a product of biological development, and all of his faculties are totally within nature. To Haeckel, there is no need for transcendental or supernatural assumptions (the independent existence of the external natural world prior to the emergence of humankind and the limited constituting ability of human consciousness are both accepted as basic facts of physical reality).

Haeckel found empirical support for his evolutionary perspective in comparative anatomy and physiology, comparative psychology, comparative embryology, and paleontology. Comparative morphology demonstrated that man is a true vertebrate, tetrapod, mammal, placental, and primate. There was even biological proof for the asserted close affinity of man with the great apes. In all important respects, the symboling human animal presented all the anatomical marks of a true ape with reason being the only prerogative that essentially distinguished him from all his lower or earlier evolutionary relatives.

Physiology and psychology demonstrated that the vital processes in man are subject to the same physical and chemical laws as those of all other animals. Likewise, man's mental faculties or psychic life (including his rational faculty) are said to differ from those of the nearest related animals only in degree and not in kind, i.e. quantitatively but not qualitatively. The human mind or soul (consciousness) evolves during the ontological development of the human individual and is dependent upon the common structure and normal functioning of the psychic organ, the human brain. As a result of this naturalist interpretation of the development of mind, personality can lay no claim to immortality.

Haeckel acknowledged that reason does distinguish man from the lower animals, but reason is not a supernatural aspect of the human mind. He held that reason is an activity of the human mind that has evolved along with the evolution of the brain. To account for the evolutionary emergence of consciousness (and eventually the emergence of reason), Haeckel adopted a position of panpsychism. His position of panpsychism resulted from the investigation of comparative psychology. However, a distinction should be made between a moderate and an extreme form of panpsychism. A moderate form of panpsychism merely maintains that all living objects manifest an aspect (and therefore a degree) of psychic development or sensitivity (this is Haeckel's own position). However, an extreme form of panpsychism advocates that ultimately everything in reality is psychic in nature, and therefore an ontological monism grounded in mind is necessary. A naturalist may support an ontological monism grounded in materialism, for the degrees of sensitivity manifested by biological organisms depend upon the development of the central nervous system and brain. As a result, one may speak of the materialist emergence of psychic activity. By maintaining that the human mind differs from the psychic activity of lower animals merely in degree, Haeckel's view agreed with those of Huxley and Darwin. His position does not imply anthropocentrism, a doctrine he rejected. He merely acknowledged that in the long series of biological organisms that have evolved man is the most complex and that degrees of sensitivity in function are directly related to the degrees of complexity in the structure of the central nervous system and brain. Therefore, since man's central nervous system and brain are the most complex in the animal kingdom, it is not surprising that he should manifest the highest degree of consciousness. In fact, man is capable of self-consciousness primarily because of his superior central nervous system and brain and their role in sociocultural interaction.

Embryology or ontogeny demonstrates a common parentage of all organisms in general and of all primates in particular. Embryology implies that all existing primates share a common ancestry; man and the great apes (orangutan, chimpanzee, and gorilla) being the latest products of primate evolution. Haeckel formulated his own "Biogenetic Law" to express the causative connection that exists between ontogeny and phylogeny (also see an earlier formulation by Karl Ernst von Baer): *"Ontogenesis is a brief and rapid recapitulation of phylogenesis,* determined by the physiological functions of heredity (generation) and adaptation (maintenance). . . . In man, as in all other organisms, the embryonic development is an epitome of the historical development of the species."[13] Haeckel's Biogenetic Law of recapitulation was regrettably exploited by extreme cultural evolutionists and condescending racists at the expense of technologico-economically little developed and nonliterate societies arbitrarily described as primitive and backward peoples. Later, this law was misinterpreted even more cynically by racist politicians in unethical ways of which Haeckel certainly would have disapproved.

Haeckel accepted Huxley's pithecometra thesis, i.e. that the differences between man and the great apes are not so great as those between the apes and the lower monkeys. Denying the different proofs that have been given for the immortality of the human soul (i.e. theological, cosmological, teleological, moral, ethnological, and ontological), he thought that there are sound scientific arguments to support this denial (i.e. physiological, historical, experimental, pathological, ontological, and phylogenetic evidence).

Paleontology supported an evolutionary interpretation of the Great Chain of Being, or *scala naturae*, previously acknowledged by the nonevolutionists. There is a hierarchy in nature represented by successive chronological developments, e.g. protozoa, protista, metazoa, invertebrates, fish, amphibians, reptiles, birds, and mammals as well as developments in plants and insects. As such, the paleontological record supported the evolutionary unity of nature and documented Haeckel's monistic view. Haeckel's own lectures on the evolution of man were very popular. His hypothesis that the "missing link" would be found in Asia inspired Eugene Dubois, his student, to travel to Java where he subsequently discovered the fossil hominid *Pithecanthropus erectus* (1890-1892). As can be seen, Haeckel's fruitfully erroneous hypothesis of the missing link led to a scientifically positive result in physical anthropology.

Haeckel's monism rested upon the Law of Substance, which in turn rested upon two great cosmic theorems: (1) the conservation of matter (Lavoisier's chemical law of the persistence or indestructibility of matter presented in 1789) and (2) the conservation of energy (Mayer's and Helmholtz's physical law of the persistence of force presented in 1842). On the basis of all this, Haeckel concluded that the sum total of force or energy in the universe remains constant, no matter what changes take place around us. Energy is eternal and infinite, like the matter on which it is inseparably dependent. The two cosmic laws are fundamentally one, their unity being expressed in the formula of the law of the persistence of matter and force. Haeckel referred to this fundamental cosmic law as the supreme Law of Substance (axiom of constancy of the universe). The law supported universal causality and unity. Haeckel's monistic ontology has benefited in the wake of Einstein's theory of relativity. The demonstrated interchangeability of matter and energy scientifically justifies a naturalist monism, for matter and energy are merely two manifestations of the one ontological stuff of the universe.

There are six essential characteristics of Haeckel's view of the cosmos: (1) the universe is infinite in space, (2) the universe is eternal in time, (3) the universe is a *perpetuum mobile* (movement is an innate property of substance), (4) there is no internal purpose whatever in the universe, (5) there is an eternal repetition in infinite time of the periodic metamorphosis of the cosmos, and (6) the idea of God is identical with that of nature or substance. Haeckel's scientifically pantheistic interpretation of reality is reminiscent of the positions held by Bruno, Spinoza, Goethe, Humboldt, Kňezević, and Einstein.[12]

Haeckel argued that evolution and theism are irreconcilable. The evolutionary perspective led to the paradoxical conception of God as a "gaseous vertebrate," or gaseous being (by the expression gaseous vertebrate, Haeckel is referring to the invisible and personal aspects of the traditional Christian conception of God). He concluded that monistic pantheism is necessarily the world-system of the modern scientist and that the modern man who has science and art (and therefore religion) needs neither a special church nor narrowly enclosed portion of space.

Haeckel thought it very probable that unicellular evolution had taken place on planets elsewhere in the universe and that higher plants and animals have evolved on these planets, very questionable that these higher plants and animals are identical in development to those on earth, wholly uncertain whether there are vertebrates, mammals, and men on other planets, much more probable that there are different plants and animals elsewhere, and perhaps higher beings transcending the intelligence of earthly men. He was also confident that future progress in astronomy, geology, physics, chemistry, biology, and anthropology would bear out his positions.

Haeckel's cosmological and evolutionary perspective is expressed in the following fundamental theorems: (1) the universe is eternal and infinite, (2) its substance, with its two attributes (matter and energy), fills infinite space and is in eternal motion, (3) this motion runs on through infinite time as an unbroken development, with a periodic change from life to death (from evolution to devolution), (4) the innumerable bodies that are scattered about the space-filling ether all obey the same Law of Substance (while the rotating masses slowly move towards their destruction and dissolution in one part of space, others are springing into new life and development in other quarters of the universe), (5) our sun is one of these unnumbered perishable bodies, and our earth is one of the countless transitory planets that encircle them, (6) our earth has gone through a long process of cooling before water, in liquid form (the first condition for organic life), could settle thereon, (7) the ensuing biogenetic process, the slow development and transformation of countless organic forms, must have taken millions of years, (8) among the different kinds of animals that arose in the later stages of the biogenetic process on earth, the vertebrates have far outstripped all other competitors in the evolutionary race, (9) the most important branch of the vertebrates, the mammals, were developed later (during the Triassic period) from the lower reptilia, (10) the most perfect and most highly developed branch of the class mammalia is the order of primates, which first put in an appearance (by development from the lowest prochoriata) at the beginning of the Tertiary period, (11) the youngest and most perfect twig of the branch primates is man, who sprang from a series of manlike apes toward the end of the Tertiary period, and (12) consequently, the so-called history of the world (i.e. the brief period of a few thousand years that measures the duration of civilization) is an evanescently short episode in the long course of organic evolution, just as this, in turn, is merely a small portion of the history of our planetary system, and as our earth is a mere speck in the infinite universe, so man himself is but a tiny grain of protoplasm in the perishable framework of organic nature.[15] Foreshadowing Carl Sagan and several others, Ernst Haeckel proposed that it is very probable that on many planets of other solar systems plants and animals have evolved similar to those on earth, perhaps even higher plants, animals, and vertebratelike beings with intelligence that may even far transcend our earthly species.

In *Monism as Connecting Religion and Science* (1892), Haeckel's science-oriented evolutionary philosophy had rejected homotheism (theoanthropomorphism), i.e. attributing human shape, flesh, and blood to the gods.[16] He held that the Christian concept of a personal God is totally erroneous in light of scientific evidence and the evolutionary perspective. Unlike Darwin, Spencer, and Huxley, Haeckel was not reticent in expounding the naturalist and atheist implications of the theory of evolution as he understood them. His philosophy of man was grounded in the special sciences informed by critical reason. He held that the evolution of man could be sufficiently accounted for by the sciences, i.e. there is no need to distort or ignore established natural facts by dogmatically holding to unverifiable philo-sophical assumptions for the sake of retaining a theological orientation. He was bold enought to clearly point out that there are no longer any legitimate reasons for believing in a super-natural origin and essence of man. Hence, for Haeckel, there is no need to construct elaborate philosophical systems in an attempt to reconcile the special sciences with theology. Theology is simply being replaced by ethically inspired and morally responsible scientific understanding and philosophical appreciation. His strong reaction to the dogmatism of the churches was reinforced by their refusal to acknowledge the implications of a naturalist interpretation of biological evolution. Few natural philosophers have been so honest and persistent in their quest for truth as Ernst Haeckle, who, even when considered solely within the limits of his time, remains a major scientific figure of undeniable historic significance.

Haeckel's *Last Words on Evolution: A Popular Retrospect and*

Summary (1905) restated his philosophical position for the last time. The work contains three essays given at the last public delivery by the great man who was rightfully referred to as the "Darwin of Germany." Rightly or wrongly, he continued to maintain the thesis that materialism (evolutionary monism) and supernaturalism are diametrically and irreconcilably opposed. He wrote that the papacy is anything but a divine institution, describing it as "the greatest swindle the world has ever submitted to" (one may or may not regret that a man of his intellectual breadth and depth did not find it possible to see this psycho-culturally influencial institution in a more critically balanced light).[17] In his essay "Charles Darwin as an Anthropologist" (1909), Haeckel acknowledged his indebtedness to the great genius for the fact of organic evolution and its mechanism of natural selection, which support a monistic view of the origin and nature of humankind, but he resisted the new theories of mutation and the gene.

Haeckel's works consistently discarded vitalism, teleology, positivism, agnosticism, and mysticism. They defended mechanism, hylozoism, a moderate form of panpsychism, pantheism, monism (matter-energy or matter-force), and the use of reason. Within his evolutionary framework, he never veered from his basic assumptions, i.e. that naturalist or antispiritualist monism supports an eternal, infinite, self-sufficient, indestructible, uncreated, and endlessly evolving universe. The empirical evidence to support the doctrine of evolution is sufficient to convince any open-minded scientist, philosopher, or even an enlightened theologian. In its general implications, Haeckel's *Weltanschauung* (worldview) remains untarnished after more than half a century of continued scientific advancement. The fact that his millions of years have become billions of years is a correction resulting in a greater perspective that he would have happily embraced. It may be charged that Haeckel neglected to establish a rigorous epistemology, avoided mental analysis, underestimated the influence of social evolution (he could have added a supplementary set of cultural theorems including a less ambiguous and more elaborate ethical stand),[18] and even deified the universe. Haeckel was primarily a zoologist, and a great one, who like Bruno and other pantheists did not experience a need for a special divinity apart from and beyond that of the evolving cosmos itself.

Haeckel's views were the source of considerable controversy in biology, philosophy, and religion. He distinguished himself in zoology, making major contributions to embryology (Biogenetic Law) and taxonomy. Unfortunately, he advocated a limited form of spontaneous generation to account for the origin of the simplest protoplasmic substance (monera) from inorganic carbonates. Nevertheless, he was convinced of the essential unity of the inorganic and organic worlds. Haeckel upheld the essential unity of the entire universe. His doctrine of substance (ontological monism) maintained that the ultimate nature of the stuff of the universe is unknown, but it rejected traditional dualism as well as atomistic materialism and idealistic monism. Yet, Haeckel's writings advocated materialism and panpsychism, strict determinism and mechanism, and a scientific rejection of the supernatural. His evolutionary monism was grounded in the law of the conservation of matter and energy (he held to the existence of ether and even supported action-at-a-distance).

Haeckel advocated the reconciliation of science (empiricism) and philosophy (rationalism and speculation) but rejected religious faith as a substitute for critical inquiry. He held that introspection must be supplemented with experimental psychology and physiology. Clearly, for him, consciousness itself is a natural phenomenon and the product of biological evolution.

Haeckel's pantheism advocates monism, a "natural equality" of egoism and altruism, the improvement of education and the human condition through more science, and is grounded in humanism, naturalism, and the evolutionary viewpoint within a cosmic perspective. With only a few important modifications, this philosophical worldview remains in step with modern thought.

NOTES AND SELECTED REFERENCES

[1] Cf. Cyril Bibby, ed., *The Essence of T. H. Huxley* (New York: St Martin's Press, 1967), Cyril Bibby, *Scientific Extraordinary: The Life and Scientific Works of Thomas Henry Huxley, 1825-1985* (New York: St Martin's Press, 1972), William Irvine, *Apes, Angels, and Victorians: The Story of Darwin, Huxley,*

and Evolution (New York: Mc-Graw Hill, 1972) and William Irvine, Thomas Henry Huxley (London: Longmans/Green, 1960).

[2]Cf. Thomas H. Huxley, Darwiniana: Essays (New York: D. Appleton, 1895).

[3]Cf. Thomas H. Huxley's "The Struggle for Existence in Human Society" (1888) in Petr Kropotkin, Mutual Aid: A Factor of Evolution (Boston: Extending Horizons Books, 1914, pp. 329-341).

[4]Cf. Charles Coulston Gillispie, Genesis and Geology: A Study in the Relations of Scientific Thought, Natural Theology, and Social Opinion in Great Britain, 1790-1850 (New York: Harper Torchbooks, 1959) and Bentley Glass, Owsei Temkin, and William L. Straus, Jr., eds., Forerunners of Darwin: 1745-1859 (Baltimore: The Johns Hopkins Press, 1959).

[5]Cf. Thomas H. Huxley, Man's Place in Nature (Ann Arbor: The University of Michigan Press, 1959).

[6]Cf. Thomas H. Huxley, On the Origin of Species or, the Causes of the Phenomena of Organic Nature (Ann Arbor: The University of Michigan Press, 1968, refer to chapter six, pp. 121-144).

[7]Cf. Thomas H. Huxley, Collected Essays (New York: D. Appleton, 1894, volume nine, p. 83). Volume eight of Huxley's Collected Essays is devoted to discourses on biology and geology, containing his lecture "On a Piece of Chalk" (1868), pp. 1-36.

[8]Cf. Ernst Haeckel, The History of Creation or the Development of the Earth and Its Inhabitants by the Action of Natural Causes: A Popular Exposition of the Doctrine of Evolution in General, and of that of Darwin, Goethe, and Lamarck in Particular (London: Kegan Paul, Trench, Trübner, 2 vols., 4th ed., 1892). This work was first published in 1868. Also refer to Haeckel's The General Morphology of Organisms (1866). Also refer to Ruth G. Rinard, "The Problem of the Organic Individual: Ernst Haeckel and the Development of the Biogenetic Law" in Journal of the History of Biology 14(2):249-275.

[9]Ibid., esp. pp. 363-446. Also refer to Ernst Haeckel, The Evolution of Man: A Popular Scientific Study (New York: G. P. Putnam's Sons, 2 vols., 5th ed., 1905). This work was first published as Anthropogenie in 1874. Also refer to Haeckel's Collected Popular Lectures on the Subject of Evolution (1878), Freedom in Science and Teaching (1878), and The Last Link: Our Present Knowledge Regarding the Origin of Men (1898).

[10]Cf. Ernst Haeckel, The Riddle of the Universe at the Close of the Nineteenth Century (New York: Harper & Brothers, 1905).

[11]Ibid., pp. 11-12.

[12]Ibid., pp. 15-16.

[13]Ibid., pp. 81, 143.

[14]Ibid., chapter fifteen, esp. pp. 288-291.

[15]Ibid., pp. 13-14.

[16]Cf. Ernst Haeckel, Monism as Connecting Religion and Science: The Confession of Faith of a Man of Science (London: Adam and Charles Black, 1895). Also refer to David H. DeGrood, Haeckel's Theory of the Unity of Nature (Boston, The Christopher Publishing House, 1965).

[17]Cf. Ernst Haeckel, *Last Words on Evolution: A Popular Retrospect and Summary* (New York: Peter Eckler, 1905, p. 177). Other works by Ernst Haeckel include *A Visit to Ceylon* (1883), *The Wonders of Life: A Popular Study of Biological Philosophy* (1904) and *Eternity* (1916).

[18]In his insightful paper "Darwin, Nietzsche, and the German Quest" delivered at the *Darwin & Evolution: A Centennial Tribute* symposium held at Canisius College (April 19, 1982) and chaired by Dr. Birx, Dr. Eugene P. Finnegan pointed out that Ernst Haeckel was a leading exponent of German Social Darwinism and, unfortunately, his views subsequently influenced Adolf Hitler. Refer to Alfred Kelly, *The Descent of Darwin: The Popularization of Darwinism in Germany, 1860-1914* (Chapel Hill: The University of North Carolina Press, 1981, esp. pp. 100-122). Of special importance is Daniel Gasman, *The Scientific Origins of National Socialism: Social Darwinism in Ernst Haeckel and the German Monist League* (New York: American Elsevier, 1971, esp. pp. xii-xiii, xx, 49-50, 115, 147-156, 169-174). A rigorous evolutionist must always be open to change and development throughout this universe, including within our human sociocultural milieu.

Chapter 10

SPENCER, FISKE, SUMNER, AND WARD

Herbert Spencer (1820-1903) was an only child in a family of teachers.[1] He had an unrestrained but unsystematic education in the natural sciences and arithmetic. Spencer's formal training was sparse; he was greatly influenced by his father, who concentrated on the intellectual side of education but did little to develop the emotional side (Spencer disliked languages, but was attracted to physics and mathematics). At the age of seventeen he was already a railroad engineer. After ten years, he became newspaper editor of the *Pilot* and eventually subeditor of the *Economist*. A reading of Charles Lyell's *Principles of Geology* (1830-1833) introduced him to the emerging idea of evolution, which became his theoretical orientation.

As an independent writer, Spencer's first book was *Social Statics* (1850).[2] It advocated a Lamarckian theory of biosocial evolution: the creative advance from homogeneity to heterogeneity is a universal law of evolution manifested in the inorganic, organic, and superorganic orders. This law of evolution is also held to be applicable to language, poetry, literature, painting, sculpture, architecture, dancing, and music. Being influenced by Kant, Malthus, Lamarck, Lyell, and especially Darwin, Spencer developed a worldview grounded in cosmic evolution. The result was his monumental work, *Synthetic Philosophy* (1860-1896), of ten volumes covering metaphysics, biology, psychology, sociology, and ethics.

For Spencer, the universe exhibited one continuous and progressive scheme of development. His epistemological view was grounded in the distinction between experience and reality, i.e. the knowable and the unknowable. He held that science studies the necessary relationships in the phenomenal manifestations of the unknowable in order to discover general laws. Space, time, force, causation, motion, substance, and matter are not innate ideas but inscrutable concepts derived from the experiential history of the human species. Education should be devoted to the sciences and art, emphasizing experimentalism and free expression respectively. Spencer rejected theism, pantheism, and atheism. He held agnosticism to be the only reasonable belief in regard to metaphysical and religious issues.

In sociology, Spencer advocated the "survival of the fittest," or progress-through-struggle, as a social principle. He believed that feelings of sociality and sympathy had led to the emergence of society. He gave an organismic interpretation of society and held to progressive evolution, the psychic unity of mankind, and the principle of laissez-faire. He was an early spokesman for industrial capitalism and advocated a strictly utilitarian system of ethics, i.e. an intelligent search for pleasure within the notion of duty. He believed that through the inheritance of acquired ethical conduct, human suffering would gradually be eliminated. Grounded in Darwinian evolution, Spencer's position is referred to as "Social Darwinism": it opted for the eventual harmony of egoism and altruism and the increase of freedom and happiness.

Herbert Spencer's "First Principles" (1862) is the opening volume of his *Synthetic Philosophy*.[3] It shows obvious influences by Kant, Lamarck, and Darwin. Like Kant, Spencer made the distinction between the phenomenal realm of knowable appearances in the cosmos and the unknowable realm of ultimate reality beyond the appearances. As a result, ontology can never be completely known. All laws are limited in application to phenomena only. Spencer also adhered to the erroneous doctrine of the inheritance of acquired characteristics developed in Lamarck's *Zoological Philosophy* (1809). Likewise, Spencer unfortunately overextended the implications of Darwin's principle of "natural selection" formulated in *On the Origin of Species* (1859): he transferred the biological concept of the survival of the fittest to the social realm.

"First Principles" is a daring attempt to establish an *a priori* law governing the phenomena of cosmology, biology, sociology, psychology, and ethics. The science-oriented work contained several major theses: (1) the cosmos is divided into a knowable realm and an unknowable realm; (2) matter, motion, space, and time are manifestations of force; (3) the cosmos passes through finite cycles, each having the three successive stages of evolution, equilibration, and dissolution. Although the work is both deductive and inductive, it finally rests upon the basic assumption of the persistence of force. Despite its errors and limitations, it represents the first comprehensive view of cosmic evolution.

Spencer attempted to reconcile the antagonisms between religion and science, establishing his own philosophical view. He maintained that if religion and science are to be reconciled the basis of reconciliation must be this deepest, widest, and most certain of all facts: that the force that the universe manifests to us is utterly inscrutable. He held that atheism, pantheism, and theism are all absolutely unthinkable. Science is simply a higher development of common knowledge: a prevision to aid the achievement of good and avoid the bad. He hoped to establish a general, rational "Theory of Things" and envisioned philosophy as a science of the conditioned, i.e. a science of that which is conditioned and limited by human thought. He maintained that religion and science are necessarily correlatives, always growing but limited. They both have as a common denominator the belief in an unknowable realm as the unconditioned: ultimate reality is behind all the appearances in the cosmos and is forever unknowable. Matter, motion, space, and time in their ultimate nature must always remain absolutely incomprehensible to the human understanding. Even force passes all human understanding. In brief, appearances are knowable while ultimate reality is forever unknowable. Spencer asserted this to be the ultimate truth that religion and science will someday admit.

For Spencer, philosophy emerges as an ultimate interpretation of the universe, i.e. knowledge of the highest degree of generality or completely unified knowledge. His view of knowledge was grounded in universal experience and the coherent theory of truth: knowledge is the synthesis of impressions and ideas. Both the nonego (impressions) and ego (ideas) are manifestations of an unknowable force. For Spencer, all knowledge can be only relative and never absolute. Space, time, matter, and motion are conscious abstractions from the experiences of force, and knowledge of them is always relative. Therefore, the first generalization of Spencer's synthetic philosophy is that force is the "ultimate of ultimates"; it is also a conditioned manifestation of the unconditioned cause. Force is the most fundamental relative reality indicating an absolute reality by which it is immediately produced.

Spencer held that the philosophy of evolution required an *a priori* law that will unite all concrete phenomena, primary truths and their implications, and express simultaneously the complex antecedents and the complex consequences that any phenomenon as a whole presents: a law of the concomitant redistribution of matter and motion from the imperceptible to the perceptible and from the perceptible to the imperceptible, as well as the continuous redistribution in nature. This synthetic generalization would apply to each detail of the cosmos (to all manifestations of the inorganic, organic, and superorganic). It would apply to all of the artificial divisions in nature that man has established to facilitate the arrangement and acquisition of knowledge within the one cosmos. Astronomical, geological, biological, psychological, sociological, and moral phenomena (as well as all manifestations of the universal and unlimited force) are subject to this same law. He assumed that there is a single metamorphosis universally progressing, wherever the reverse metamorphosis has not set in. Both inductively and deductively, Spencer considered the process of evolution to be one in principle and one in fact. After examining the empirical sciences, he gave the following formulation of his *a priori* law of evolution: "Evolution is an integration of matter and concomitant dissipation of motion; during which the matter passes from an indefinite, incoherent homogeneity to a definite, coherent heterogeneity; and during which the retained motion undergoes a parallel transformation."[4]

To Spencer, in both its simple and compound manifestations, the cosmic process of evolution advances in a geometrical progression along with advancing heterogeneity toward a stage of equilibration and quiescence in which the most extreme multi-formity and most complex moving equilibrium are established. The temporality of this state is limited, and complete equilibrium or the arresting of evolution is followed by universal dissolution. Spencer held that universal dissolution is an inevitable complement or countermovement of universal evolution. As a result, his cosmology is necessarily cyclical, there being an eternal alternation of evolution and dissolution in the totality of things. This Spencerian idea had a profound influence on the eminent Serbian cosmologist and philosopher of history Božidar Knežević who, however, never embraced Spencer's Social Darwinism (Nietzsche also taught a cyclical view of the universe).

In general, "First Principles" established truths, inferences, and an *a priori* law of evolution seemingly applicable to the appearances of things, i.e. the knowable phenomena of nature manifested by the unknowable force. Science and religion meet at a common agreement: the cosmic force behind the totality of things is eternal, persistent, unperceivable, and inconceivable. The connection between the conditioned phenomenal order and the unconditioned ontological order is forever inscrutable. Spencer also claimed that the implications are no more materialist than they are spiritualist, and no more spiritualist than they are materialist.

In "Principles of Biology" Spencer rejected the theories of spontaneous generation and special creation: the phenomenon of life is interpreted according to the cosmic law of evolution as formulated in "First Principles."[5] Life is the continuous adjustment of internal relations to external relations, a moving equilibrium toward increasing complexity, greater unity, and ultimate equilibration. Spencer's concept of the survival of the fittest (doctrine of indirect equilibration) is an early restatement of Darwin's Malthusian principle of natural selection. As Darwin's, Spencer's biological evolution held to the erroneous Lamarckian conception of the inheritance of acquired characteristics (Darwin, Wallace, Spencer, Huxley, Haeckel, Kovalevskii, and Gray were all probably quite ignorant of Gregor Johann Mendel's work, which established the foundation for modern genetics). Spencer derived arguments for the evolutionary hypothesis from classification, embryology, biogeographical distribution, and a detailed account of anatomical and physiological developments in plants and animals. He concluded that man's natural and social equilibrium will never remain complete in the cyclic nature of this inscrutable cosmos.

In Spencer's "Principles of Sociology," the superorganic or human society resembles an individual organism: (1) it increases in mass, (2) it increases in complexity, (3) it increases in integration or interdependence, and (4) its form continues independent of its units, which are born, develop, and die.[6] This organic nature of society is the foundation of Spencerian sociology. In agreement with the universal law of evolution, the development of the superorganic aims toward equilibration. During this development, society progresses from a militant stage to an industrial stage where state control is reduced to the minimum and individual liberty is expanded to the maximum possible. Spencer, an archconservative, rejected all forms of socialism and extended Darwin's biological concept of natural selection by adopting it as a social principle: for the improvement and perfection of the human race, the physically and mentally imperfect are not to be preserved. Spencerian sociology is clearly too directly grounded in biology (survival of the fittest is assumed best for the human species).

Spencer's "Principles of Psychology" recognized the potentiality of mind in life, just as the potentiality of life is in inorganic matter itself.[7] He made a distinction between the nonego (modes of motion) and the ego (modes of feeling). The ego develops by infinitesimal steps and is continuously in a process of relating and adjusting to the external environment. As with life, psychical evolution displays modes of consciousness in hierarchical order: reflex action, instinct, memory, reason, feeling, and will. Like Kant, Schopenhauer, and Nietzsche, Spencer gives priority to the will but neglects the unconscious. The Lamarckian influence resulted in his maintaining that the ideas of space, time, and moral precepts have arisen through the immense accumulations of experience in the historical development of man and are transmitted (innately inherited) as *a priori* intuitions that have been consolidated into modifications of the central nervous system (this is a further attempt to reconcile Kantian epistemology with an evolutionary cosmology). Spencer held that the evolution of consciousness increases by degrees and that there is no manifestation of consciousness different in kind. There is a parallel evolution of outer physical relations and inner mental relations: they continuously interact and adapt to each other while increasing in number, complexity, and heterogeneity. Spencer also concluded that the substantial ego behind the states of consciousness is unknowable, being a part of the unknowable force.

Spencer's "Principles of Ethics" presented a form of hedonism.[8] Correlated with the evolution of structures and functions, the evolution of conduct is by degrees, accumulated and transmitted genetically, and increases in complexity. The evolution of conduct is toward the greater fulfillment of an individual life within a social context. His ultimate ethical principle or supreme moral law is the "law of equal freedom": the *summum bonum* is pleasure. The distinction is made between relative and absolute ethics. As long as any degree of pain or evil is present in those affected by good conduct, the conduct is only relatively right; as long as there is any degree of pain or evil, conduct cannot be absolutely right or good. The goal of ethical evolution is the removal of war, pain, and evil, and the increase of happiness and social harmony. Spencer hoped that eventually moral conduct would become purely natural and instinctive. Like his sociology and psychology, Spencerian ethics is grounded in evolutionary biology.

Although Spencer's *Synthetic Philosophy* is a tremendous achievement of integrative thought, there are in it some glaring faults. His law of evolution is a synthetic descriptive generalization, admittedly assumptive: to call it a universal and necessary law of the cosmos is to reject the principle of verification. The distinction between the knowable and the unknowable is inadvertently dogmatic and could limit future inquiry. There is no empirical evidence to substantiate the doctrine of the inheritance of acquired characteristics in biology, epistemology, or ethics. The mistaken overextension of the Darwinian concept of natural selection into sociology is unwarranted and has aptly been called by Marvin Farber "pseudoevolutionism."[9] Lastly, the whole system has an air of long-range pessimism because of the inevitability of dissolution, although the concept of evolutionary cycles must be taken seriously by speculative philosophers as a considerable cosmological possibility. Despite its errors, biases, and other limitations, its major contribution to scientific philosophy is its attempt at a systematic introduction to evolution, both terrestrial and celestial.

John Fiske (1842-1901) wrote *Outlines of Cosmic Philosophy, Based on the Doctrine of Evolution, with Criticism on the Positive Philosophy* (1874).[10] This volume was a reinstatement, with additions, of Herbert Spencer's *Synthetic Philosophy*. In it, Fiske referred to the latter's discovery of a cosmic law of evolution as the greatest intellectual achievement since Newton's discovery of the law of gravitation. Fiske was essentially a disciple and expositor of Spencer's philosophy (nevertheless, the scientific theory of biological evolution was formulated principally by Charles Darwin, although the general idea of evolutionary development was first extended to the cosmos by Herbert Spencer).[11]

Fiske was dissatisfied with orthodox Christianity, being drawn to the philosophical and theological implications of the modern evolutionary sciences. Like Spencer, he was concerned with the relationship between philosophy and religion.[12] He originally made a Kantian assumption that there is an independent reality beyond consciousness that is essentially unknowable.

Fiske's most important contribution to evolutionary philosophy was his doctrine of the significance of the prolongation of the period of infancy as a factor in the biosocial development of human nature.[13] His theory was grounded in the biological fact that the human infant's brain continues to develop after its birth (he acknowledged the differences in growth patterns between man's brain and those of the great apes). He taught that the need for the prolonged care and education of dependent and vulnerable human infants explained the origin of the family, clans, and society, made the transmission and accumulation of knowledge and therefore the improvement of society possible, and contributed to the growth of altruism. Fiske held to the unity of all knowledge, the inevitability of progress, and the ultimate harmony of science and religion. His views were grounded in evolutionary idealism and cosmic theism. It is interesting to note that Fiske had overextended into the religious realm Darwin's concern for human adaptation and survival. He held that established religions are means of human adaptation to the unknowable. In *The Unseen World* (1876), he now believed in the personal immortality of the human soul (his position was influenced by New England liberal Protestantism).

Fiske's later works took an increasingly spiritualist and conservative orientation. There had been a slow break from the Spencerian philosophy and a movement toward a theistic interpretation of the universe, e.g. *The Destiny of Man* (1884) illustrates Fiske's change of attitude.[14] Spencer's unknowable force is now held by Fiske to be the Christian God. His view of evolution acquired a theistic, teleological, and anthropocentric orientation: the goal of evolution is the creation and perfection of man. Evolution is God's process by which He perfects His creation. In *Through Nature to God* (1899), Fiske presented his final position.[15] Reacting to the writings of Huxley, he held that man differs in kind from the apes. In fact, the universe is dichotomized between man and all other things in it. He argued that evolution exists purely for the sake of moral ends and that science offers confirmation of the existence of God and immortality.

At Yale University, sociologist and political economist William Graham Sumner (1840-1910) used Herbert Spencer's three-volume work *The Principles of Sociology* (1876-1896) as a text in his very popular and successful courses.[16] Although he had studied theology and became an ordained Episcopalian minister, Sumner eventually turned his intellectual interests to sociology (he was devoted to both Darwin's scientific theory of biological evolution primarily by means of natural selection or the survival of the fittest and Spencer's philosophy of sociocultural evolution grounded in unmitigated individualism).

Sumner became the most vigorous and influential of the Social Darwinists and taught that competition is the law of nature while the survival of the fittest is the law of civilization. He was greatly influenced by the writings of Malthus, Darwin, Huxley, Haeckel, and Spencer (as well as Gumplowicz, Lippert, Ratzenhofer, and Ricardo). His philosophy of human social evolution is based upon three major concepts: (1) the prudence and diligence of the "Protestant Ethic" (1904) of the German sociologist Max Weber, (2) the doctrines of classical economics, and (3) the Darwinian explanatory principle of natural selection.

Sumner took a scientific view of the human being: man is not a divine special creation of God but an intelligent animal brought forth on earth by the same natural evolutionary forces that had produced other planetary forms of life. Like Spencer, he saw human society as a superorganism gradually changing at a geological tempo. In fact, Sumner extended the explanatory principles of both the struggle for existence and the survival of the fittest from Darwinian biology to apply, as concepts, to the hardships and perils of the intense competitive life of social man in the continuing improvement of the human world and ongoing battle against nature itself. He taught that there is a direct analogy between the natural selection of fitter organisms through their superior adaptability in animal struggle and the social selection of fitter citizens through their economic virtues in human competition for cultural advancement and the conquest of nature: "The law of the survival of the fittest was not made by man and cannot be abrogated by man. We can only, by interfering with it, produce the survival of the unfittest. . . . The whole retrospect of human history runs downwards towards beastlike misery and slavery to the destructive forces of nature. . . . Every successful effort to widen the power of man over nature is a real victory over poverty, vice, and misery, taking things in general and in the long run."[17]

A staunch conservative, Sumner was a champion of the industrialized capitalist society and adherent to social determinism (in his somber vision of the future, optimism and pessimism are alike impertinent). He rigorously defended the *status quo* by attacking equality, democracy, reformism, socialism, communism, nihilism, and government interventionism as if there were no value differences among these. Clearly, his Social Darwinism advocated and defended a dogmatic laissez-faire view of the human predicament. Sumner claimed that the fundamental problem of all societies is the population-land ratio, argued that wars and revolutions that overthrow useless institutions are a comparative good, and maintained that the two most important virtues are self-denial and hard work. In the human social struggle for existence, he taught that money is the essential token of success: moral progress is represented especially in the accumulation of economic virtues and, as such, millionaires are the bloom of a competitive civilization. As a side effect, he realistically foresaw the depletion of our planet's natural resources and "earth hunger" as the inevitable result.

Sumner's two major works are *What Social Classes Owe to Each Other* (1883) and *Folkways* (1906). His ambitious attempt at a comprehensive synthesis was never achieved, although this project was finished by Albert Galloway Keller and published in four volumes as *The Science of Society* (1927). In *Folkways*, generally acknowledged to be his most valuable contribution to theoretical sociology, Sumner adopted the framework of absolute ethical relativism and taught that humans act to satisfy four basic types of needs: hunger, sex, vanity, and fear of ghosts[18] (this emphasis on human needs anticipates the organismic interpretation and functionalist approach of the eminent Polish-born British cultural anthropologist Bronislav Kasper Malinowski, whose several major works rank with the most seminal contributions to the science of human behavior). For Sumner, folkways are both the habits of the individual and the customs of the society: those folkways necessary for the survival of the group become the mores and values of its community.

Paradoxically enough, despite his intellectual stature, the theologically trained William Graham Sumner displayed much less of the compassion characteristic of the Sermon on the Mount then did the scientifically trained geopaleontologist turned sociologist Lester Frank Ward, who was equally opposed to Social Darwinism in both its rightest and leftist forms.

Lester Frank Ward (1841-1913) was a distinguished geologist, paleontologist, botanist, and subsequent major sociologist who boldly advocated deliberately organized and ethically guided social reforms; not until 1906 did he teach sociology on a full-time basis when invited to Brown University for life.[19] In his two-volume masterpiece *Dynamic Sociology* (1883, 1897), he viewed human society from a cosmic perspective: there are references to astronomy, physics, chemistry, biology, psychology, and anthropology before he focused upon sociological material itself.[20] Special attention is given to a survey of the positive philosophy of Comte and the synthetic philosophy of Spencer.

Ward's sweeping perspective is refreshing and surprisingly modern: planet earth is a mere fragment of cosmic history, life originated from creative chemical syntheses, and man is an animal whose mind is an emergent product of organic evolution (human mental powers include feelings or emotions, reason, foresight, memory, and imagination). Although there is no cosmic purpose in physical reality as such, there is at least human purpose in the realm of civilization: purposeful activity is a proper function of both the individual and the society as a whole.

In his six-volume work *Glimpses of the Cosmos* (1913-1918), Ward praises science, which he referred to as the great iconoclast, and writes,

> It has led us into the arena of the infinite universe, and taught us to contemplate the wonders of nature, from the vast firmament of revolving spheres to the infinitesimal world of moving atoms; from the sparkling crystal to the living organism, the contemplation of which sublime truths yields to the mind a holier ecstasy than any reflections upon the character and attributes of anthropomorphic deities, or any selfish hopes of a future eternity of bliss. And when we remember that which is the crowning glory of science, that with all these blessings she has never cost one human life, one drop of human blood, one pang of human suffering, how long will the world hesitate to pronounce its decision?[21]

In his understandable enthusiasm for science in general, Ward depicts its progress here in a somewhat excessively idyllic and insufficiently realistic manner. The progress of science, in fact, has not been entirely free from human tragedy and suffering (some of these were caused by its religious and political opponents, while others resulted from the initially destructive impact of its not always wise and humane pursuit and application). For Ward, in any event, human progress is achieved by the scientific triumphs over the laws of physical nature and the errors of theology, metaphysics, and superstition. As such, he advocated the prudent engineering of human society and its creative energies.

Ward was a socially compassionate democrat, naturalist, and humanist. He optimistically envisioned a planned society through the constructive legislative intervention of a representative government: while environment transforms the animal, man transforms the environment by applying his educated intellect (scientific intelligence) to his own improvement and goals so as to increase human happiness and reduce human misery. Man's task is not to imitate the laws of nature but to observe them, appropriate them, and direct them to his own needs and desires. Likewise, if nature progresses through the destruction of the weak, man progresses through the protection of the weak. In these significant insights, Ward clearly anticipated the modern concepts of democratic planning and the welfare state.

Ward attacked the natural-law and laissez-faire individualism of Spencer and the other Social Darwinists, especially Sumner. He made a sharp distinction between the purposeless evolution of the physical earth with its plant and animal worlds on the one hand and the purposive actions of human mental evolution on the other. Like Kropotkin, and in complete sympathy with the newer currents of social thought, Ward saw unchecked human competition, to the exclusion of mutual aid and social cooperation, as prodigiously wasteful and dangerously destructive of human values.[22] In brief, his aim was to demolish the tradition of biological sociology (i.e. the unmodified extension of Malthus and Darwin to the human realm) as well as the excessive tendency toward mathematical quantomania. For Ward, *genetic* phenomena encompass the natural results of mechanical but blind physical and biological forces, while *telic* phenomena are those artificial results governed by the teleological calculations and conscious controls of dynamic human purpose and will. Somehow or other, telic behavior is an outgrowth of antecedent genetic evolution.

Although seeing no continuity between the wasteful genetic processes of nature and the purposive actions of human conduct, Ward did present an organismic view of society. He also stressed the importance of human feelings as forces behind ethical and rational motivation in social dynamics conducive to social improvement. Admiring Comte, influenced by Spencer and Hegel, uninterested in Marx, and rejecting Sumner's Social Darwinism, Ward foresaw the future formation of a "sociocracy" as the result of collective telesis (i.e. the purposeful democratic control of society by society as a whole).

During the American Gilded Age, conservative thinkers generally claimed that Darwin's theory of evolution gave scientific legitimacy and justification to untrammeled economic and political competition. In truth, however, Darwin himself never made such socially ruthless and reactionary statements and probably never dreamed that his somewhat one-sidedly competitive picture of raw nature would be mistransposed to human society as a would-be rationalization for man's inhumanity to man.

NOTES AND SELECTED REFERENCES

[1]Cf. Paul F. Boller, Jr., *American Thought in Transition: The Impact of Evolutionary Naturalism, 1865-1900* (Chicago: Rand McNally, 1971, esp. pp. 47-56); John C. Greene, *Darwin and the Modern World View* (Baton Rouge: Louisiana State University Press, 1981, esp. chapter three, pp. 88-128); Richard Hofstadter, *Social Darwinism in American Thought* (Boston: Beacon Press, rev. ed., 1955, esp. chapter two, pp. 31-50); William Henry Hudson, *Herbert Spencer* (New York: Dodge Publishing Company, 1908); Don Martindale, *The Nature and Types of Sociological Theory* (Boston: Houghton Mifflin, 1960, pp. 174-175, 165-168, 187-189, 207, 211, 277, 345, 444, 461, 532); Jay Rumney, *Herbert Spencer's Sociology* (New York: Atherton Press, 1965); Richard L. Schoenwald, ed., *Nineteenth-Century Thought: The Discovery of Change* (Englewood Cliffs, New Jersey: Prentice-Hall, 1965, pp. 129-166); Nicholas S. Timasheff, *Sociological Theory: Its Nature and Growth* (New York: Random House, 3rd ed., 1967, esp. chapter three, pp. 32-44); and Philip P. Wiener, *Evolution and the Founders of Pragmatism* (New York: Harper Torchbooks, 1965, esp. pp. 70, 83, 94, 103, 192, 195, 197). Also refer to Philip Appleman, ed., *Darwin* (New York: W. W. Norton, 2nd. ed., 1979, part four, "Darwin and Society" pp. 389-510).

[2]Cf. Herbert Spencer, *The Man Versus the State* (1884), Caldwell, Idaho: Caxton Printers, 1940.

[3]Cf. Herbert Spencer, *First Principles* (New York: The De Witt Revolving Fund, Inc., 1958). Also refer to F. Howard Collins, *An Epitome of the Synthetic Philosophy* (New York: D. Appleton, 1889).

[4]Cf. Herbert Spencer, *First Principles* (New York: The De Witt Revolving Fund, Inc., 1958, p. 394).

[5]Cf. Herbert Spencer, *Synthetic Philosophy*, volumes two and three, "Principles of Biology" (1864-1872).

[6]Cf. Herbert Spencer, *Synthetic Philosophy*, volumes six, seven, and eight, "Principles of Sociology" (1876-1896). Also refer to Herbert Spencer, *The Study of Sociology* (1873), Ann Arbor: University of Michigan Press, 1966, esp. pp. 282, 341-342, 346, 350, 358, 364-365.

[7]Cf. Herbert Spencer, *Synthetic Philosophy*, volumes four and five, "Principles of Psychology" (1870-1872).

[8]Cf. Herbert Spencer, *Synthetic Philosophy*, volumes nine and ten, "Principles of Ethics" (1879-1893). Also refer to Herbert Spencer, *The Data of Ethics* (New York: A. L. Burt, 1879).

[9]Cf. Marvin Farber, *Basic Issues of Philosophy: Experience, Reality, and Human Values* (New York: Harper Torchbooks, 1968, p. 215).

[10]Cf. John Fiske, *Outlines of Cosmic Philosophy, Based on the Doctrine of Evolution, with Criticisms on the Positive Philosophy* (New York: Houghton/ Mifflin, 4 vols., 1874). Also refer to Paul F. Boller, Jr., *American Thought in Transition: The Impact of Evolutionary Naturalism, 1865-1900* (Chicago: Rand McNally, 1971, esp. pp. 40-41); Richard Hofstadter, *Social Darwinism in American Thought* (Boston: Beacon Press, rev. ed., 1955, esp. pp. 13-15, 19-22, 24, 31-32, 48); and Philip P. Wiener, *Evolution and the Founders of Pragmatism* (New York: Harper Torchbooks, 1965, esp. 129-151).

[11]Cf. John Fiske, *Darwinism and Other Essays* (New York: Houghton/Mifflin, rev. ed., 1885).

[12]Cf. John Fiske's "Herbert Spencer's Service to Religion" (1902) in R. J. Wilson, ed., *Darwinism and the American Intellectual: A Book of Readings* (Homewood, Illinois: The Dorsey Press, 1967, pp. 71-78).

[13]Cf. John Fiske, *A Century of Science and Other Essays* (New York: Houghton/ Mifflin, 1899, "The Part Played by Infancy in the Evolution of Man" pp. 100-121).

[14]Cf. John Fiske, *The Destiny of Man Viewed in the Light of His Origin* (New York: Houghton/Mifflin, 1884).

[15]Cf. John Fiske's "Through Nature to God" (1899) in R. J. Wilson, ed., *Darwinism and the American Intellectual: A Book of Readings* (Homewood, Illinois: The Dorsey Press, 1967, pp. 112-123).

[16]Cf. Paul F. Boller, Jr., *American Thought in Transition: The Impact of Evolutionary Naturalism, 1865-1900* (Chicago: Rand McNally, 1971, esp. pp. 56-62); Maurice R. Davie, *William Graham Sumner* (New York: Thomas Y. Crowell, 1963); Richard Hofstadter, *Social Darwinism in American Thought* (Boston: Beacon Press, rev. ed., 1955, esp. pp. 51-66); Don Martindale, *The Nature and Types of Sociological Theory* (Boston: Houghton/Mifflin, 1960, esp. pp. 165-168, 187-189); Nicholas S. Timasheff, *Sociological Theory: Its Nature and Growth* (New York: Random House, 3rd ed., 1967, esp. pp. 68-71); and Philip P. Wiener, *Evolution and the Founders of Pragmatism* (New York: Harper Torchbooks, 1965, pp. 67, 260).

[17]Cf. William Graham Sumner's "Sociology" in R. J. Wilson, ed., *Darwinism and the American Intellectual: A Book of Readings* (Homewood, Illinois: The Dorsey Press, 1967, pp. 143, 147).

[18]Cf. William Graham Sumner, *Folkways: A Study of the Sociological Importance of Usages, Manners, Customs, Mores, and Morals* (New York: Mentor Books, 1940).

[19]Cf. Paul F. Boller, Jr., *American Thought in Transition: The Impact of Evolutionary Naturalism, 1865-1900* (Chicago: Rand McNally, 1971, esp. pp. 64-69); Israel Gerver, *Lester Frank Ward* (New York: Thomas Y. Crowell, 1963); Richard Hofstadter, *Social Darwinism in American Thought* (Boston: Beacon Press, rev. ed., 1955, esp. pp. 67-84); Don Martindale, *The Nature and Types of Sociological Theory* (Boston: Houghton/Mifflin, 1960, esp. pp. 69-72); and Nicholas S. Timasheff, *Sociological Theory: Its Nature and Growth* (New York: Random House, 3rd. ed., 1967, pp. 74-82).

[20]Cf. Lester F. Ward, *Dynamic Sociology, or Applied Social Sicence as Based upon Statical Sociology and the Less Complex Sciences* (New York: Greenwood Press, 2 vols., 1968).

[21]Cf. Lester F. Ward, *Glimpses of the Cosmos* (New York: G. P. Putnam's Sons, 6 vols., 1913-1918, volume one, "Science vs. Theology" p. 55). Also refer to Lester F. Ward's "Mind as a Social Factor" in R. J. Wilson, ed., *Darwinism and the American Intellectual: A Book of Readings* (Homewood, Illinois: The Dorsey Press, 1967, pp. 149-161).

[22]Cf. Petr Kropotkin, *Mutual Aid: A Factor of Evolution* (Boston: Extending Horizons Books, 1914).

Chapter 11.

NIETZSCHE

F riedrich Wilhelm Nietzsche (1844-1900) was a brilliant philologist, process philosopher, insightful psychologist, engaging poet, and the antichrist.[1] He was both a critical observer and a thought-provoking interpreter. His views were greatly influenced by the Presocratic Heraclitus, the brutally honest Schopenhauer, and the incomparable Wagner (he eventually emancipated himself from the two German masters).[2] In his rigorous philosophical quest for truth, Nietzsche emerged as the major scathing critic of the mediocrity throughout European society and culture in the last century.

Perhaps the most significant and influential philosopher since Kant and Hegel, Nietzsche was the most complex and fascinating thinker of recent times. His intellectual creativity triumphed over half-blind and painful eyes, severe migraine headaches, manifold physical agonies (including violent vomiting and continuous stomach disorders), and long periods of depression. His writings are rich with ideas and insights and were often presented through the ingenious use of symbols, aphorisms, and epigrams.[3]

Nietzsche was the founder of atheistic existentialism (whereas Kierkegaard, who turned away from science, was the father of theistic existentialism).[4] The German thinker was a man of flesh and blood who thrived on provocative ideas and challenging perspectives. The protean nature of this great philosopher eagerly both sought and explored the contradictions and paradoxes within the flux of human existence.

Philosophizing with a hammer and writing in blood (as he put it), Nietzsche gloried in the shock value of deliberate but often artful exaggerations. This intellectual giant (but at the same time emotional cripple) constructed no rational conceptual system. This pathetic and pitiful figure was a victim of his own isolation and self-induced loneliness. He longed for a loyal friend, but was his own worst enemy (his associations with Paul Deussen, Erwin Rohde, Franz Overbeck, Peter Gast, and Paul Rée were far from ideal). Nietzsche was unappreciated, yet this wandering German of the Alps saw further and deeper than most other philosophers. He dedicated his life to fathoming the concrete universe and human nature within it. His remarkable genius both analyzed the problems of the modern world and synthesized a new worldview in order to give meaning and purpose to the totality of existence.

Nietzsche was a creative scholar, desiring truth and loving necessity. He rashly claimed that "God is dead" and despised the "human, all too human" values and attitudes of his own time. To avoid nihilism, he developed his own conceptual framework of penetrating insights and bold speculations: the master and slave moralities, the pervasive will-to-power, the coming superman as creative intellect beyond good and evil, and the awesome cosmic doctrine of the eternal recurrence of the same. This philosopher of rejection and overcoming was both visionary and futurist. The German held that Christianity is responsible for the entrenched bad conscience in the decadent European society and culture. Since man is the only evaluating animal, Nietzsche called for a rigorous reevaluation of all values and ideals. His own process philosophy was concerned with life: it was both an optimistic and futuristic view of things. The universe consists of physical energy and ceaseless activity (there is no other world of spirit). Man must grow by overcoming the struggles and sufferings within the human condition.

Nietzsche's intellectual development passed through three major stages: from original insights into Greek culture and a love for music (especially the early operas of Richard Wagner) through a ruthless criticism of western values (particularly democracy, socialism, nationalism, and utilitarianism) to an interest in science and the presentation of his own philosophy of cosmic becoming. He abhorred conformism and mediocrity, despised supernaturalism (especially Christianity) and philosophical idealism, and rejected the major philosophers of western intellectual history (Plato, Aristotle, Augustine, Aquinas, Kant, Hegel, and finally even Schopenhauer). He claimed that these thinkers and their values had reduced humankind to the lowest common denominator.

Nietzsche called for a rigorous reevaluation or transevaluation of all values *(Umwertung aller Werte)*. He brashly rejected the theistic beliefs in a personal God, personal immortality of the human soul, freedom of the human will, a moral world order, and a divine destiny for the future of mankind, as well as Kant's noumenal thing-in-itself *(Ding an sich)*. He claimed these beliefs to be grounded in false values that suppress the masses as well as restrict the creative activity of the superior intellects.

Nietzsche's penetrating insights into the psychological makeup of man represent a transitional stage between the works of Schopenhauer and Freud. He investigated guilt, pity, resentment, suffering, compassion, sublimation, and bad conscience. Rejecting all religions, his philosophy emphasized the individual in nature and stressed that conflict is necessary for growth. He did respect the special sciences (particularly chemistry, physiology, psychology, and medicine). Sigmund Freud, who also owed much to Darwin and his theory of evolution, said of Nietzsche that he had a more thorough knowledge of himself than any other man who ever lived or was ever likely to live.

Nietzsche's academic interests passed from traditional theology and classical philology through nihilism to process philosophy and optimistic fatalism. His mature philosophy of overcoming taught the will-to-power *(Der Wille zur Macht)*, prophesied the superman *(Der Übermensch)*, and established the awesome cosmic doctrine of eternal recurrence *(Die Ewige Wiederkehr)*.

As a rigorous naturalist, Nietzsche saw man as a product of and totally within nature. Although accepting the Darwinian theory of biological evolution, he saw the future development of man in terms of advances in psychology and morality. He distinguished between the slave-morality of the masses and the master-morality of the superior individuals free from the false values of the modern world. Unlike the common man, the superior individual creates his own values beyond good and evil (there is no fixed ethical framework). As such, Nietzsche found even Wagner to be "human, all too human," as he put it. He claimed that *Der Ring des Nibelungen* (1876) and *Parsifal* (1882) are works for the masses and not the products of a superior artist (this resulted in the break between the two geniuses in May, 1878).[5]

What is the ultimate nature of reality? For Nietzsche, everything is the manifestation of the will-to-power: reality is an eternal becoming, and will is the essence of all things. The flux of the world is grounded in the primordial will-to-power and the awesome idea of the eternal recurrence of the same events. In the struggle of life from worm through ape to man, the will continues to overcome the obstacles of survival in order to grow and develop toward greater independent creativity. The will-to-power as the active creative force pervasive in the cosmos is ever-seeking self-aggrandizement through overcoming resistance in general and striving for human excellence in particular: one has the voluntarist affirmation and enhancement of life beyond good and evil.

In early August of 1881, near Sils-Maria while walking alone through the wooded Swiss Alps of Upper Engadine "6000 feet beyond man and time" along the lake of Silvaplana, the restless Nietzsche claimed to have experienced a new vision far superior to that of all other men. This remarkable event took place on a mountain pass when the solitary wanderer, deep in thought, came upon a powerful pyramidal rock not far from Surlei. An instant result was Nietzsche's intuitive grasp of the idea of the eternal recurrence of the same events as a joyful interpretation of ultimate reality. Clearly, he was euphoric over this new paradigm.

The eternal recurrence of the same is Nietzsche's most comprehensive, challenging, and obscure idea concerning reality.[6] His metaphysics as exemplified in his argument for the eternal recurrence (or general speculations on reality) was grounded in several basic and interconnected *a priori* propositions assumed to be true: the total sum of energy or force in this universe is finite (*Kraft* is held to be the essential stuff of the cosmos); the number of states of energy is finite; all energy is conserved (neither created nor destroyed); time is infinite, cyclical, and irreversible; all energy has infinite duration; change is eternal (there never has been, nor will be, a state of maximum entropy, disorder, or equilibrium); and the principle of sufficient reason is invoked.

In a cosmic cycle, the number of states (changes, combinations, and developments) of force is finite, and a sequence of all such states represents a limited series of events. That is, there is a limit to the number of possible states of the whole universe. The number of distinct irreducible events being finite in an eternal universe, a series absolutely and completely identical to that which now exists must at some future time be reproduced. Therefore, what now is will be repeated an infinite number of times (eternity stretches both into the past and into the future). The entire sequence of such states is repeated in the same series endlessly in this great roulette game of existence (grounded in a pervasive strict determinism), unknown and seemingly unknowable in principle to the human animal. Briefly, each full cycle is a circular movement of absolutely identical actual events and relationships.

Nietzsche regarded his cosmic theory of the eternal return as the most scientific of hypotheses and, in fact, sought in the disciplines of chemistry and mathematics for an empirical and a formal confirmation of this engaging repetitive "law" of the process universe respectively. He even referred to the eternal recurrence as the mightest thought of thoughts and the highest formula of the supreme Dionysian affirmation of life that is at all attainable. References to the eternal recurrence are primarily found in *The Joyful Wisdom* (1882), *Thus Spake Zarathustra* (1885), and *The Will to Power* (1888); the second volume is considered to be his major work.

Since both time and space are circular, all events and relationships are within the closed wheel of brute becoming as a neverending maze of vital force. In this vicious circle, either all returns because nothing has ever made any sense whatsoever or else things never make any sense except by the return of all things in the same series without beginning or end. From this dynamic view of nature, everything in reality consists of a finite pattern of doings (or arrangement of experiences) that is eternally repeatable: one has a closed philosophy of innumerable perspectives. Essentially, the universe is a recurring cycle of activity. It is caught in an exclusive ring of recurrence, since again and again the same finite series of events must follow a repeated sequence that always leads back to it and so on again forever (all experiences and relationships, including various destructions and recreations, are locked in this precise eternal repetition); that is, everything (variations and repetitions) is trapped within the exclusive fabric of an eternal but finite universe.

Within the framework of the first law of thermodynamics (the principle of the conservation of energy), Nietzsche maintained that the universe as unceasing activity or perpetual motion lasts forever. Neither time nor the world ever had a beginning or will ever have an end: backwards and forwards, change and reality are both eternal. Time is an eternal knot, reality is an endless cycle, and the will-to-power is a creative force permeating all things. Time is eternity itself (creative, noncreative, convoluted, and involuted). In brief, there is no last cycle or finale of nothingness (only the endless return of finite becoming): the universal ring of eternal becoming is the ineluctable necessary law of reality.

Nietzsche placed a tremendous amount of importance on his concept of the eternal recurrence, which held for him great cosmological and existential (psychological, moral, and personal) significance. The German philosopher claimed that the entire course of cosmic history has no transcendent meaning, purpose, or direction: there is no ultimate aim, absolute end, or final goal to the cyclical universe.

It may be argued that Nietzsche's *a priori* law of return is compatible with strict determinism (personal freedom is an illusion). Nietzsche found ecstatic joy (*amor fati* or the love of absolute necessity in the universe) in his philosophy of over-coming, rooted as it is in his cosmic doctrine of the eternal recurrence with each finite cycle exactly the same in both all general features and all specific details.

Nietzsche's idea of eternal recurrence had for him far-reaching metaphysical and metaethical importance. This concept is crucial to his entire thought-system, having both cosmological and existential significance for man within nature beyond any possible psychological application. The eternal recurrence of the same is a metaphysical attempt to justify reality, especially human existence with all its physical pain and moral suffering (it is also a psychological crutch for the human awareness of finitude in the face of everlasting oblivion). Yet, in the Nietzschean worldview, every moment is of infinite value (the perpetuation of momentaneity). All the moments of any man's life, then, must return forever. This idea will bring devastation to some and ecstacy to others. Therefore, Nietzsche's categorical imperative is *once is forever!* One faces the supreme challenge to make a decision for all eternity (a decision that has already been made an infinite number of times but is unknown).

Nietzsche taught that humankind desires eternity, and he longed for personal immortality. Despite his own loneliness and despair, Nietzsche loved the world and could not bear the thought of passing completely out of existence never to live again (better to suffer and endure repeatedly the same agonies of this life over and over for all time than never to exist again): his ardent thirst for actual eternity preferred unending misfortune to the utter nothingness of nonexistence.

Nietzsche's life is a part of the completely determined series of events that constitutes an entire cosmic cycle and therefore, from the suprahistorical perspective of the eternal recurrence, it is inevitable that he will live the completely identical life in each of the infinite number of repeated cycles. Yet there is no continuity of consciousness, memory, or record that transcends a single cycle (no recollection of the earlier recurrences). However, one may even surmise that this idea of return actually had its origin in a *déjà vu!*

The Nietzschean concept of recurrence is a positive doctrine of an eternally repeating afterlife identical to each previous beforelife, of which there have already been an infinite number (one has the exaltation of each and every moment). This awesome view of reality gives immortality to all things, including Friedrich Nietzsche. It attributes cosmic order to chaos, presents an existential challenge to man, and justifies the suffering and finitude of human existence within each cycle. Reminiscent of the myths of Tantalus and Sisyphus, the suffering Nietzsche found personal satisfaction in what he thought to be the absolute certainty of the return of the same throughout the whole of eternity. The personal implication of this cosmology is that Nietzsche will actually live forever as a recurring finite event throughout infinite time. His own existential maxim or categorical imperative beyond good and evil is a duty to the eternal recurrence: So act (or so be) that you would be willing to act in just that manner (or be just this way) an infinity of times.

Taking the principle of increasing variation seriously, one could have an eternal series of finite cycles with each differing in content so that no two cycles would ever be absolutely identical in the cosmic ring of events (in fact, Herbert Spencer did propose this view of things). Should the active universe be eternal in time and infinite in both space and energy, one can even imagine the possibility of an infinite number of identical (or very similar) finite cosmic cycles occurring simultaneously. As such, Nietzsche could live an infinitely infinite number of distinct times as his actual experience of this universe on earth extends again and again beyond his life on our planet to come up again and again as the same (or slightly different) experiences on an infinite number of other worlds. In fact, this infinite number of simultaneous cycles could be staggered so that Nietzsche would always exist sometime somewhere!

According to the theory of return, when this author is again writing this chapter on Nietzsche in the next finite cycle of eternal recurrence, it will once more be August 1982 but he shall have no memory of having written it an infinite number of times before (or even just once before, not to mention that he shall be writing this identical chapter an infinite number of times in the future).

In scope and implications, Nietzsche's awesome doctrine of the eternal recurrence of the same was the central idea and metaphysical foundation of his entire conceptual framework. It justified human existence in general (especially his own tragic life) and, in particular, gave philosophical support to his other two interdependent and compatible ideas of the will-to-power and the coming of the overman. Nietzsche had cleverly replaced the traditional theological belief in a transcendent personal god with the unique metaphysical assumption of an indifferent this-worldly return of the same throughout eternity. He regarded his discovery of the eternal recurrence of the same as his greatest creation: a triumphant hymn in praise of the dynamic universe, our earth, all life, and human existence. It is a metaphysical substitute for the religious god (a comforting "theology" of tragic and/or joyful destiny). In brief, one has an atheistic theology of the eternal recurrence of indiscernible replicas, which remains an incredible thesis with fascinating extrapolations (even if it is a product of intuition not linked to modern science or mathematics).

Nietzsche had seriously planned to eventually write a systematic and definitive volume to be titled *The Eternal Recurrence,* a book devoted solely to the proof and implications of this incredible idea. Unfortunately, however, this plan was never fulfilled.

It may be argued that Nietzsche's doctrine of the return of the same is merely a profound thought-experiment of human existential significance that could radically alter one's entire view of the whole world rather than a true scientific statement as the ultimate eternal confirmation of all reality. One may have complete indifference to such a doctrine of eternal monotony and subsequent cosmic boredom, since it clearly rules out any true novelty or creativity. However, Nietzsche's cosmic doctrine of the eternal recurrence of the same is not a self-contradictory hypothesis, and therefore it remains, in the broadest sense, a possible conceptual explanation of the ultimate scheme of things. This grand assertion is both engaging and intriguing to the human imagination. For many, the theory of return is fascinating but unconvincing. At this time, empirical evidence and logic neither negate nor support the cosmic doctrine of eternal recurrence (but in terms of mathematics, it does seem an extremely unlikely interpretation of the universe).

It has been claimed that Darwin aroused Nietzsche from his dogmatic slumber. As such, Nietzsche did hold that sea animals had evolved into land animals (and apes had evolved into men) but, like Darwin, he neither saw this biological process as necessary progress nor assigned to it teleological value. Yet Nietzsche's dynamic cosmology was the expression of a profound philosopher, whereas Darwin's organic evolution was the discovery of a critical naturalist.

Nietzsche writes about the becoming of life as a part of the flux of reality, although he saw Darwin's emphasis on endless struggle as true but lethal. Both Nietzsche and Darwin disregarded the distinction between the one and the many (particularly in terms of value and survival): Nietzsche only celebrated the one as the creative superman (the goal of our species cannot lie in the end but only in its highest specimens, for humankind of today, seen as the image of the animal, is but a transition between the past or inferior ape and the future or superior overman), while Darwin emphasized the many as a large biological population or species. The German philosopher stressed the odds against the survival of vulnerable complex living structures or unique specimens, whereas the English naturalist saw the adaptation and evolution of species or varieties brought about primarily by natural selection operating on chance favorable (because useful) variations in great numbers of individuals. In short, for Nietzsche, what does not break one under makes one even stronger.

Although Nietzsche consistently emphasized the pervasive conflict in life, he did not see selection working in favor of the progress of species (e.g. the last man on earth will be inferior to the modern man, just as the organized instincts of any herd are inferior to the intellectual superiority of the exceptional individual). Sheer numbers and mere survivals are an anathema to Nietzsche's anti-Darwinian value of life on earth, since the highest specimens must struggle against the lower masses (clearly his position is a form of Social Darwinism). As an evolutionist, Nietzsche saw present man as a link between the past ape and the future overman. He envisioned the coming of the superior individual: the superman yet to come will be as advanced over the man of today as the human animal has evolved beyond the common worm! Briefly, Nietzsche did not maintain that the human species would improve through biological evolution, but on the contrary, he seems to have thought that on the whole it would in fact degenerate to a despicable race of last men (of course, there will be a sharp contrast between the biological last man and the intellectual overman).

In general, one may argue that Nietzsche's metaphysical doctrine of the eternal recurrence is not inconsistent with Darwin's scientific theory of biological evolution (at least in accounting for the struggle and development of life throughout earth history from the worm through ape to man within a single process). However, there are crucial differences.

Like Darwin, Nietzsche rejected all teleological interpretations of nature. Although the German philosopher acknowledged the conceptual contribution of the English naturalist, he never embraced the former's mechanist/materialist explanation for the creative development of life on earth. Instead, Nietzsche's cosmic vision clearly upheld a vitalist interpretation of cyclic process and necessary determinism (although when extended to life this vitalism is neither Lamarckian inheritance nor Darwinian selection). Unlike Darwin's mechanist/materialist interpretation of biological evolution, Nietzsche's vitalist view of life was grounded in the primordial impulse of the will-to-power pervasive throughout the development of the entire world. Nietzsche advocated a vitalist interpretation of life emphasizing the individual, the survival of the weakest, and especially the will-to-power; Darwin had presented a mechanist/materialist view of life grounded in the principle of natural selection (or the survival of the fittest) as applied to the history of populations.

Unlike Nietzsche, Darwin's entire writings delve into those rocks and fossils and living species that give the theory of evolution its scientific foundation. In fact, the English naturalist never clearly disclosed his opinions on the philosophical and theological implications of his evolutionary framework. Unlike Darwin, however, Nietzsche presented a view of life that challenged the very foundations of philosophy and theology. Yet, the German thinker never seriously considered the findings of those special sciences that, from geology and paleontology to biology and anthropology, actually give the empirical evidence needed to support the claims of the theory of evolution.

Also unlike Darwin, who concentrated on the origin of new plant and animal species and established biological evolution on a firm scientific foundation, Nietzsche investigated the history of western values and emphasized the coming of the higher man (overman or superman) within a process metaphysics. Differing from Darwin, Nietzsche sees the distance between the individual overman and the rest of humankind as far greater than the distance that separates the ordinary man from the other animals, e.g. the ape or the worm. For the German thinker, quantity usually imperils quality while excellence issues from deprivation and catastrophe in the exceptional individual rather than from the slow accumulation of favorable traits in a social unit (as argued in the Darwinian interpretation of the evolution of life).

Finally, it is intriguing to contemplate a public debate on these matters between Nietzsche and Darwin. Similarly, one wonders what the German genius would have to say about the new science of exobiology and those recent advances in genetic engineering?

In his philosophical quest of truth, Nietzsche became the major critic of the modern human situation. He was neither a systematic thinker nor a rationalist. Yet, within the framework of the reevaluation of all values and the enormous importance of the idea of the recurrence doctrine, the German philosopher stressed the struggle for personal excellence in this life (particularly emphasizing the significance of intellectual creativity in the arts, e.g. music and poetry, as well as philosophy); certainly, he taught the courageous suffering unto heroic joy.

Nietzsche's value judgments have both cognitive and non-cognitive elements; however, his Dionysian affirmation of self and life was ultimately grounded in reason and the special sciences.[7] He held that one must create one's values: the strong man knows the most radical perspectivism and rejoices in it (he is a creator, an immoralist, and a yes-sayer to life with its assumed freedom and responsibility). Nietzsche was a yes-sayer to all experience and thereby advocated the total affirmation of life.[8]

Disaster finally struck Friedrich Nietzsche in Turin on January 3, 1889. Sobbing, he collapsed in the street of Piazza Carlo Alberto after having thrown himself on the neck of a horse that had been brutally mistreated by its cruel owner. The tragic Nietzsche had become insane (probably due to tertiary syphilis). For this "artistic Socrates" the rest was silence: he lingered in his madness for over ten years. Nietzsche died in Weimar on 25 August 1900 and, with ironic appropriateness, was buried in the cemetery at Röcken in the Prussian province of Saxony (he had returned to the place of his birth). Although his genius was still unrecognized by the masses and unappreciated by the intellectuals, within a few years the fame of this great philosopher began to spread around the world. Suffice it to say that Nietzsche was a prophet of the shape of things to come.

NOTES AND SELECTED REFERENCES

[1]Cf. Ivo Frenzel, *Friedrich Nietzsche: An Illustrated Biography* (New York: Pegasus, 1967); Ronald Hayman, *Nietzsche: A Critical Life* (New York: Oxford University Press, 1980, esp. pp. 167, 199, 302); R. J. Hollingdale, *Nietzsche: The Man and His Philosophy* (Baton Rouge: Louisiana State University Press, 1965, esp. pp. 88-91, 96, 120-125, 167, 176-178, 191-202, 229, 239-240); Janko Lavrin, *Nietzsche: A Biographical Introduction* (New York: Charles Scribner's Sons, 1971, esp. pp. 27, 83-84). Also refer to Eric Heller, *The Disinherited Mind: Essays in Modern German Literature and Thought* (Philadelphia: Dufour and Saifer, 1952, esp. pp. 51-140), Eric Heller, *The Artist's Journey into the Interior and Other Essays* (New York: Random House, 1959, esp. pp. 171-226), and M. S. Silk and J. P. Stern, *Nietzsche on Tragedy* (Cambridge: Cambridge University Press, 1981, esp. pp. 228, 335) as well as J. P. Stern, *A Study of Nietzsche* (Cambridge: Cambridge University Press, 1979).

[2]Cf. Friedrich Nietzsche, *Schopenhauer as Educator* (Chicago: Henry Regnery Company, 1965). Concerning Schopenhauer, Nietzsche was at first very influenced by his *The World as Will and Representation* (part one 1818, part two 1844) but later rejected its pessimism and metaphysics. Concerning Wagner, see Nietzsche's four works: *The Birth of Tragedy, or Hellenism and Pessimism* (3rd ed., 1886), whose first edition of 1872 subtitled *Out of the Spirit of Music* contained a "Preface to Richard Wagner" and devoted its final sections to the great musician; *Richard Wagner in Bayreuth* (1876); *The Case of Wagner: A Musicians' Problem* (1888); and *Nietzsche Contra Wagner: Out of the Files of a Psychologist* (1877-1888).

[3]Nietzsche's works include: *The Birth of Tragedy: Out of the Spirit of Music* (1872, 1874, 1886), *Thoughts Out of Season* (1873-1874, 1876), *Human, All-Too-Human: A Book for Free Spirits* (1878-1880, 1886), *The Dawn of Day: Thoughts on the Prejudices of Morality* (1881, 1887), *The Joyful Wisdom* (1882, 1887), *Thus Spake Zarathustra: A Book for Everyone and No One* (1883-1885, 1892), *Beyond Good and Evil: Prelude to a Philosophy of the Future* (1886), *On the Genealogy of Morals: A Polemic* (1887), *The Case of Wagner: A Musicians' Problem* (1888), *The Twilight of the Idols, or How to Philosophize with a Hammer* (1889), *Nietzsche Contra Wagner: Out of the Files of a Psychologist* (1895), *The Antichrist* (1895), *The Will to Power* (1901, 1906, 1911), and *Ecce Homo* (1908). Also refer to Friedrich Nietzsche, *Philosophy in the Tragic Age of the Greeks* (Chicago: Regnery Gateway, 1962) and Christopher Middleton, ed., *Selected Letters of Friedrich Nietzsche* (Chicago: The University of Chicago Press, 1969).

⁴Cf. H. J. Blackham, *Six Existentialist Thinkers* (New York: Harper Torch-
books, 1959, pp. 23-42), Bernd Magnus, *Nietzsche's Existential Imperative*
(Bloomington: Indiana University Press, 1978), and Kurt F. Reinhardt, *The
Existentialist Revolt: The Main Themes and Phases of Existentialism* (New
York: Frederick Ungar, 1960, pp. 59-120).

⁵Cf. *The Nietzsche-Wagner Correspondence* (New York: Liveright, 1949). Also
refer to Dietrich Fischer-Dieskau, *Wagner and Nietzsche* (New York: The
Seabury Press, 1976) and Frederick R. Love, *Young Nietzsche and the
Wagnerian Experience* (New York: AMS Press, Inc., 1966). Nietzsche vehemently
objected to the theatrically oriented *Der Ring des Nibelungen* (1876) and the
Christian mysticism of *Parsifal* (1882). Yet, he loved Wagner the man (his
father image) until the end of his life.

⁶For primary references to the eternal recurrence, refer to *The Portable
Nietzsche* translated and edited with a critical introduction, prefaces, and notes
by Walter Kaufmann (New York: The Viking Press, 1968, pp. 98, 101-102, 257,
269-270, 327-333, 339-340, 364, 430, 434-436, 459, 563). Also see: Harold
Alderman, *Nietzsche's Gift* (Athens: Ohio University Press, 1977, esp. pp. 83-
112); David B. Allison, ed., *The New Nietzsche: Contemporary Styles of
Interpretation* (New York: Delta, 1977, esp. pp. 107-120, 232-246); Crane
Brinton, *Nietzsche* (New York: Harper Torchbooks, 1965, esp. pp. 76, 81, 145-
149, 138-141); Milič Čapek, "Eternal Return" in *The Encyclopedia of
Philosophy*, ed. by Paul Edwards, (New York: Macmillan, 1967, vol. 3, pp. 61-
63); Geoffrey Clive, ed., *The Philosophy of Nietzsche* (New York: Mentor
Books, 1965, esp. pp. 544-591); Arthur C. Danto, *Nietzsche as Philosopher*
(New York: Macmillan, 1968, esp. pp. 21, 34, 70, 76, 187-188, 201-215, 223-224);
R. J. Hollingdale, *A Nietzsche Reader* (New York: Penguin Books, 1977, esp.
pp. 249-262); Richard Lowell Howey, *Heidegger and Jaspers on Nietzsche: A
Critical Examination of Heidegger's and Jaspers' Interpretations of Nietzsche*
(The Hague: Martinus Nijhoff, 1973, esp. pp. 85-97, 148-159); Karl Jaspers,
*Nietzsche: An Introduction to the Understanding of His Philosophical
Activity* (Chicago: Henry Regnery, 1965, esp. pp. 48, 103, 297, 350-367, 375,
445); Walter Kaufmann, ed., *Nietzsche: Philosopher, Psychologist, Antichrist*
(Princeton, New Jersey: Princeton University Press, 1974, 4th ed., esp. pp. xiii,
8, 11, 53, 66-68, 87-88, 96, 114, 118, 121-122, 131, 136-137, 142-143, 149, 150-151,
154, 161, 167, 175, 188-189, 205, 246-247, 264, 267, 285-286, 294, 306-307, 309,
311, 313, 316-333, 363, 402, 457, 500); Hans Küng, *Does God Exist? An Answer
for Today* (New York: Vintage Books, 1981, esp. pp. 341-424, 495, 659); F. A.
Lea, *The Tragic Philosopher: A Study of Friedrich Nietzsche* (New York:
Philosophical Library, 1957, esp. pp. 171-172, 211-226, 254-288, 316, 321-322);
Karl Löwith, *From Hegel to Nietzsche: The Revolution in Nineteenth-
Century Thought* (Garden City, New York: Anchor Books, 1967, esp. pp. 59,
217-218, 428); George A. Morgan, *What Nietzsche Means* (New York: Harper
Torchbooks, 1965, esp. pp. 59-65, 77-83, 122, 163-167, 254, 279-290, 293, 302-
307, 310-315, 356, 366); Robert C. Solomon, ed., *Nietzsche: A Collection of*

Critical Essays (Notre Dame, Indiana: University of Notre Dame Press, 1973, esp. pp. 316-357); John Stambaugh, *Nietzsche's Thought of Eternal Return* (Baltimore: The Johns Hopkins University Press, 1972); and J. P. Stern, *Freidrich Nietzsche* (New York: Penguin Books, 1979, esp. pp. 56, 79-81, 116, 119-120, 131, 143-144, 157).

[7]Cf. John T. Wilcox, *Truth and Value in Nietzsche: A Study of His Metaethics and Epistemology* (Ann Arbor: The University of Michigan Press, 1974). Also refer to Lawrence M. Hinman, "Nietzsche, Metaphor, and Truth" in *Philosophy and Phenomenological Research,* 43 (2): 179-199.

[8]Cf. Rose Pfeffer, *Nietzsche: Disciple of Dionysus* (Lewisburg, Pennsylvania: Bucknell University Press, 1972, esp. pp. 129-198, 256-258, 266). Recent books include: Frederick Copleston, S.J., *Friedrich Nietzsche: Philosopher of Culture* (New York: Harper & Row, 1975, pp. 16, 18-19, 61-62, 151, 157-158, 160, 170, 196, 208, 212, 216, 218-219, 223, 227-229, 231, 233-234, 242, 245), Gilles Deleuze, *Nietzsche and Philosophy* (New York: Columbia University Press, 1983, esp. pp. 27-29, 47-49, 68-72) trans. by Hugh Tomlinson, Lawrence J. Hatab, *Nietzsche and Eternal Recurrence: The Redemtion of Time and Becoming* (Washington, D.C.: University Press of America, 1979, esp. pp. 93-116), Roger Hollinrake, *Nietzsche, Wagner, and the Philosophy of Pessimism* (Edison, New Jersey: Allen Unwin, 1982), and Richard Schacht, *Nietzsche* (Boston: Routledge and Kegan Paul, 1983, esp. pp. 245-247, 253-266, 346, 380-383). Early works of general interest are Janko Lavrin, *Nietzsche and Modern Consciousness: A Psycho-critical Study* (London: W. Collins Sons & Co., 1922), Anthony M. Ludovici, *Nietzsche: His Life and Works* (New York: Dodge, 1910, esp. pp. 58-74), and Heinrich Mann, *The Living Thoughts of Nietzsche* (London: Cassell, 1939).

KNEŽEVIĆ

Around the turn of the century, the literature on the theory of evolution received a singularly significant contribution from a highly gifted and original thinker writing in Serbo-Croatian (a Slavic language unfortunately inaccessible to most Western scholars then and even today). This lonely genius was Božidar Knežević (1862-1905), who not only accepted the principles of both Darwinian and, with reservations, Spencerian evolutions but also combined and extended these frameworks along with his own penetrating insights into a cosmic vision that focused upon planetary history from a pananthropic perspective. His philosophy of evolution represents a unique system of ideas and speculations that are scientifically plausible, strikingly modern, and worthy of serious attention. Although this thought-provoking and even at times disturbing message was formulated almost a century ago, it is especially relevant today.

Božidar Knezević was born in Ub, a small town in rural Northwestern Serbia. His father was a shopkeeper who died when the boy was only one year old. He was brought up by his loving mother, a woman of notable intelligence, and a cruel stepfather who treated him without either understanding or affection. During his formative years, the youngster was often forced to look after his stepfather's icy shop while the coarse and insensitive man spent his winter hours drinking at the nearby inn. Nevertheless, these considerable hardships did not prevent the boy's inquisitive mind (with its powerful imagination) from speculating on the nature of the universe and wondering about the meaning, purpose, and ultimate destiny of humankind within the cosmic scheme of things.

With probing intellect and soaring genius, Knežević continuously explored and devoured volumes of world literature in numerous languages (including English, German, French, and Russian). Besides these basic tools of research, he was also required to learn the natural and social sciences along with the arts and humanities. As a gymnasium pupil in Belgrade and later as a university student in the same city, he earned his living performing various menial tasks. Throughout his extensive studies, he demonstrated special interests in history and philosophy.

Although a thinker of exceptional erudition, Knežević was denied a university professorship by jealous colleagues of incomparably inferior qualifications. He was thereby forced to spend most of his short career as a gymnasium director or teacher in several dull and isolated semiurban centers of nineteenth-century Serbia, far from Belgrade with its bookshops and libraries that he loved so dearly.

In sharp contrast to the tragic drabness and even misery of his external existence (further burdened with the crushing responsibilities of caring for his wife and eleven children as well as several nephews and nieces), Knežević lived a rich and glorious internal life of remarkable loftiness and intensity.

Only three years before his death at the age of forty-three, Knežević finally succeeded in obtaining a teaching position in Belgrade; at long last he received a handsome salary, which alleviated his poverty, a royal decoration for intellectual merit, a prestigious position as assistant editor of the official *Serbian Gazette,* and a membership-at-large in his country's Supreme Council for Education. Unfortunately, all this deserved recognition came too late to be rightfully enjoyed. His never robust health soon deteriorated into tuberculosis that, combined with a possible heart ailment, ultimately ended his physical life. Yet, the products of his creative mental activity still live on in his own seminal works and the scholarly writings of others.[1]

As a naturalist philosopher of cosmic evolution, Božidar Knežević was both a rationalist and an empiricist. His two major books are *Succession in History* (1898) and *Proportion in History* (1901): these two works are jointly combined in a single volume entitled *The Principles of History*.[2] A third important book consists of his 876 aphorisms published as *Thoughts* (partially in 1902 and later in their entirety in 1931).[3] The rest of his scholarly activity includes many articles on outstanding ideologists of the last century along with voluminous translations from English, German, French, and Russian (including Henry Thomas Buckle's *History of Civilization in England*, which first appeared in 1891).

For his major concepts, like many other evolutionists, the boldly speculative Knežević was indebted to the ideas of Comte and Hegel as well as the writings of Darwin, Marx, and Spencer. But as an eclectic of remarkably creative genius, this visionary achieved an original worldview that deftly synthesized both positivism and historicism within a cosmic perspective of events and things. The result was a vast, dynamic, and unique interpretation of humanity's place and destiny within the determining laws of a first evolving and then devoluting universe. Yet, as a new intellectual edifice, this comprehensive philosophy amounts to a great deal more than a mere sum of its borrowed ingredients. It is a provocative and profound statement on the breadth and depth of the cosmos in general and the history of our earth in particular.

Knežević's encyclopedic excursions ranged from astronomy and geology through biology and anthropology to sociology and psychology. He was firmly commited to science, reason, and the brotherhood of humankind. His thoughts were free from religious bias, ethnic partiality, and academic dogmatism. As a futurist, he envisioned a worldwide sociocultural system as the outgrowth of human progress grounded in the contributions of technology and increased historical understanding.

Knežević's views on man, history, and nature were elegantly and skillfully interwoven into a holistic and intelligible scheme of singular cosmic ascent and subsequent singular universal descent. They incorporated the following major ideas: an astronomical perspective, an organismic orientation, planetary history as evolutionary teleology and materialistic determinism, human history as scientific advancement and sociocultural convergence, and the growing consciousness, rationality, and ever-increasing freedom within the progressive development of our global species.

Knežević rejected all views supporting geocentrism, zoocentrism, and anthropocentrism. He also refuted astrology, alchemy, and phrenology in favor of scientific inquiry and rational reflection upon empirical facts and human experiences. Never losing sight of the causal and temporal priority of the physical universe, he wrote, "Nature is prior to man, and all that is natural precedes the human and the humane. Nature was completed in all its important aspects when man appeared; he arrived at a far advanced moment of cosmic development, and for this reason we assert the priority of nature."[4] In another context, he observed, "Now that astronomy has bound the earth to the cosmos, biology the animals to the earth (and man to the animals), and psychology the soul of man to the soul of the animal, it remains for scientific history to bind man to man."[5]

Like Kant, Knežević speculated about the immensity of our universe without ever leaving the narrow confines of his homeland. The great Serbian naturalist of cosmic history taught that the earth is not the center of the cosmos nor is the human animal a divinely chosen creature. He recognized three major stages of planetary evolution: inorganic, organic, and psychic levels of development. Despite its antiquity, mankind is seen as a recent product of (and totally within) natural history; its destiny is bound to the general laws of material reality.

Knežević held that cosmic reality is an infinite series of finite evolutionary cycles, all broadly analogous to, but each uniquely different from, both one another and this particular unfolding round containing the ephemeral history of the earth and its human species. He recognized forward evolution and subsequent backward devolution as essential aspects of both cosmic and planetary development (material reality is a single semicircle).

To understand and appreciate humankind's true place within universal history, Knežević wrote,

> The real significance of Darwin's theory lies in his endeavor to transpose to organic nature the mechanical concept which Newton applied to the cosmos (to consider organic life in terms of cosmic laws), to reduce all the laws in all fields of human thought to the exactness of the laws of astronomy would be to reach the ideal stage of reason. . . . The entire

progress of the modern era starts with the separation of state from religion, because, when the church lost its secular power, it could no longer keep science in its service and prevent free inquiry."[6]

Kneževič's single semicircular scheme of earth history consists of an inexorable evolutionary ascent from the original primordially pervasive unconscious chaos to the fulcrum of a temporary intellectual and moral convergence and unification of a free mankind on a global scale in a strictly secularized and naturalistic sense, followed by an implacable reverse process of devolutionary descent from this orderly but ephemeral equilibrium back into the disproportion and necessity of an equally pervasive and ultimate unconscious chaos.

To Kneževič, the material universe was utterly indifferent to the fleeting incident of human existence: were human beings suddenly to disappear from the face of this earth, the stars and planets of this galaxy and others would undoubtedly endure (the assessment of the proper place of mental activity and of the human species in the totality of nature is a necessary precondition for any intellectually defensible understanding of, and axiologically acceptable appreciation for, the overall cosmic scheme of things). Since all things must die, the human species with its society and civilization is merely a luxury of nature that will inevitably disappear within the endless flux of material reality (as will the organic and inorganic realms).

Kneževič did envision other universal processes within the infinity of superspace and the eternality of supertime, each more rather than less analogous to the semicircular history of our earth and its sun (characterized by the inexorable movement from chaos through order to chaos). He conceived of the histories of other planets, stars, cosmic systems, and even entire universes as phases and episodes of the immeasurable and incomprehensible universal process of God or Nature.

Kneževič did not rule out the rare occurrence of alternative forms of life and intelligence existing elsewhere on other celestial bodies. The Serbian wrote,

> Looking at many a human being, it inadvertantly occurs to one that, if the earth had stopped in her evolution at the creation of the ape, she would have remained more consistent with her strength and skill. With

the creation of humankind, she showed herself as an inept dilettante, a rather unfortunate imitator of some other more fortunate planet. Even in the case of better earthly creatures, the earth's basic simian disposition frequently breaks through. . . . Man is only at particular points in the cosmos : . . he is confined only to particular moments of universal life.[7]

Yet, he did not foresee man's ever escaping the finite bonds of this earth. In fact, he held that the eventual doom of our own species is inseparably linked to the fatal destiny of this perishable planet as an integral part of an ephemeral solar system.

Knežević maintained that time is the force that reduces eons to moments, galaxies to atoms, and all life (including human splendor) to a handful of ashes. In his own words, "The entire life of today's universe, from its initial to its ultimate chaos, is only one sound in the enormous general harmony of the cosmos; only a clock on which the hands meet again in the immensity of universal time. . . . the life of the earth is only a clock on which humankind's entire historical life is only a second, organic life a minute, and the inorganic life of things all the rest of time. . . . In nature, in the cosmos, there is neither past nor future; all is past and all is future."[8]

Knežević held that no particular science, ideology, philosophy, or religion could be a complete and definitive repository of the whole truth. For him, even the ultimate historically accessible truth available to man could be only relatively true in the sense of being the highest approximation to an agreement with reality. It is no wonder that he regarded even his own intellectual system, like all the others before and after it, as one of a series of perishable palaces from whose debris grow new structures of thought. But concerning reason itself, the wise Serbian wrote, "Only with maturity does reason succeed in taking its place by reducing fantasy and pleasure to their proper limits . . . it is only through reason that man is acquiring greater and greater control over nature and over himself; it is only reason that introduces justice and freedom among men . . . Reason is the highest point the earth can ever reach through man."[9]

Knežević sought to explain the apparent incomprehensibility of the humanly accessible universe by pointing out that it can, in fact, be comprehended only if we realize that the entire seemingly eternal and infinite cosmos (of which we are presently aware) is not the ultimate whole but only a small part of the superspatial and supertemporal process of material reality. He refers to this ongoing process as the unfinished biography of God, seeing humankind within the natural history of life on this planet without any religious (let alone denominational) preferences or presuppositions: no personal god or mystical apotheosis will save humankind from cosmic devolution and universal entropy. However, reminiscent of Spinoza and Bruno, the free-thinking Kneževic as pantheist and humanist held God to be Nature.

Kneževic was neither encumbered nor shackled by a myopic and superannuated natural theology or archaic and obscure mysticism; he viewed science and reason as the instruments of a natural faith. The Serbian scholar did recognize the value of religion as an important formative influence on early man. He extensively discussed its mixed historical impact on the emergence of the arts, philosophy, and the special sciences (religion offers emotional comfort, psychological reassurance, moral guidance, and social orientation). However, its richly ambivalent effect on humanity's evolving maturity and wisdom became especially evident in more recent periods of history as religion changed, not always necessarily for the better. In many instances, it became a yoke that repressed and sought to thwart humankind's curiosity, exploration of the unknown, and ongoing intellectual development. Thus, for social progress and cultural development, it was essential to separate the Church from the State in order to assure the very preconditions necessary for rational thought and free inquiry.[10]

Although he accepted the Darwinian theory of biological evolution, Kneževic was never a Social Darwinist. He was above all forms of ethnic and racial prejudice, religious and ideological intolerance, intellectual narrow-mindedness, and emotional bigotry. He remained deeply concerned with human rights, dignity, and brotherhood. The futurist favored scientific and technological advances, economic and moral development, and political emancipation in the context of an increasingly democratic and just social order. As a humanist, he firmly believed in the moral responsibility and value of individual as well as collective efforts. Concerning the future of our own species, he claimed that (at the peak of its maturity) a united mankind will be able to reconcile within itself all the successive errors of its earlier phases of evolution.

There is also a timely and grimly practical aspect to Knežević's predictions about the future. Anticipating our energy crisis due to the earth's finite and therefore exhaustible resources, he foresaw a cult of the sun as the final religion of nature, marking the end of conscious science on earth, just as it had marked the beginning of unconscious religion in the remote past (he did not consider energies other than those of solar origin).

Knežević was neither a laborious writer nor an abstruse thinker. On the contrary, as appropriately noted by one critic, "What makes Knežević such a pleasure to read is his accessibility; one need not be a philosophical scholar to grasp his ideas, for they are not an intricate web of abstractions; rather, they are so fluid and soundly structured that it seems as though they were easily rendered; yet this is the mark of genius: the ability to produce simplicity (be it through pen, brush, or musical instrument) after absorbing the complexities of the human struggle."[11]

Božidar Knežević was primarily a historian who regarded astronomy as the quintessential model of scientific exactness and mathematical precision. He has given us an intellectually daring and dispassionately objective cosmology. It contains an important message for the modern world and, as such, deserves to be read and studied by all thoughtful persons who are concerned with the survival and further evolution of humanity at least on our earth if not perhaps elsewhere in the material universe.[12]

NOTES AND SELECTED REFERENCES

[1]Dr. George V. Tomashevich, the most authoritative Knežević scholar in the English-speaking world, has contributed the following works: "Božidar Knežević and His Aphorisms" in *Serbian Studies* (Chicago, 1981), 1 (2): 25-44, "Božidar Knežević: A Yugoslav Philosopher of History" in *History, The Anatomy of Time: The Final Phase of Sunlight* by Božidar Knežević (New York: Philosophical Library, 1980, pp. 1-31), "Božidar Knežević: Jugoslovenski filosof istorije" in *Kultura* (Beograd, 1972), 16: 70-90, "Božidar Knežević: A Yugoslav Philosopher of History" in *The Slavonic and East European Review* (London, 1957), XXXV (85): 443-461, and "Povodom

Pedesetogodišnjice Smrti Božidara Kneževica" in *Naša Reč* (Paris, 1955). Other references to Božidar Knežević can be found in: Ksenija Atanasijević, *Penseurs Yugoslaves* (Belgrade: Bureau Central de Presse, 1937, pp. 136-184), Howard Becker and Harry Elmer Barnes, *Social Thought From Lore to Science*, vol. 3, 3rd ed. (New York: Dover, 1961, p. 1085), H. James Birx, "Knežević and Teilhard de Chardin: Two Visions of Cosmic Evolution" in *Serbian Studies* (Chicago, 1982), 1 (4), pp. 53-63, Kosta Grubačić, *Božidar Knežević: Monografija o znamenitom srpskom filozofu istorije* (Sarajevo: Veselin Masleša, 1962), M. Dragan Jeremić, *Čovek i istorija* by Božidar Knežević (Novi Sad/Beograd: Matica srpska/Srpska književna zadruga (SKZ), 1972, pp. 5-39), Mihailo Marković, "Yugoslav Philosophy" in *The Encyclopedia of Philosophy* edited by Paul Edwards (New York, Macmillan, 1967, vol. 8, p. 361), Dennis Quick, "The Unraveling of Man" in *Serb World* (Milwaukee, 1981), 2 (2): 22, Joseph S. Roucek, "Sociology in Yugoslavia" in *Twentieth Century Sociology*, edited by Georges Gurvitch and Wilbert E. Moore (New York: Philosophical Library, 1945, pp. 740-754), Andrija B. Stojković, *L'Evolution de la philosophie serbe: Aperçu* (Beograd: Radiša Timotić, 1977, pp. 43, 45-47, 51, 63-64, 85-88), Vladimir Vujić, *Zakon reda u istoriji* by Božidar Knežević (Beograd: Geca Kon, 1920, pp. i-xxiii), and W. Warren Wagar, "Preface" in *History, The Anatomy of Time: The Final Phase of Sunlight* by Božidar Knežević (New York: Philosophical Library, 1980, pp. xiii-xvii).

[2]Cf. Božidar Knežević, *History, The Anatomy of Time: The Final Phase of Sunlight* (New York: Philosophical Library, 1980), translated by George V. Tomashevich in collaboration with Sherwood A. Wakeman.

[3]Cf. George V. Tomashevich, "Božidar Knežević and His Aphorisms" in *Serbian Studies* (Spring, 1981), 1 (2): 25-44. Contains an essay, eighty-five aphorisms, and critical comments.

[4]Božidar Knežević, *History, The Anatomy of Time: The Final Phase of Sunlight*, p. 35.

[5]*Ibid.*, pp. 99-100.

[6]*Ibid.*, pp. 68, 71.

[7]*Aphorisms 642* and *59* by Božidar Knežević, translated by George V. Tomashevich.

[8]*Aphorisms 13, 24*, and *44* by Božidar Knežević, translated by George V. Tomashevich.

[9]Božidar Knežević, *History, The Anatomy of Time: The Final Phase of Sunlight*, pp. 103, 159, 229.

[10]Cf. George V. Tomashevich, "Reflections on Science and Religion" in *Free Inquiry* (Summer 1981), 1 (3): 34-36.

[11]Dennis Quick, "The Unraveling of Man" in *Serb World* (Milwaukee, 1981), 2 (2): 22.

[12]In addition to Božidar Knežević, several other nineteenth- and twentieth-century Serbian intellectuals are worthy of mention in connection with various aspects of the theory of evolution: Nikola Tesla, materialist philosopher and electrical engineer whose inventive genius envisioned both organic forms and

intelligent beings on other planets elsewhere in the universe; Jovan Cvijić, world-renowned geomorphologist and anthropogeographer who studied tectonics and glaciation; Branislav Petronijević, famous paleontologist and impressive mathematical philosopher who reconstructed the fossil remains of *Archaeopteryx* and produced an evolutionist metaphysics; Milutin Milanković, internationally recognized paleoclimatologist and theoretical astrophysicist who provided the mathematical foundation for the Ice Age periodization of the Pleistocene epoch; Nedeljko Divac, outstanding research entomologist who translated Darwin's *Origin* and *Descent* into Serbo-Croatian; Ivan Djaja, widely esteemed physiologist of both fermentation and animal energetics who made pioneering contributions to the science of hypothermia; and Ksenija Atanasijević, distinguished philosopher and philologist whose extensive scholarship included critical studies of process thinkers from Empedocles, Aristotle, and Bruno to the more recent works of Hegel, Schopenhauer, and Nietzsche. Among Croatian intellectuals worthy of mention are Roger Boscovich, eminent astrophysicist and process naturalist, as well as Andrija Mohorovičić and Stjepan Mohorovičić, both prominent for their contributions to seismology, meteorology, and physics (the latter gave his name to the term Mohorovičić Discontinuity in earth geology) and especially Dragutin Gorjanović-Kramberger, internationally recognized paleoanthropologist and geologist who discovered the Krapina fossil remains of the Neanderthal phase of human evolution. Valuable contributions to the natural sciences (with special bearing on the theory of evolution) were also made by numerous scholars of other Yugoslav nations and nationalities, too numerous to be identified by name. (This footnote was suggested by and prepared with Dr. George V. Tomashevich.)

Chapter 13

BERGSON AND ROSTAND

Henri Louis Bergson (1859-1941) was an internationally known French philosopher of Polish Jewish ancestry. He was a professor at the Collège de France (1900-1921), and spent his life teaching, lecturing, and writing on his novel philosophical worldview. He opposed materialism and mechanism, grounding his own system in the application of intuition to the process of evolution. In 1927, he received the Nobel Prize in literature for his major work *Creative Evolution* (1907). Bergson was greatly influenced by the theories of Darwin and Spencer. He rejected Democritus's materialist atomism, Plato's eternally fixed forms, Aristotle's immutable species, and Kant's rigid epistemological categories. His mystical orientation is similar to that of Plotinus (they both advocated a process metaphysics).

Bergson's conceptual framework is ultimately dualistic, distinguishing between the special sciences and metaphysics, intellect and intuition, scientific or mathematical time and experiential or "pure" time, mechanistic determinism and the spontaneous freedom of the whole personality, the material body and the spiritual mind united by perception in real duration, sensorimotor or "habit" memory of animals and the "pure" spiritual memory of man, materialist evolution and creative evolution, and a closed society and an open society. According to Bergson, one has the following dualistic scheme: *Science* is concerned with mathematics and logic, appearances, space, matter (extension), fixity, mechanism, static nature, body, relative knowledge, and a closed society; whereas *metaphysics* is concerned with intuition and mysticism, reality, time, spirit (life), change, vitalism, creative evolution, mind or spirit (consciousness), absolute knowledge, and an open society.

Bergson was a seminal thinker concerned with space, time, matter, life, evolution, consciousness, creativity, free will, intuition, and mysticism. He presented an original insight into the phenomenon of real duration, continuous becoming, and creative evolution as a cosmic process. For Bergson, God is love and the object of love, and the evolutionary process has a divine purpose (this was his final interpretation). As a systematic thinker, he is impossible to understand without referring to the whole spectrum of his thought. Nevertheless, his development of the significance of intuition can easily be taken as the central point of his entire philosophical structure. His views have led to the development of existentialism in one direction and process philosophy in another. Ultimately, Bergson's rejection of reason in favor of intuition must be critically evaluated in light of the rapid advancement of the special sciences. In the last analysis, his dualistic orientation is grounded in religious beliefs and metaphysical assumptions not subject to empirical verification or rational inquiry.

Bergsonian metaphysics is incomplete without taking into account knowledge from biology, physiology, and psychology. His view is not mathematically oriented but rather time oriented: science and reason (mathematics and logic) are replaced by intuition. Intuition is seen as the bridge between science and metaphysics: metaphysics is knowledge of things-in-themselves or absolute knowledge of reality. Therefore, where Kant's work resulted in a critique of "pure" reason, Bergson's work resulted in a critique of "pure" intuition. As such, it rejected mechanism and materialism, and established a metaphysics (however incomplete) embracing planetary evolution, diverging creativity, emerging novelty, and freedom. According to Bergson, only intuition can disclose the inner workings of reality.

Bergson focused on man's freedom, consciousness, and place in the cosmos. His view was at the same time mystical and scientific, resulting in unavoidable ambiguities. His attempt was a bold one: the reconciliation of subjectivity and mysticism with rational thought and objective empirical science. However, priority was given to intuition and mysticism. The major problems result from his basic presuppositions. For Bergson, there is a sharp distinction between time and space, as well as two categorically distinct realms: the organic and the inorganic. What results is a metaphysical dualism between life (spirit) and matter (stuff). Life is active and dynamic, while matter is passive and static. When matter is infused with life, the result is creative evolution. But unlike Darwinian mechanism, which offered "natural selection" as the primary principle for explaining biological evolution, Bergsonian vitalism is an attempt at a systematic approach to the creative aspect of all evolution that had been allegedly ignored by the prior mechanistic materialists.

What method did Bergson use to become aware of the creative flux of reality? He did not appeal to science (mathematics) or reason (logic). His metaphysics rested on intuition. Intuition not only gives an awareness of duration or time (metaphysics) but also illuminates the process of memory (epistemology). Bergson's emphasis on intuition moved from a rejection of metaphysics based upon reason to an exaltation of intuition and mysticism as the true source of ultimate knowledge. Only intuition grasps creative reality as time and change or pure duration.

In *Time and Free Will* (1889), Bergson's philosophy asks one to turn from the fragments of science and immerse oneself in the living stream of things as the flux of reality.[2] As already noted, Bergson makes a sharp distinction between space and time. Science, being objective, is concerned with space and arrives at its concepts through the analysis of the quantitative multiplicity of space. But for Bergson, this analysis distorts the continuous flux of reality. Therefore, intuition, being subjective, is concerned with time and arrives at the concept of *durée réelle* through sympathetic insight into the qualitative multiplicity of consciousness in continuous creative flux. Science, using mathematics, can never obtain absolute reality; absolute reality (metaphysics) is known only through intuition. Number belongs to space, and therefore mathematics is the tool of the sciences (Bergson had already asserted the belief that space cannot contain or reveal absolute reality, and therefore neither can mathematics or logic). According to Bergson, one is misled if one believes that time is homogeneous. Time is not homogeneous, and it is distinct from space; as pure duration *(durée réelle)*, it is time that holds the key to absolute reality.

Time and duration can be known only intuitively, never through the application of science, mathematics, or logic within space. Time is intuitively grasped by the flux of consciousness. Pure duration, that which endures, exists only in the mind. Unlike in pure space where there are merely simultaneities, the mind with its memory contains pure duration; for in memory, things endure. Space is external and homogeneous, while duration is internal and heterogeneous. Bergson established in this early work (1) the limitations of science (mathematics and logic) and (2) the significance of intuition. Thus in Bergson's first major volume, he clearly exalted intuition over science. It is immediate intuition that allows one to see time and motion. Real duration is intuitively known within the enduring memory of the flux of consciousness.

In *Matter and Memory* (1896), Bergson intuitively distinguished among matter, perception, and memory.[2] For Bergson, immediate intuition continues to be the method by which one may discover reality. Memory adds to intuition simply because memory is the collection of previous intuitions (pure memory is unconscious). Intuition shows one the distinction between body and memory and their union. Although Bergson was neither a materialist nor an idealist, he still desired to make a distinction between the body (matter) in space and the soul (memory or spirit) in time. He tried to do this by interjecting perception as a third and intermediate principle. Thus, he avoided an extreme form of ontological dualism or spiritual monism. In brief, perception links space with duration and matter with memory.

Bergson's *An Introduction to Metaphysics* (1903) is nothing short of a critique of intuition.[3] It is a significant work in the development of Bergsonian philosophy. A crucial distinction is made between the intellect and intuition. Intellect, the method of the sciences, can never give us a feeling of absolute reality. The latter is obtained only through the proper method of metaphysics, i.e. intuition or "intellectual sympathy." What does one discover when one applies intuition to one's inner experiences? One discovers the flowing of personality through time: the ego is a ceaselessly changing yet continuous process. Duration is in sharp contrast to the mathematically "spatialized" time of the intellect. Duration is (1) a heterogeneous flux or becoming, (2) an irreversible flowing always toward the future, (3) a continually creating newness or novelty that is hence intrinsically unpredictable, (4) an inexhaustible source of freedom, and (5) its living reality can never be communicated by images or concepts but rather must be directly intuited. Since both time (duration) and motion are apprehended intuitively, reality can only be known intuitively. However, metaphysics does not oppose the sciences but rather complements them.

Bergson insists that logic, mathematics, and the sciences (e.g. physics, biology, and the social sciences) give us but relative knowledge as they necessarily demand the use of symbols, but metaphysics, by definition, is the search for absolute knowledge and does not require symbols. It is intuition that gives absolute knowledge, and it is therefore the method of metaphysics. Bergson believed that "intellectual sympathy" is not difficult: it is a simple act to intuit space, time, motion, matter, and memory. They seem to have always been "given" to one as data of perception as well as absolute metaphysical certainties (such knowledge is said to require no symbolization).

To summarize, there is one absolute reality that one seizes through intuition (and not by simple analysis or the use of symbols) and that is one's own personality in its flowing through time, i.e. one is immediately aware that the self endures and thus one experiences pure duration *(durée réelle)*. Intuition, then, is the metaphysical investigation of what is essential and unique in an object, and duration is essential and unique to the human self. This pure duration excludes all ideas of juxtaposition, reciprocal externality, and extension. It is likewise impossible to represent by concepts or symbols, i.e. by abstract general or simple ideas.

Bergson strongly believed that both empiricism and rationalism fail in obtaining absolute knowledge of the ego. Only intuition reveals the absolute knowledge of the ego, i.e. pure duration. Pure duration is within the ego, as only the human ego has extensive and specific memory. Space, being exterior to the human ego, has no memory and therefore is not durable (does not endure in the same sense that memory endures). Space has no duration but only instantaneity. Only time as memory has duration. In the final analysis, intuition attains the absolute while bringing science and metaphysics together through perception.

Philosophy cannot ignore the knowledge given to it by the sciences. Bergson himself was greatly interested in biology and psychology and became an avid evolutionist (an emergent evolutionist to be exact). What he was reacting against was the alleged misconception that the methods of science, which symbolize and spatialize, give one absolute knowledge. Science gives one facts, but it is the intuitive method of metaphysics that reveals to one the ultimate reality of things. That is to say, through intuition one is aware of the flux of one's own consciousness in time and similarly by analogy becomes aware of the flux of things in the external world. In short, reality is the flow of time (whether internal or external), and this is only appre- hended through intuition, which does not spatialize or symbolize.

Thus far in the development of his thought, Bergson's use of intuition (immediate experience, integral experience, pure per- ception, or intellectual sympathy) has established two "types" of time (objective and subjective) and a mind-body "dualism" (memory-matter dualism). His intuitive method is a deliberate reaction against the use of understanding (science and reason) in the attainment of absolute knowledge, since the method of understanding itself is claimed to distort the ultimate flux of things by symbolizing and spatializing them. Reality is not space but rather time, and there are two "types" of time: (1) objective time (natural scientific time), which is an abstract mathematical conception, extended, homogeneous, and passively unital (it is, in essence, an illusion as the product of rational thought) and (2) subjective time (conscious time), which is a concrete, flowing (flux), irreversible, heterogeneous, indivisible, active, creative, accumulative, endless, immediately experienced (intuited) pro- cess (it is, in essence, real time or "pure time," *durée réelle*, grasped intuitively within the continuous flux of our conscious- ness).

Once again, the real time *(durée réelle)* of consciousness and the abstract time of science are said to unite through perception: the mind is real duration. It is continuous and irreversible; it is a heterogeneous flux of activity creating novelty and manifesting the intuition that gives man his freedom. Through perceptions, mind is united with matter, which is relatively stable in comparison to the flux of consciousness. Mind and body are related by a convergence in time through perception. In summary: (1) intellect uses symbols and conceptualizes, giving relative external knowledge (in Bergson's view, such static abstractions of time and motion falsify real becoming because, for him, the special sciences, logic, and mathematics can never give an adequate and complete account of the dynamic universe) while (2) intuition is immediate, nonconceptual, and nonsymbolic (intellectual sympathy), yielding absolute knowledge of the becoming of reality.

Creative Evolution (1907) is Bergson's masterpiece.[4] In it, he rebels against the mechanism and materialism of Darwinian evolution and the fixities and rigidities of logical and mathematical interpretations of reality. He further develops his conception of intuition, illustrating its significance for a true understanding of emergent evolution. Bergson is still time-oriented. *Creative Evolution* gives him the opportunity to present his interests in intuition, matter, spirit, and evolution within a system metaphysical and scientific. To be sure, he never entertained the possibility that the evolutionary process could be reduced to mathematical predictability. Bergson rejects radical finalism and radical mechanism, claiming that each distorts the free creativity of the evolutionary process. It is because of the use of intellect that the essence of emergent evolution is missed, but if the intellect cannot give a true view of the evolutionary process, intuition can, for just as intuition is able to show one the absolute reality of his own internal ego (i.e. make one aware of subjective duration, or *durée réelle*), an "intellectual sympathy" with the evolution of life will reveal to one the objective duration of external nature.

By going "counter to the natural bent of the intellect" (i.e. by using intuition) one discovers the irreducible, irreversible, creative, divergent, and durative nature of the evolutionary process. Evolution is creative: the universe is in a continuous flux of emergent novelty. In order to give an adequate philosophical account of this intuitive fact, Bergson established a metaphysical concept: the *élan vital* or vital principle as the force, impetus, or "spirit" within matter. It causes the creative divergence that is scientifically obvious. But alas, the *élan vital* is a metaphysical conception that is not scientifically verifiable. For Bergson, the *élan vital* is, however, metaphysically sufficient to account for the holistic, functional, divergent, and creative coadaptation of an organism through successive alternations of form toward greater and greater complexity, consciousness, and emergent novelty. It is also the *élan vital* that gives evolution its continuation and duration. This impetus, although continuous, manifests itself in three divergent lines: plants, insects, and animals.

For Bergson, metaphysics supplements the findings of biology through the use of intuition. Intuition makes one aware of a vital impetus, the *élan vital*, within the evolutionary process itself. Intuition reveals to one with absolute certainty the cosmic process of becoming. Bergson argues that the *élan vital* is responsible for (1) a current of cumulative and irreversible consciousness, (2) evolutionary continuity, (3) novelty (the creative "leaps" in evolution), (4) the increased complexity through time in the evolutionary process, (5) "spirit" attempting to transcend matter, (6) vitalism (the life-force, spirit, or creative agency that is the original impetus from which life springs as well as the force that maintains the continuity of evolution and its continuous emergent novelty), and (7) God as unceasing life, action, and freedom. Hence, as mind is the spirit of the material body, so the *élan vital* is the spirit within the matter of the cosmos. In the process of divergent evolution, plants manifest fixity and torpor, insects reveal instinct, and animals display increasing degrees of consciousness culminating in human self-consciousness. Man displays reflection, reason (intellect), and intuition and is thus the apex of creative evolution.

From Bergson's point of view, intelligence and instinct both have innate knowledge. Even at the extremities of the two principal lines of evolution (the insects and the vertebrates), instinct and intelligence do not occur in pure form. In general, Bergson believed that instinct is innate knowledge of things (matter) while intelligence is innate knowledge of relations (form). Instinct is sympathy for things, i.e. sympathy for matter as is illustrated especially in the Hymenoptera and Lepidoptera. In contrast, intellect is knowledge of relations or static states and thus cannot "know" continuity, mobility, or the essence of evolution as creative becoming. Intuition is "purified" instinct (remembering that the sympathy of instinct is free from the use of symbols, where intellect is not) that accompanies human intellect and intuits becoming or evolution, i.e. the continuity of a change, which is pure mobility. Instinct, under the influence of intellect, has become disinterested in things (matter). Being self-conscious, it reflects on the interpenetration of subjective duration and objective life as the flux of reality. There is, then, a conscious unit between mind, with its duration, and matter, which is in flux. Because of evolution, there necessarily must be within the unity of the cosmos a relationship between pure duration and creative evolution (i.e. between *durée réelle* and the *élan vital*). This relationship is "known" through the use of intuition.

A continual coming and going is possible between nature and mind, because the spirit in nature, the *élan vital*, emerges in man as intuition. Intuition is the *élan vital* conscious of itself: the *élan vital* has consciously become aware of itself through intuition within the self-consciousness of man. Intuition is the pivotal concept of Bergson's philosophy, for it alone can link the consciousness of mind *(durée réelle)* with the stream-of-consciousness *(élan vital)* in matter. This is knowledge that the sciences, logic, and mathematics can never provide. Only through intuition is absolute reality known as time, whether subjective *(durée réelle)* or objective *(élan vital)*.

Bergson deplores the tendency in science and philosophy to mistake its conceptualizations (products of the understanding) for absolute reality. Time is ruled out by intelligence. Intelligence, looking for fixity, masks the flow of time by conceiving it as a juxtaposition of "instants" on a line. Time is not oriented to science or mathematics because it is an indivisible flow and therefore has no measurable parts. If one thinks one is measuring time, in reality one is measuring space. True time is only the intuitive experience of duration. Therefore, one can distinguish between real lived time and its "spatialization" and "symbolization" into the external objects and events of scientific inquiry.

Duration and Simultaneity: With Reference to Einstein's Theory (1922) is an attempt to refute the thesis of the existence of the multiple real times in the theory of relativity.[5] Bergson advanced arguments to save the absolute character of time from the problem of "asymmetrical aging" known as the clock paradox of the identical twins (the theory of multiple times challenged Bergson's major philosophical intuition about the absolute unity of true time). As such, he defended his intuitive view of duration as a single, indivisible flow of true time against Einstein's thesis. For Bergson, the duration of the universe includes both external time or the *élan vital* and internal time or *durée réelle*. There is only one continuous time in the universe and only one stream of consciousness. He held that there is no dichotomy between time external and time internal. However, the "leap" from *durée réelle* to the *élan vital* is unclear. One is told that it is done intuitively, of course, involving memory and perception.

There is no reason to limit duration to consciousness. According to Bergson, evolution is only intelligible when one extends spirit to it. Duration exteriorizes itself as spatialized time: it is space, rather than time, that is measurable. Real duration is experienced and as such contains the very stuff of the existence of all things. Time remains absolute reality and is learned as well as retained only in the realm of intuition. Furthermore, there is only one continuous time in the universe. This is Bergson's philosophical commitment, and it is ultimately a mystical commitment.

In Bergson's *The Two Sources of Morality and Religion* (1932), the discovery of God is made only by that sort of intuition that is the mystical experience of exceptional persons.[6] This intuition (mysticism) leads to intense activity in order to ensure man's future evolution toward a mystical "open society" instead of the merely intellectually-oriented "closed society." By extending intuition to mysticism, Bergson hoped that in the future more mystics would change the earth into an "open society," i.e. a society with unlimited possibilities, progressive and flexible in its moral and religious beliefs, and mystical in nature.[7] The intellectually routine and mechanical society is conservative, authoritarian, self-centered, and closed both morally (static and absolute) and religiously (ritualistic and dogmatic) and as such negates change, spontaneity (intuition), and freedom (free will).

Bergson's far-reaching influence is especially noticeable in the evolutionary synthesis of Pierre Teilhard de Chardin, who, consciously or not, never explicitly admitted an obvious indebtedness to his predecessor. Despite this curious point, the major positive contribution of both is their recognition that one way or another the fact of evolution must be taken seriously.

French philosophers of evolution from Lamarck through Bergson to Teilhard de Chardin, not to forget the German zoologist and philosopher Hans Adolf Eduard Driesch (1867-1941), have emphasized a vitalist interpretation of the creative history of emerging life on our earth: a special place is given to the appearance of humankind within the living scheme of things, and there is usually an appeal to teleology, spiritualism, and even a personal God as the creator of the universe.

In sharp contrast to this vitalist tradition, the eminent French biologist and philosopher Jean Rostand (1894-1977) argued for a critical materialist view of the world. For him, there is no metaphysical explanation of the problems of life: nothing but facts and a series of logical reasonings that follow from these facts. His early interest in biology had been stimulated by a reading of Charles Darwin's *On the Origin of Species* (1859) and through a correspondence with the great entomologist Jean Henri Fabre (1823-1915).

Internationally acclaimed for his laboratory experiments in artificially induced parthenogenesis in frogs and toads (as well as sea urchins, silkworms, and rabbits), Rostand has boldly expressed his opinions on a range of subjects, some very controversial, from biochemistry and the origin of life to animal evolution and moral philosophy. As a dedicated scientist and passionate humanist, he stressed the need for the human person to strive to retain the value of individuality in spite of the incredible advances in twentieth-century science and technology: the threat of depersonalization will be especially acute in the future as a result of the awesome prospects of test-tube life and genetic engineering.

Focusing on the biological nature of man, Rostand points out in his poetic prose that the human animal of modern civilization is, in terms of genetics, substantially identical with those ancestral cave dwellers who made and used paleolithic tools/weapons to survive and thrive during the Pleistocene Age; as such, our species (a grandchild of fishes and a greatnephew of snails) represents a living anachronism whose flesh is contemporary with that of the mammoth, mastodon, and woolly rhinoceros. He claims that man's uniqueness is essentially his reflective consciousness, which is rooted in the cerebral cortex and allows for thoughts ranging from Darwinian evolution to Einsteinian relativity. Nevertheless, for this great biologist, ephemeral humanity can sober itself abruptly by giving rational attention to its dwarfed significance among the innumerable galaxies of this immeasurable cosmos.[8]

Rostand's prolific writings have distinguished him as a skilled and profound popularizer of modern science. He is the author of over fifty books (most on scientific subjects), as well as numerous experimental research reports and collections of critical essays. As a convinced evolutionist concerned above all with the biological and psychological potentials of humankind, he championed the affirmation of and respect for life and free inquiry while offering a glimpse at the possible future of our own species.

In one of his major works, *Humanly Possible: A Biologist's Notes on the Future of Mankind* (1970), Rostand has expressed a curious thought related to his own biological investigations: "If we knew everything about the frog, we would know everything about life, including human life."[9] In his frogs, Rostand saw the entire universe. Be that as it may, he was always aware of the irreplaceable value of each person as a unique individual. One certainly hopes that future scientists and philosophers will be as concerned with the ethical and moral dimensions of human existence as they undoubtedly will be for the theoretical implications and practical consequences of their own research into the nature of things.

Reminiscent of Bernard de Fontenelle (1657-1757), whose famous work *Conversations on the Plurality of Worlds* (1686) and other writings popularized the science of the age and offered speculations in the field of astromony, Rostand as an imaginative biologist wrote that most probably life exists on many other planets in our universe: although life on other celestial bodies is unlikely to have evolved plant and animal forms similar to those we know here on earth, there may be elsewhere something much superior to man (although not something just exactly the same as man).[10]

As an independent researcher and scholar, Jean Rostand became a leading authority on genetics and embryology as well as crustaceans and the theory of evolution. His books include *Can Man be Modified?* (1959) and *Evolution* (1962). As a moral philosopher, he was deeply disturbed by the scientific implications of his own work in experimental heredity. Ironically, Rostand disavowed contact with the modern world, the future of which he may so profoundly affect.

NOTES AND SELECTED REFERENCES

[1]Cf. Henri Bergson, *Time and Free Will: An Essay on the Immediate Data of Consciousness* (New York: Harper Torchbooks, 1960).

[2]Cf. Henri Bergson, *Matter and Memory* (London: George Allen & Unwin, 1962).

[3]Cf. Henri Bergson, *An Introduction to Metaphysics* (New York: Bobbs-Merrill, 2nd rev. ed., 1955).

[4]Cf. Henri Bergson, *Creative Evolution* (New York: Modern Library, 1944). In chapter one, Bergson writes: "The 'vital principle' may indeed not explain much, but it is at least a sort of label affixed to our ignorance, so as to remind us of this occasionally, while mechanism invites us to ignore that ignorance" (p. 48).

[5]Cf. Henri Bergson, *Duration and Simultaneity: With Reference to Einstein's Theory* (New York: Bobbs-Merrill, 1965).

[6]Cf. Henri Bergson, *The Two Sources of Morality and Religion* (Garden City, New York: Doubleday, 1935). Also refer to Henri Bergson, *An Introduction to Metaphysics: The Creative Mind* (Totowa, New Jersey: Littlefield/Adams, 1965). This work was first copyrighted in 1946 by the Philosophical Library, Inc.

[7]Cf. Karl R. Popper, *The Open Society and Its Enemies* (New York: Harper Torchbooks, 2 vols., 1963). Refer to volume one, pp. 202, 294, 314; and volume two, pp. 229, 258, 307, 315-316, 361.

[8]Cf. Jean Rostand, *The Substance of Man* (New York: Doubleday, 1962).

[9]Cf. Jean Rostand, *Humanly Possible: A Biologist's Notes on the Future of Mankind* (New York: Saturday Review Press, 1973).

[10]Cf. Jean Rostand and Paul Bodin, *Life, the Great Adventure* (New York: Charles Scribner's Sons, 1956, pp. 11-15).

Chapter 14.

ALEXANDER, SELLARS, SMUTS, AND MORGAN

F ollowing the publication of Charles Darwin's *On The Origin of Species* (1859), many philosophers accepted the evolutionary perspective and adopted it within their systematic interpretations of man and the cosmos, while some reacted against a crude mechanist and materialist worldview.

The emergent evolutionists presented a new interpretation of the universe in general and the history of living things on earth in particular. Some presented an emergent cosmology, while others focused only upon planetary history. This school is primarily represented in the works of Alexander, Sellars, Smuts, and Morgan. These four emergent evolutionists were concerned with three major concepts: emergence, levels, and novelties. They rejected mechanist materialism, vitalism, teleological finalism, preformationism, and reductionism. Their differing schemes of emergent evolutionism emphasized the unpredictable emergence of variety, diversity, and complexity within the discontinuous history of creative advancement: each new emergent quality engendered by evolution is held to represent a new level of existence (there may even be sublevels within levels). Emergent qualities are novel and unpredictable. The emergent evolutionists presented speculative descriptions of cosmic and biological evolution rather than explanatory demonstrations. However, it is interesting to speculate that perhaps somewhere else in the vast cosmos an evolutionary process identical with (or very similar to) our own planetary history may have taken place, is taking place, or will take place. If so, the emergent evolutionists would be hard put to explain the identical or nearly identical occurrence of emergent novelties within their own limited scheme of things.

Samuel Alexander (1859-1938) was a realist interested in biology and physiological psychology. In his major work, *Space, Time, and Deity* (1920), he gave a comprehensive metaphysical analysis of the emergence of qualities within a space-time continuum.[1] His work distinguished between Space-Time or Motion as an objective, continuous, infinite whole and the finite homogeneous units of space-time or motion that fill this infinity (the stuff or substance of the infinite universe in Space-Time and Motion of which finite space-times or motions are merely fragments). It referred to these cosmic units as point-instants, finite extensions, durations, or pure events: they are not homogeneous units of psychic energy. In brief, this metaphysical scheme holds that reality is a four-dimensional continuum of interrelated complexes of instances of spatiotemporal motions.

Alexander taught that within this plenum there is a nisus, or creative trend. As a result of this innate tendency, Space-Time or Motion generates a chronological hierarchy of new, finite, "empirical" qualities. The successive order of emergent qualities is said to be as follows: motions, materiality with primary or constant qualities, materiality with secondary or variable qualities, life (organic), mind (consciousness), and deity.

As an emergent evolutionist, Alexander saw the cosmos as a hierarchy of qualities differing in kind. He distinguished between emergent qualities in general and the various levels of reality differing in degree, i.e. there are levels of degree within each of the major levels that differ in kind. He emphasized that mental functions succeeded, but are dependent upon, biological structures (just as living functions succeeded but are dependent upon physical structures).

Alexander also attempted to reconcile theism and pantheism within an evolutionary framework. He gave priority to Time, holding that Time generated Space-Time. In turn, Space-Time generated an order of successive, finite, empirical qualities; the latest emergent quality is consciousness. He then made a distinction between God and the deity within the irreversible creative process or nisus of Space-Time. God is the infinite world as Space-Time or Motion (this is the pantheistic aspect of God). But God is also dynamic, for the infinite universe is evolving toward deity, i.e. God is "pregnant" with deity. What is deity? Deity is the next higher, finite, empirical quality that has yet to emerge from the productive movement or nisus of Time. Within the infinite series of levels of existence, deity is always that angelic quality next to emerge in the scheme of things. God is the infinite universe itself, while the deity is only a portion of God's nature. The deity to emerge will differ in kind (not merely in degree) from mind or spirit or personality; deity is God's divine quality yet to emerge ahead of the present state of emergent evolution. In short, God is infinitely immanent here and now while the next deity is finitely transcendent in the time to come.

This attempt to reconcile theism with pantheism (transcendence with immanence) is grounded in religious sentiment or natural piety. Since the quality of mind or consciousness emerged from the quality of life, the divine quality of deity will emerge out of consciousness. Religious experiences are feelings for the vague future quality of deity. To use an analogy, the universe is the infinite body of God while the future deity is His finite "mind" yet to emerge (this view is organismic and teleological). God is the whole universe engaged in process toward the emergence of this new quality, and religion is the sentiment in one that draws one toward it. One is caught in the movement of the world toward this higher level of existence, the next emergent quality as deity.

Roy Wood Sellars (1880-1973) adopted a position of critical realism or evolutionary naturalism. His doctrine of emergent evolution was a reaction against Platonism and Kantianism. Epistemologically, it advocated a position of critical and physical realism. Metaphysically, it held to emergent materialism. Sellars took time, evolution, creative synthesis, novelty, and accumulative growth seriously. He asserted that there is a discoverable orderliness, massiveness, and immanent executiveness in nature. Nature itself represents a four-dimensional manifold of space-time and manifests four primary categories: space, time, thinghood, and causality. Evolution implies novelty and genetic continuity and has resulted in distinct levels of development but not levels of reality. Matter is active and capable of high levels of organization and accumulation resulting in the emergence of new properties by degree.

Sellars's major work is *Evolutionary Naturalism* (1922).[2] In it, he wrote that the logical structure of nature or emergent evolution reveals a hierarchy of a pyramid or tierlike construction representing stages or levels of complexity of organization and degrees of freedom (as well as new laws and categories). Historically, the general levels are represented by matter, life, mind, and society. But Sellars does not imply teleology, i.e. nature is neither evolving according to a predetermined design nor moving toward an end or goal. However, it is unfortunate that he never developed his principle of organization within a systematic explanation of emergent evolution.

As a process philosopher, Sellars held that new properties emerge as a result of the evolution of matter into new organizations. Man has emerged out of organic evolution, and mind has emerged by degree as a new property or category of highly complex organization. As a consequence of his critical realism, he maintained a naturalist or psychobiological view of the mind-body problem. The whole of man must be included in nature and nature so conceived that this inclusion is possible; thought is retrospective and supervenes upon reality. He argued that consciousness is an activity totally dependent upon the proper functioning of the brain. Therefore, human immortality is clearly impossible since there cannot be personal consciousness without a brain; all emergent properties are grounded in matter. Sellars's evolutionary cosmology and nonreductive materialistic ontology are in line with the advancements of the special sciences, and his democratic socialism and humanism are in the best interests of the human condition.

As did Marvin Farber (1901-1980), Roy Wood Sellars upheld a humanist philosophy advocating a materialist ontology and an evolutionary cosmology. He grounded knowing in biology and saw the human organism existing totally within the flux of emergent nature. His other books include *Critical Realism* (1916), *The Philosophy of Physical Realism* (1932), and *Philosophy for the Future: The Quest of Modern Materialism* (1949).

Jan Christiaan Smuts (1870-1950) proposed a philosophy of holism as an attempt to reconcile the conflicting views and implications of mechanism and vitalism. In *Holism and Evolution* (1925), he presented a creative and holistic interpretation of planetary evolution within a non-Euclidean universe, i.e. within a curved or warped space-time continuum.[3] Within the creative synthesis of evolution, he taught that matter, life, mind, and personality represent the successive major advances of cosmic energy. He stated that the universe is ultimately intensely active energy or action. Matter is concentrated structural energy or action that is active, plastic, and transmutable. It is capable of manifesting creative forms, arrangements, or patterns and values.

Smuts taught that the fundamental factor of the universe is its wholeness. Wholeness is the basic characteristic and tendency of events within the geometrical progression of cosmic development from necessity to greater degrees of freedom; holism is the ultimate category from which are derived the physical, chemical, organic, psychic, and personal categories of nature. Therefore, Smuts spoke of the evolution of wholes (creative reality is caused by the continued production of new wholes from preexisting wholes). Thus arise the physical, chemical, organic, psychic, and personal categories that are all expressive of holistic activity at its various levels and reducible to terms of holism. Thus mind structures presuppose life structures, and life structures presuppose energy structures, which are themselves graded according to the various forms of physical and chemical grouping.

Smuts held that the quality of mind represents a new product of emergent evolution. The superstructure of mind is immeasurably greater than the brain or neural structure on which it rests and is something of a quite different order that marks a revolutionary departure from the organic order out of which it originated. He viewed man as a wholistic structure in keeping with his philosophy of holism. His view of evolution is naturalistic with no resort to teleology, spiritualism, or an absolute. Although his pervasive metaphysical factor of holism remains inscrutable, he has provided a naturalist framework.

C. Lloyd Morgan (1852-1936) was a biologist and philosopher who presented a unique metaphysical system in which he described the historical emergence of new kinds of intrinsic relatedness among evolving spatiotemporal events. As a result of his comparative psychological studies, he rejected both radical behaviorism and teleology. His conceptual framework was neither mechanistic nor finalistic; his metaphysical system supported a naturalist version of evolution.

In his major work, *Emergent Evolution* (1927), Morgan distinguished among three kinds of relatedness or natural systems: matter, life, and mind (i.e. physico-chemical, vital, and conscious events).[4] Matter, life, and mind represent distinct emergent levels within which there are qualitative differences. For him, evolution is a progressive advance with novelty, or the actualization of enfolded possibilities: there is an ascending hierarchy of kinds or orders of relatedness from those in the atom to those in reflective consciousness (this hierarchy of emergent relations presents itself as a pyramid with an atomic base and reflective consciousness near its apex).

Morgan acknowledged an independent evolving physical world as a system of events and a God as the ultimate source and directing activity on which the continuity and progress of emergent evolution are ultimately dependent: by acknowledgment, he meant a judgment whose verification lies beyond the range of such positive proof as naturalist criticism rightly demands. Morgan taught that such an acknowledgment of God was necessary to supplement a scientific interpretation of evolution (an ultimate synthesis of interpretation and explanation would result). God is presented as directive activity, i.e. the creative source of emergent evolution. God is efficiency, causality, and dependence. God explains not only the directive activity in evolution but also that which from above draws all things and all men upwards. That is, God is immanent as the all-embracing activity and transcendent as the goal of emergent evolution. Morgan claimed that only through intuition may one enjoy the creative activity in evolution, as well as feel at one with existence.

Morgan held that emergent evolution manifests a threefold relatedness of involution, dependence, and correlation. Involution refers to the position that higher events involve the existence of lower events from which they have emerged, i.e. conscious events (mind) presuppose physiological events (life), and organic events (life) presuppose physico-chemical events (matter). In short, no mind without life and no life without matter. Dependence refers to the position that the existence of events also depends upon supervenient events: involution and dependence supplement each other within the space-time continuum of emerging events, i.e. as one descends the pyramid of evolution, one is concerned with dependence, but as one ascends the pyramid of evolution, one is aware of involution (as such, the coherency of evolution is guaranteed from above and below, respectively). Correlation refers to the inseparable union of the physical and the psychical. That these three categories are *a priori* is a judgment of natural piety or an acknowledgment. Emergent evolution recognizes creativity as the novelty of new kinds of events or relations in space-time, but the immanent activity and transcendent source are not susceptible of scientific proof.

In *The Emergence of Novelty* (1933), Morgan held that the historical advance of all cosmic events may be plotted on an ascending curve with God as its Final Cause.[5] However, he never held to a mystical union of the activity and source of emergent cosmic evolution. Instead, he stressed the significance of organization. Hence, evolution is the pervasive advancement of creative organization rather than merely natural aggregation. He is the only emergent evolutionist to resort to the influence of a transcendent God to account for the creative advance in the evolutionary process.

The emergent evolutionists rejected mechanist, crude materialist, and reductionist interpretations of the process of nature. Instead, they emphasized the dynamic creativity and unpredictable novelty pervasive throughout cosmic and/or earth history. However, these views are aligned more closely with metaphysics than with the special sciences and reason (Sellars's naturalist interpretation of evolution is a welcome exception to this predominantly speculative orientation in recent process thought).[6]

NOTES AND SELECTED REFERENCES

[1]Cf. S. Alexander, *Space, Time, and Deity* (New York: Dover, 2 vols., 1966). Also refer to S. Alexander, *Philosophical and Literary Pieces* (London:

Macmillan, 1939, esp. "Theism and Pantheism" pp. 316-331 as well as "Artistic Creation and Cosmic Creation" pp. 256-278).

[2]Cf. Roy Wood Sellars, *Evolutionary Naturalism* (Chicago: Open Court, repr. 1969). Also refer to Roy Wood Sellars, ed., *Philosophy for the Future: The Quest of Modern Materialism* (1949) and *The Philosophy of Physical Realism* (1966).

[3]Cf. Jan Christiaan Smuts, *Holism and Evolution* (New York: Viking Press, 1961).

[4]Cf. C. Lloyd Morgan, *Emergent Evolution* (New York: Henry Holt, 1927).

[5]Cf. C. Lloyd Morgan, *The Emergence of Novelty* (London: Williams & Norgate, 1933).

[6]Cf. Preston Warren, "Roy Wood Sellars (1880-1973)" in *Philosophy and Phenomenological Research*, XXXIV(2): 300-301. As the first critical realist, Roy Wood Sellars dedicated his intellectual life to a reformed materialism which treated the fundamental questions in ontology, epistemology, and axiology: he developed his emergent evolutionism as a substantive naturalism, attempted to clarify the semantic issues of the relativity theory, held to an identity theory of the mind-body relationship, and drafted the "Humanist Manifesto" (1933) within his own philosophy of critical naturalism. Also refer to Archie J. Bahm, "Evolutionary Naturalism" in *Philosophy and Phenomenological Research*, XV(1): 1-12, as well as Norman Paul Melchert, *Realism, Materialism, and the Mind: The Philosophy of Roy Wood Sellars* (Springfield, Illinois: Charles C Thomas, 1968).

Chapter 15

PEIRCE, JAMES, AND DEWEY

Pragmatism has been the most influential philosophy in America, representing both an evolving philosophical movement and a science-oriented attitude. This school of thought is best represented in the works of Peirce, James, and Dewey. Pragmatism was founded in 1878 by Charles Sanders Peirce, who had established the Metaphysical Club at Old Cambridge in 1871. At Harvard, Peirce's *pragmaticism* grew out of his study of the phenomenology of human thought and the uses of language. He hoped to work out a general theory of signs: a method for clarifying the meanings of intellectual signs (words, ideas, or concepts) used in communication for the purpose of helping to solve scientific or philosophical problems. For Peirce, a sign must have some conceivable practical consequence; he presented a method of inquiry grounded in a pragmatic theory of truth. However, William James's "Philosophical Conceptions and Practical Results" (1898) offered a new view of *pragmatism* emphasizing religious beliefs, morality, and values within lived human experience. Adopting a psychological orientation, he proposed that the meaning or truth of ideas, beliefs, or values resides in their "cash value" or usefulness. Therefore, James defended the "will to believe" within his own interpretation of pragmatism. Yet, John Dewey was concerned with "warranted assertions" derived from scientific inquiry grounded in a naturalist logic. His *instrumentalism* was concerned with science and values (education and morality).

Charles Saunders Peirce (1839-1914) was, as noted, the founder of pragmatism.[1] He was interested in the physical sciences (especially astronomy), philosophy, mathematics, and logic. His philosophical development passed through four distinct systems. The first system (1859-1861) was a form of extreme post-Kantian idealism grounded in a threefold ontological classification of all reality into matter, mind, and God (the three categories of It, Thou, and I, respectively). Following this early idealism, Peirce began the serious study of logic. The second system (1866-1870) argued for his three ontological categories of matter, mind, and God being necessary for signhood within converging inquiry (as such, he gave a doctrine that was both phenomenalistic and "realistic"). The third system (1870-1884) presented a modern logic of relations supporting utilitarian pragmatism and the doubt-belief theory of converging inquiry: pragmatism was presented as a theory of meaning particularly attributable to scientific definitions concerning the real as a permanent possibility of sensation, and the doubt-belief theory of converging inquiry was set in the context of biological evolution. The fourth system (1885-1914) involved a complete revision of the earlier three categories, which are now rendered as simply three classes of relations: monadic, dyadic, and triadic. He held that all thought is in the form of signs and therefore irreducibly triadic.

Peirce taught that a sound philosophy requires rigorously acknowledging the theory of evolution. Writing at the close of the nineteenth century, he recognized and incorporated three basic theories of biological evolution: Lamarckian, Cataclysmic, and Darwinian. They represent evolution by creative love, mechanical necessity, or fortuitous variations respectively (i.e. agapastic, anacastic, or tychastic evolution). Peirce favored the agapastic interpretation of biological evolution.[2] He presented a worldview supporting both an evolutionary cosmology and an evolutionary epistemology. In 1905, Peirce referred to his doctrine of pragmatism as pragmaticism to distinguish it from William James's religious form of pragmatism.

Peirce presented an intriguing if unconvincing evolutionary cosmology ultimately rooted in pervasive idealism. In fact, his interpretation of cosmic evolution was based upon the Lamarckian doctrine of the inheritance of acquired characteristics. To begin with, Peirce held to the historical continuity of reality and taught that the universe as a whole is a living and evolving organism of pervasive feelings and habits: the laws of nature describe the habits of the universe. He argued that the human mind contains innate adaptive knowledge accumulated through centuries of experience as a result of the inheritance of acquired characteristics. Judgments of common sense are more likely than not to be a source of true hypotheses.

To account for the origin of the universe, Peirce held that there had been an initial condition in which the whole cosmos itself was nonexistent, i.e. a state of absolute nothingness. From this state of pure zero or unbounded potentiality, he taught that there originated by evolutionary logic (the logic of freedom or potentiality) a state of potential consciousness or an undifferentiated continuum of pure feelings completely without order. From this primordial chaos of unpersonalized feelings and pure chance, the cosmos has evolved arbitrary brute force or action followed by living intelligence. Cosmic evolution has manifested three successive but different modes of metaphysical being: feeling, antirational action, and rational thought. This sequence represents the evolutionary perfecting of habits toward greater order or harmony. This cosmic view supported both the temporal continuity (doctrine of *synechism*) and absolute chance (doctrine of *tychism*) within the universe. Even now, there is chance within the universe and, as such, natural laws are not yet completely exact.

For Peirce, cosmic evolution is a general continuous tendency of increasing habit and decreasing chance developing toward an inevitable, rational, absolutely perfect, and symmetrical goal. In short, the cosmos continues to evolve from its past state of homogeneity (chaos and absolute chance) to a future state of heterogeneity as complete order of "concrete reasonableness" and maximum beauty. Ontologically, Peirce clearly held that the entire universe is mind rather than matter and that human life and scientific inquiry must adapt to (and remain in harmony with) the direction of cosmic evolution.

William James (1842-1910) was influenced by the New England transcendentalists and romantic humanitarians.[3] Nevertheless, he remained, for the most part, a science-oriented thinker. As a young man, James traveled widely and developed a profound sensitivity for the breadth and richness of human experience (especially aesthetic and religious experiences). But abandoning his desire to become an artist, James devoted himself to the study of the natural sciences, e.g. chemistry, comparative anatomy and physiology, and medicine. During his scientific training, he adopted an evolutionary and naturalist orientation. In June 1869, he received his medical degree from Harvard and then began teaching anatomy and physiology (1873), psychology (1875), and philosophy (1879).

Turning from the natural sciences, William James now devoted his attention to the relationship between psychology and philosophy. He was particularly concerned with questions about the human condition. He offered a philosophical defense of morality, free will, and religious beliefs. His form of pragmatism is primarily a theory of meanings (rather than truths) grounded in the world of pure experience. His pragmatism rejected the possibility of a pure phenomenological description of inner experience, but it supported naturalism and the possibility of an *a posteriori* metaphysics, both scientific and theoretical. James's pragmatism is an open-ended method for settling metaphysical disputes within a practical context. It is grounded in an empiricist attitude recognizing the relationship between science and philosophy. Pragmatism is concerned with consequences (theories thus become instruments): *the truth is the name of whatever proves itself to be good in the way of belief, and good, too, for definite, assignable reasons* (*Pragmatism*, 1907, p. 59).

As an evolutionary naturalist, William James rejected the traditional mind-body dualism in the philosophical literature. However, he did distinguish between mental states and the underlying physical states of the brain. He saw psychology as providing an adequate introspective description of mental states and pragmatic philosophy as a critical analysis of the meanings in human experience. Above all, he focused upon spontaneous religious experience. An evolutionist, he held that there must be a meaningful adaptive function to religious feelings. He developed his doctrine of the will to believe, a vague philosophical defense of religious beliefs within his otherwise naturalist orientation. He rejected traditional supernaturalism, merely holding that religious experience suggests that there is a higher part of the universe. His pragmatism was an attempt to formulate a metaphysics grounded in natural processes, the special sciences, and humanism. His vision recognized both the breadth and depth of human experience, and he remained a mediator between tough-mindedness and tender-mindedness.

As the most recent and major figure in the school of American pragmatism, John Dewey (1859-1952) had been influenced by Kantian and Hegelian idealism.[4] However, he became greatly inspired by the works of Charles Darwin and William James. As a result, he abandoned his early idealism for a dynamic biologico-anthropological orientation within a naturalist evolutionary framework. Although adopting a dynamic organismic perspective, he developed no philosophical system in the traditional sense. Unlike Kant and Hegel, he left speculative philosophy in order to concentrate on a practical social consideration of psychology and education within a democratic community of creative intelligence.

Dewey emerged as America's most influential philosopher. His form of pragmatism differed from both Peirce's and James's, and was referred to as instrumentalism. His philosophy advocated naturalism and humanism within a practical context uniting thought and action. Influenced by Darwinism (though not Social Darwinism), he taught that ideas are instruments or tools to solve problematic or indeterminate situations (ideas allow for human adaptation and survival within a social context). For Dewey, ideas are guides to action; true learning takes place only when the individual is free to explore and test his world. In brief, he advocated individual self-realization within a harmonious democratic society.

As a rigorous naturalist, Dewey held the natural world to be prior to the emergence of human existence: nature is in continuous change, process, and interaction. Human experience is a product of and totally within dynamic nature (human experience is both reflective and nonreflective). Human growth is through experience and education. Education results from an interaction (dialectical relationship) between the human knower and the knowable events of nature. Action is central to Dewey's philosophy and results in the creative self-realization of the individual within the community.

Dewey emphasized both the scientific method and rigorous reflection. He called the empirical method the denotative method. His naturalist philosophy rejected traditional dualisms: the knowable and the unknowable (becoming and being), philosophy and science, rationalism and empiricism (intellectualism and sensationalism), mind and body, and the theoretical and practical (thought and action). Also, he rejected both supernaturalism and innate ideas. For Dewey, there are only methodological truths. Ideas are meaningful, valuable, and true if they result in fruitful inferences and/or consequences. The results of thought and action determine pragmatic significance. Unfortunately, Dewey's metaphysics (ontology and cosmology) was never clearly developed.

As an evolutionist, Dewey recognized three major natural levels of development within the continuity of planetary history: the physicochemical, psychophysical, and human experience levels. Five major points emerge from his instrumentalism or experimentalism: (1) the intellect or reason is on a higher plane than the other human qualities, (2) ideas emerge out of man's need to establish an equilibrium between himself and his environment (this point illustrates Darwin's influence on Dewey's form of pragmatism), (3) ideas are meaningful if and only if they are applicable to problematic or indeterminate situations, (4) ideas are true if and only if they solve problematic or indeterminate situations, and (5) the scientific method is applicable to ethics and morals (human principles and social behavior) as well as to nature, and there is no need to have recourse to any form of supernaturalism.

In summary, Dewey's pragmatism held that (1) metaphysics should be descriptive and not merely speculative (he called for a naturalist philosophy), (2) subjectivism results in an unwarranted dualism (he rightly rejected the traditional mind-body dichotomy), (3) philosophy must recognize the facts and values of the special sciences, e.g. physics, chemistry, biology, anthropology, and psychology (he taught that the study of anthropology is an ideal preparation for philosophy), (4) knowledge and the awareness of problems are obtained through experience, (5) there are no absolute truths, (6) ideas are *meaningful* if they are applicable to human problems and *true* if they eliminate such problems, (7) education does not necessarily liberate a person, but it does develop his creative intellect, (8) there is no distinction between knowing and doing: to know is to act, and (9) it is the aim of future philosophy to clarify man's ideas, i.e. to determine the *meanings* and *truths* of human ideas and human actions. Dewey's pragmatism was essentially a theory of inquiry.

As a doctrine and movement, pragmatism dominated American philosophy in the early part of this century. Its three major exponents were Peirce, James, and Dewey. Each of these philosophers had his own interpretation of pragmatism. For the most part, they aligned themselves with the advances in the natural and social sciences (especially biology, anthropology, and psychology).

NOTES AND SELECTED REFERENCES

[1]Cf. Justus Buchler, ed., *Philosophical Writings of Peirce* (New York: Dover, 1955, esp. "Evolutionary Love" pp. 361-374), Richard Hofstadter, *Social Darwinism in American Thought* (Boston: Beacon Press, rev. ed., 1955, esp. chapter seven, "The Current of Pragmatism" pp. 123-142), Vincent Tomas, ed., *Essays in the Philosophy of Science* by Charles S. Peirce (New York: Bobbs-Merrill, 1957, esp. "The Order of Nature" pp. 105-125), Philip P. Wiener, ed., *Charles S. Peirce: Selected Writings (Values in a Universe of Chance)* (New York: Dover, 1966), and Philip P. Wiener, *Evolution and the Founders of Pragmatism* (New York: Harper Torchbooks, 1965, esp. chapter four, "The Evolutionism and Pragmaticism of Peirce" pp. 70-96).

[2]Cf. Charles S. Peirce, "Evolutionary Love" In *The Monist*, 1893, 3: 176-200.

[3]Cf. William James, *Pragmatism and Four Essays from 'The Meaning of Truth'* (New York: Meridian Books, 1955) and *The Will to Believe and Other Essays in Popular Philosophy and Human Immortality* (New York: Dover, 1956). Also refer to Alburey Castell, ed., *Essays in Pragmatism* by William James (New York: Hafner, 1948), Richard Hofstadter, *Social Darwinism in American Thought* (Boston: Beacon Press, rev. ed. 1955, esp. chapter seven, "The Current of Pragmatism" pp. 123-142), John J. McDermott, ed., *The Writings of William James* (New York: Random House, 1967), and Philip P. Wiener, *Evolution and the Founders of Pragmatism* (New York: Harper Torchbooks, 1965, esp. chapter five, "Darwinism in James's Psychology and Pragmatism" pp. 97-128).

[4]Cf. John Dewey, *The Influence of Darwin on Philosophy* (1909) and other

essays in contemporary thought (Bloomington: Indiana University Press, 1965, esp. title essay pp. 1-19) and *On Experience, Nature, and Freedom: Representative Selections* (New York: Bobbs-Merrill, 1960). Also refer to Richard Hofstadter, *Social Darwinism in American Thought* (Boston: Beacon Press, rev. ed., 1955, esp. chapter seven, "The Current of Pragmatism" pp. 123-142) and Philip P. Wiener, *Evolution and the Founders of Pragmatism* (New York: Harper Torchbooks, 1965).

Chapter 16

WHITEHEAD

Alfred North Whitehead (1861-1947) was interested in the new facts and concepts within relativity physics, quantum mechanics, and evolutionary biology. No longer were space, time, and motion considered independent absolutes, simple causality and determinism applicable to alleged material atoms, or biological species fixed entities within a static nature. Whitehead's intellectual life passed through three successive but distinct stages: (1) mathematics and logic,[1] (2) history and philosophy of science,[2] and (3) process metaphysics.[3] Although he was both empiricist and rationalist in his philosophical quest, he emerged as a major speculative and systematic thinker of this century devoted to an aesthetic appreciation for the creative advance of the Cosmos. He always held that philosophy is an attempt to express the infinity of the universe in terms of the limitations of language. Yet, for a major philosopher emphasizing experience and activity, Whitehead's personal life seems to be remarkably uneventful. His satisfaction came from the inner world of creative and rigorous reflection.[4]

Whitehead's *Science and the Modern World* (1925) is a series of lectures devoted to the development of western science and philosophy during the last three centuries (consideration of epistemology is admittedly entirely excluded).[5] From Copernicus and Bruno to modern ideas in physics and biology, it emphasizes that the growth of the special sciences directly influences conceptual frameworks. He held that the creative development or evolution of the Cosmos represents a fundamental unity and that cosmology must do justice to objective and subjective realms of inquiry.

Whitehead's major work is *Process and Reality: An Essay in Cosmology* (1929).[6] It is concerned with speculative philosophy as an endeavor to frame a comprehensive, coherent, logical, and applicable scheme of things. It relies upon knowledge gained through reason and experience, stressing the elucidation of immediate experience and resultant feelings of satisfaction. Whitehead's "philosophy of organism" gives a metaphysical description of process and reality. It treats the fluidity and plurality of the world in a most general way, as well as the eternal structures assumed necessary to give unity and rational order to this particular cosmic system.

Whitehead presented the "fallacy of dogmatic finality": no permanently fixed, sufficient explanation for the universe is ever possible. Similarly, he presented the "fallacy of misplaced concreteness": a crucial distinction is required between the ontological status of ideas and the objective reality that they purport to reflect. He also upheld the "ontological principle": there is a rational explanation for the existence of all objects or events; all existing things emerge from preexisting things; something comes from something; all existence has an ontological source. The ontological principle can be summarized as, If there is no actual entity, then there is no reason. Whitehead presented a categorical scheme covering the ultimate, existence, explanation, and obligation. For him, the world is composed of actual entities and actual occasions or events within the becoming of things. Each entity manifests a dipolarity (a physical and a mental pole). Likewise, there are eternal objects or potentials or possibilities (forms, qualities, and values). God's primordial nature is the realm of eternal objects, while God's consequent nature is the realm of events. God experiences the world in a reciprocal relationship, deriving feelings and understanding (God's superjective nature).

For Whitehead, God is an actual entity necessary for a comprehensive explanation of Nature.[7] God relates to the World and gives it existence, stability, order, creativity, novelty, purpose, and value. God is held to be the chief exemplification of all metaphysical principles: He is eternal, immanent, self-consistent, and transcends any finite cosmic epoch. As an actual entity, God is also dipolar. The primordial nature or conceptual experience of God is the "Object of desire" of the eternal urge of the universe. This aspect of God is His unlimited potentiality, i.e. it is the realm of the Eternal Objects, while the consequent nature or physical experience of God is the finite, fluent multiplicity of the World. As such, God and the World are ontologically interrelated through an eternal reciprocal relationship of everlasting creativity seeking a perfect unity. Clearly, this is a form of panentheism that to a rigorous scientific naturalist remains less than completely convincing despite its impressive intellectual loftiness and unquestionable personal integrity.

For Whitehead, the World is an extensive continuum of internal and external interconnectedness. To reject this position is to commit the "fallacy of simple location." The World is a hierarchy of societies or enduring objects (nexus). The eternal objects are ingressed or participate in the enduring objects. Societies evolve into new societies, giving objective immortality to actual entities (however, this does not imply personal immortality). The preservation of new aggregations of events in novel societies is referred to as the theory of objectification.

The ultimate category of the World is creativity, the process by which the one and the many teleologically develop toward novelty. In describing universals and particulars, no sharp distinction is possible (Whitehead's doctrine of universal relativity). The present cosmic epoch manifests its own peculiar laws and creative order. Feelings (conformal, conceptual, simple comparative, or complex comparative) or prehensions (physical or conceptual, positive or negative) are pervasive throughout the actual world. They are experienced by the human knower through the mixed mode of symbolic reference (both through presentational immediacy and causal efficacy).

Each entity is a subject-superject within the process of becoming. Likewise, each entity manifests an innate subjective aim or appetition (subjectivist principle). The World is a preestablished harmony of societies moving toward new creative unities or concrescences (principle of satisfaction). In short, each object or event of the World is at the same time an external subject and a subjective form for every other experiencing actual entity. Transmutation refers to the act of deriving macrocosmic perceptions from microcosmic prehensions.

In *Process and Reality,* Whitehead exhibited an analytic and synthetic mind.[8] His original views called for new perspectives and terminology. He gave priority to experience and feelings, resulting in an idealist cosmology. Nevertheless, he attempted to systematically reconcile the temporal realm of becoming with the eternal realm of Being. He gave methodological preference to mathematics and ontological preference to experience and resulting feelings. Whitehead desired to construct a modern process cosmology. His dynamic system has the advantage of being open-ended to infinite possibilities (potentialities). The universe is an infinite series of unique cosmic epochs, each cosmic epoch having its own rational order: Nature is a creative continuum. God and the World are never complete. There is no final achievement (goal or end); no final cosmic satisfaction resulting from the creative urge of the World toward a novel synthesis. In general, Whitehead's cosmology is remarkably open but idealistic as a result of his residually anthropocentric and anthropomorphic orientation. By going beyond the basic facts of empirical experience, his speculations are open to criticism. However, he will be remembered for his contributions to logic, mathematics, and the history and philosophy of science as well as his emphasis on the creative development of nature, principle of causal efficacy, and method of extensive abstraction.[9]

NOTES AND SELECTED REFERENCES

[1]Cf. Alfred North Whitehead and Bertrand Russell, *Principia Mathematica* (Cambridge: Cambridge University Press, 1910, paperback ed. 1962).

[2]Cf. Alfred North Whitehead, *The Concept of Nature* (Cambridge: Cambridge University Press, 1920, paperback ed. 1964), *Science and the Modern World* (New York: Macmillan, 1925, Free Press paperback ed. 1967), *Symbolism: Its Meaning and Effect* (New York: Macmillan, 1927, Capricorn Books paperback ed. 1959), *The Aims of Education and Other Essays* (New York: Macmillan, 1919, Free Press paperback ed. 1967), *Modes of Thought* (New York: Macmillan 1938, Free Press paperback ed. 1968), and *The Interpretation of Science: Selected Essays* (New York: Bobbs-Merrill, 1961). Also refer to Laurence Bright, *Whitehead's Philosophy of Physics* (New York: Sheed and

Ward, 1958). Although Whitehead rarely wrote about Darwin or evolution, see *Science and the Modern World* pp. 93, 100-103, 107-112, and *Modes of Thought*, pp. 112-113.

[3]Cf. Alfred North Whitehead, *Process and Reality: An Essay in Cosmology* (New York: Macmillan, 1929, Free Press paperback ed. 1969) and *Adventures of Ideas* (New York: Macmillian, 1933, Free Press paperback ed. 1967, esp. part two, pp. 103-129).

[4]Cf. Lucien Price, *Dialogues of Alfred North Whitehead* (New York: Mentor Books, 1956). Whitehead is quoted as saying: "I have been rereading Huxley's *Letters*, especially the second volume. He strikes me as one of those men who fall just under the first rank, immensely able but not great. Darwin, on the other hand, is truly great, but he is the dullest great man I can think of. He and Huxley had grasped the principle of evolution in material life, but it never occurred to them to ask how evolution in material life could result in a man like, let us say, Newton" (pp. 228-229). It is worth pointing out that Whitehead had an early interest in historical archaeology. Also refer to "Autobiographical Notes" in Paul Arthur Schilpp, ed., *The Philosophy of Alfred North Whitehead* (New York: Tudor, 2nd ed., 1951, pp. 3-14).

[5]Cf. *Science and the Modern World* (New York: Free Press, 1967).

[6]Cf. *Process and Reality: An Essay in Cosmology* (New York: Free Press, 1969).

[7]Cf. *Ibid.*, "God and the World" pp. 403-413, as well as the earlier work *Religion in the Making* (New York, Macmillan, 1926, Meridian Books paperback ed. 1960, esp. pp. 66-78, 91-101, and 143-154).

[8]Refer to the following studies of Whitehead's thought: William A. Christian, *An Interpretation of Whitehead's Metaphysics* (New Haven: Yale University Press, 1959), Dorothy Emmet, *Whitehead's Philosophy of Organism* (New York: St. Martin's Press, 2nd ed., 1966), A.H. Johnson, *Whitehead's Theory of Reality* (New York: Dover, 1962), Ivor Leclerc, *Whitehead's Metaphysics: An Introductory Exposition* (New York: Macmillan, 1958), Ivor Leclerc, ed., *The Relevance of Whitehead* (New York: Macmillan, 1961), Victor Lowe, *Understanding Whitehead* (Baltimore: The Johns Hopkins Press, 1962), W. Mays, *The Philosophy of Whitehead* (New York: Collier Books, 1962), Norman Pittenger, *Alfred North Whitehead* (Richmond, Virginia: John Knox Press, 1969), Edward Pols, *Whitehead's Metaphysics: A Critical Examination of Process and Reality* (Carbondale: Southern Illinois University Press, 1967), Paul F. Schmidt, *Perception and Cosmology in Whitehead's Philosophy* (New Brunswick, New Jersey: Rutgers University Press, 1967) and Donald W. Sherburne, ed., *A Key to Whitehead's Process and Reality* (New York: Macmillan, 1966). Also refer to Paul Arthur Schilpp, ed., *The Philosophy of Alfred North Whitehead* (New York: Tudor, 2nd ed., 1951) as well as Charles Hartshorne, *A Natural Theology for Our Time* (LaSalle, Illinois: Open Court, 1967, esp. pp. 23, 26, 74, 92, and 115) and R.G. Collingwood, *The Idea of Nature* (Oxford, The Clarendon Press, 1945, last chapter, pp. 158-177, for references to Alexander, Whitehead, and modern cosmology). Also refer to Dean R. Fowler, "Alfred North Whitehead" in *Zygon: Journal of Religion and Science*, 11(1):50-69.

[9]Whitehead's organic philosophy is admittedly not a form of naturalism, and there are very few references to Charles Darwin in his writings. Refer to Ruth Nanda Anshen, ed., *Alfred North Whitehead: His Reflections on Man and Nature* (New York, Harper & Brothers, 1961, esp. pp. 10-31) and Alfred North Whitehead, *Nature and Life* (New York: Greenwood Press, 1977).

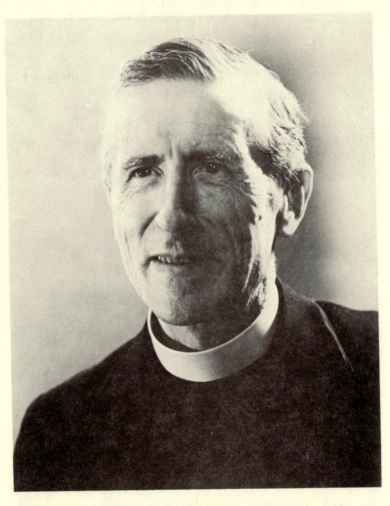

Pierre Teilhard de Chardin (1881-1955). Courtesy of American Museum of Natural History.

Chapter 17

TEILHARD DE CHARDIN

Pierre Teilhard de Chardin (1881-1955) was a very remarkable man, who dedicated his life to scientific research and the service of a personal God. His poetic and inspiring writings, especially *The Phenomenon of Man* (1940), are an attempt to reconcile within an evolutionary perspective the special sciences, a naturalist process philosophy, and a profoundly personal and mystical interpretation of traditional Catholic theology; matter and spirit, thought and action, personalism and collectivism, plurality and unity, pantheism and theism as process panentheism; and the empirically documented facts, far-reaching philosophical implications, and religious consequences of planetary evolution with certain elements of Christian supernaturalism and cosmic mysticism.

As a result of Teilhard's explicit but unsuccessful request for the publication of his own slightly revised edition of *The Phenomenon of Man* in 1948, Pope Pius XII issued an Encyclical Letter (*Humani generis*, 12 August 1950) in which he gave preference and priority to a Thomist interpretation of Divine Revelation as contained in the Holy Scriptures over the evidence of the empirical sciences and the arguments of logic. In this document, the Pope warns that opinions on the theory of evolution, which he was willing to leave an open question, may be erroneous, i.e. fictitious or conjectural.

This authoritative Papal pronouncement held that evolutionism results from a desire to be novel, and therefore the doctrine is merely a question of hypotheses or possibly even a false science. It warned that evolutionism leads to the formless and unstable tenets of a new philosophy and sterile speculation, claiming the major error to be too free an interpretation of the historical books of the Old Testament. Likewise, the Encyclical Letter held the doctrine of evolution to be not only plainly at variance with Holy Scriptures but even to be false by experience as well.

That such a position was a direct attack against Teilhard's new philosophy of evolution is fairly obvious. In fact, it was against such a dogmatic position as expressed by the Pope that the Jesuit scientist had attempted to establish a better Christianity, i.e. a Metachristianity within a cosmic, naturalist, and mystical evolutionary framework. It is worth noting that although his naturalist writings on evolution were duly sealed in a supernaturalist envelope they never received the *Nihil Obstat* and *Imprimatur*, which declare that a book or pamphlet is considered to be free from doctrinal or moral error in the eyes of the Roman Catholic Church. Moreover, not only was the publication of the Papal Encyclical *Humani generis* an implicit attack against Teilhard's less than orthodox thoughts, but a *Monitum* decree (March, 1962) issued by the Holy Office on his works went even as far as to warn bishops and heads of seminaries of the doctrinal errors said to be inherent in his interpretation of mankind within nature. It is painfully embarrassing and profoundly regrettable that in the middle of this century a gentle, devout, and brilliant intellect like Teilhard should have had to suffer silencing by the Roman Catholic Church because of his bold curiosity and daring originality in areas clearly lying beyond the Church authorities' scientific competence, let alone their expertise.

Nevertheless, the sudden enthusiasm elicited by Teilhard's posthumously published works was nothing less than phenomenal. Yet, a proper analysis and evaluation of his unique synthesis was not immediately forthcoming. Scientists referred to it as primarily religious and mystical, theologians objected to its unorthodoxy, and philosophers, for the most part, chose to ignore it. The truth is that the French scholar went far beyond any other philosopher of evolution in trying to reconcile the particular sciences, dynamic philosophy, and mystical religion within a holistic view of humankind's history and destiny in a divinely generated and God-embraced universe.

It is unfortunate that during his lifetime Teilhard never had the opportunity to submit his writings to the sort of professional criticism that could have eliminated at least some of their many ambiguities. More than a quarter of a century after his death, Teilhard's thought still remains both attractive and controversial. Evaluations of its importance continue to vary enormously.[1] Strongly negative assessments include those made from different viewpoints by George Gaylord Simpson, P.B. Medawar, and Jacques Maritain. At the other extreme are those made by Theodosius Dobzhansky and Sir Julian Huxley who, despite certain reservations, highly praised Teilhard's attempt at an evolutionary synthesis.

Simpson did refer to Teilhard as a phenomenal man, an accomplished scientist and profound theologian who uniquely combined in one well-integrated personality an evolutionary naturalist and a religious mystic. However, Maritain (one of the leading Thomists of this century) held that the Jesuit priest had, in fact, given to science a dazzling primacy and thereby committed an unforgivable sin against the intellect. As such, Maritain referred to the thought-system as theological fiction. Yet, the strongest rejection of Teilhard's worldview came from Medawar, who held it to be an antiscientific, unintelligible, and impractical philosophical fiction containing a feeble argument abominably expressed.

Although Teilhard's work has been referred to as a false science beyond demonstration, a subjectively contrived piece of theology-fiction, and even sheer nonsense to both science and theology, nevertheless, there were a few who were favorable to it. Sir Julian Huxley wrote a sympathetic and admiring introduction to the English translation of *The Phenomenon of Man*, thus giving this work philosophical respectability and scientific prestige. Huxley had been attempting an evolutionary synthesis, and therefore he welcomed Teilhard's bold and holistic conception of our species's place in the history of nature.

Similar to earlier systematic thinkers (e.g. Aristotle, Spinoza, Leibniz, and Hegel), Teilhard had desired to present an overview of man's position in the universe. Where some others before him had merely been concerned with change and development, Teilhard, as did Herbert Spencer, saw the whole cosmos (including the human species) in evolution. Because he was at once a scientist, philosopher, theologian, and poetic mystic, his unique achievement is very difficult to place, assess, interpret, and evaluate.

As will become evident, Teilhard's shortcomings are due to the planetary and theological orientations he needed to take in order for his daring but unsuccessful effort to achieve a reconciliation between the natural universe and a supernatural God within a self-personalizing and converging evolutionary panentheism of mystical involution.

It is a safe generalization to say that Teilhard's Roman Catholic philosophy of mystical evolution rests upon four basic interrelated concepts: (1) evolutionary panpsychism or spiritual monism, (2) the accelerating cosmic law of increasing centro-complexity/consciousness, (3) critical thresholds among transitional phases of terrestrial evolution, and (4) the Omega Point.

Teilhard was not restricted to one method or field of inquiry. What resulted from this approach was a Christian natural theology favoring vitalism, teleology, and spiritualism. As such, it is still awaiting the sort of comprehensive rigorous assessment that it undoubtedly merits in light of the established facts of the special sciences and critical reflection. For better or worse, the historical materialist may attempt to reinterpret Teilhard as Marx had attempted to reinterpret Hegel.[2] For the naturalist, the essential contribution of Pierre Teilhard de Chardin is to be found in his unequivocal commitment to the evolutionary framework with all its implications, including almost certainly some to which he never gave explicit thought or attention.

Teilhard wrote three books, each a major contribution: *The Divine Milieu* (1927), *The Phenomenon of Man* (1940, with a postscript and appendix in 1948), and *Man's Place in Nature: The Human Zoological Group* (1950).[3] His ideas are further developed in voluminous collections of expert essays, scholarly articles, and both insightful and revealing letters to family, friends, and scholarly associates.[4]

Human inquiry is never free from conscious or unconscious

predilection and consequent preferential judgments: scientists, philosophers, theologians, artists, statesmen, and politicians all have their unavoidable vested interests. This is why no thought-system springs forth unaffected by its author's personality or outside of the economic, sociopolitical, and cultural conditions of his or her national background and general place in the process of history. Teilhard's worldview is no exception. To remove the Jesuit scientist and his system from such a context of history would amount to an injustice not only to him personally but also to the facts and principles of evolution. Therefore, in order to understand and appreciate the development of Teilhard's philosophy of evolution from his early World War I writings and first book, *The Divine Milieu,* to his major systematic work, *The Phenomenon of Man,* and the later, scientifically oriented *Man's Place in Nature: The Human Zoological Group,* it is necessary to consider the significant events in his life that influenced the growth and development of his thought as recorded in his books, essays, articles, and letters.

On 1 May 1881, Marie-Joseph-Pierre Teilhard de Chardin was born at Sarcenat in the province of Auvergne, near Orcines and Clermont-Ferrand, France.[5] He was the fourth of eleven children, and his ancestry could boast of such luminaries as Pascal and Voltaire. The influence of his mother, Berthe-Adèle de Dompierre, kindled in him a deep life-long love of and dedication to Christian mysticism. As for his father, Emmanuel Teilhard de Chardin, he encouraged an active and ardent interest in natural history. Therefore, even as a child, Teilhard was immersed in science and religion equally and at the same time.

Teilhard had an irresistible need for one essential thing, a supernatural entity that would be simultaneously necessary and sufficient in itself. As a result, his developing interests in astronomy, mineralogy, biology, and entomology were emotionally (though not intellectually) secondary to his profound personal yearning for an all-embracing supercosmic absolute.

Leaving behind his most precious possessions (a boyhood collection of rocks and pebbles), Teilhard, at the age of eleven, entered the Jesuit secondary school of Notre-Dame de Mongre in 1892. He studied Latin, Greek, German, and the natural sciences as well as developed an interest in philosophy. On 20 March 1899, he entered the Jesuit novitiate in Aix-en-Provence, for at the age of seventeen his desire to be "most perfect" determined his choice of future vocation as a Jesuit priest. As a junior, he continued his studies in geology and philosophy on the channel island of Jersey (1902-1905). After finishing his scholasticate in Jersey, he was sent to teach chemistry and physics at the Jesuit College of the Holy Family in Cairo, Egypt (1905-1908). When free from teaching the natural and social sciences, which included botany, the young instructor preferred to research in the deserts.

Teilhard's first publication was an article entitled "A Week at Fayum" (1908), followed by an early professional publication under the title of "The Eocene Strata of the Minieh Region" (1908). But despite his continuing accomplishments in both geology and paleontology, his religious beliefs remained orthodox and free from scientific influences. He next was sent to Ore Place, the Jesuit house in Hastings on the Sussex coast, England, for further theological studies (1908-1912). While remaining interested in numerous scientific subjects (rocks, fossils, plants, insects, and birds), Teilhard's religious orientation was always present in his thoughts and writings.

On 24 August 1911, Teilhard was ordained priest in the chapel at Ore Place. The following year, he successfully passed his final exam in theology, which resulted in his earning the equivalent to a doctorate in that discipline. He then pursued his scientific researches at the Museum of Natural History in Paris under Marcellin Boule, a noted professor of paleontology. The Jesuit priest joined the Geological Society of France and established a life-long friendship with the Abbé Henri Breuil, a distinguished specialist in prehistory.

Sometime during this period, Teilhard had read Henri Bergson's major and definitive work, *Creative Evolution* (1907). The book had a profound influence on the Jesuit scientist's own incipient thoughts. No longer able to hold to a literal interpretation of the traditional scriptural account of divine creation as recorded in the sacred myth of Genesis as found in the Old Testament of the Holy Bible, he shifted from orthodoxy to an evolutionary perspective. Within a scientific and religious framework, he now viewed the entire universe as a dynamic process, and as such he even referred to it as a cosmogenesis. Unlike Bergson, however, who saw planetary evolution as an essential creative and diverging movement of life, Teilhard formulated a Catholic interpretation of terrestrial evolution that stressed the converging and involuting aspects of spirit as it moves to unite a collective humanity with a personal God at the future Omega Point.

Bergson's view of evolution had been founded upon an ontological dualism between inert matter and animated spirit as consciousness, or the *élan vital*. He held the latter lifeforce to be responsible for the irreversibility, historical continuity, and increasing novelty or creativity as well as complexity, diversity, and consciousness manifested throughout planetary evolution. Teilhard, however, adopted a monistic ontology giving a privileged position to consciousness or spirit. In brief, one will see that Teilhardian evolution is vitalistic, cumulative, irreversible, accelerating, personalizing, teleological, and converging or involuting toward ever-greater improbabilities, complexity/consciousness, value, freedom and unity as creative equilibrium or spiritual synthesis, and the perfection of being itself at the Omega Point.

For Teilhard, the theory of evolution provided the necessary framework within which he could synthesize his many interests in science and philosophy with his deepest commitments to theology and mysticism. This doctrine could also do justice to his cosmic interpretation and eschatological vision of Christ (he was already publically advocating the validity and even necessity of an evolutionary perspective).

Teilhard's first encounter with human paleontology (physical anthropology) was a rather unfortunate one. In 1909, he had met the amateur geologist Charles Dawson. In 1912, joined by Sir Arthur Smith-Woodward, they visited Piltdown, England, where Dawson unearthed a new fragment of a human skull and Teilhard found an elephant molar. The three returned to the same site in 1913, and this time Teilhard found an additional canine tooth attributed to the lower jaw of the now infamous Piltdown Man *(Eoanthropus dawsoni)*. This incident is significant, however, because it kindled in him a lasting interest in paleoanthropology. Fortunately, the cautious young scientist remained skeptical of the skull fragments and teeth of Piltdown Man, for this fossil find presented a structural anomaly in the then emerging picture of human evolution. In 1953, Dr. Kenneth P. Oakley, using the then newly discovered fluorine dating method, revealed that the specimen was, in fact, a fraudulent miscombination of a relatively recent human cranium with an orangutan mandible.[6] Teilhard remarked that "Nothing seemed to 'fit' together—it's better that it all fell through."

The admired Professor Stephen Jay Gould of Harvard University has made an impressive attempt at the detective reconstruction of this whole unfortunate footnote to the history of evolutionary science; however, this recent attempt seems to be more imaginative and brilliant than convincing and sound. If "guilty" at all, the highly ethical Teilhard could be so only by circumstantial association and perhaps omission but not by deliberate conspiracy and/or conscious commission.[6]

Nevertheless, Teilhard's future scientific research took him to every major human paleontological site known in his lifetime. In fact, the removal of Piltdown Man from the legitimate record of fossil evidence for human evolution actually favored the Jesuit priest's preferred interpretation by supporting the hypothesis of a continuous evolution of the human brain and its cultural outcome. (Charles Darwin had erroneously maintained that the modern cranial capacity had evolved before man became a tool maker; one may now safely assume that the human animal was a tool maker over two million years ago with only about a third of its present cranial capacity. Human intelligence and cultural creativity have evolved together in a feedback process of interaction of structurally and functionally interchangeable causes and effects.)

In 1913, Teilhard joined the Abbé Breuil on an expedition to the prehistoric sites in the Pyrenees.[7] Early mammalian fossils from Quercy and the Rheims areas in the southwest of France provided Teilhard with the basis for his dissertation in geopaleontology. He worked on his thesis at the Institute of Human Paleontology in the Museum of Natural History, Paris.

Teilhard's work was interrupted when he was drafted in December 1914 and assigned to the thirteenth division of the medical corps. He served at the front as a stretcher-bearer (1914-1919) and was cited three times for distinguished service: he was made Chevalier of the Légion d'Honneur, and held both the Croix de Guerre and the coveted Médaille Militaire. He had shown great courage and humility, and his letters to his cousin Marguerite Teillhard-Chambon (Claude Aragonnès) illustrate his continued optimism and developing mysticism even during the grimmest years of the war.

Teilhard's experiences during this tragic conflict did not deaden his spirit.[9] On the contrary, he emerged more dedicated to the two objects of his passion: a personal God and the evolving universe. His belief in a theistic cosmogenesis prevented him from adopting the intellectually tempting but emotionally unsatisfying position of pantheism. His mystical interpretation of a converging and involuting universe allowed for the future culminating union of a spiritualized material cosmos with an all-enclosing personal God, i.e. just as he had temporarily and discretely replaced traditional theism with a dynamic but provisional pantheism, he now reconciled and transcended both of these antithetical viewpoints in a synthetic and processual panentheism. He focused on this last position and dedicated his life to a scientific demonstration and philosophical clarification of his belief in the validity and necessity of this, for him, definitive viewpoint concerning the survival and fulfillment of humankind on earth. For his unique theological perspective and inspiration, the Jesuit priest relied upon the writings of Saints Paul and John of Patmos to substantiate the seemingly Roman Catholic orthodoxy of his cosmic and mystical view of a finalistic panentheism in terms of God and humankind.

After the war, with renewed strength and hope in the future progress of mankind, Teilhard took his solemn vows on 26 May 1918 at Sainte-Foy-Les-Lon, France. His early writings during the war had been primarily devoted to the explication of the immanence and transcendence of a personal God, i.e. the converging and involuting evolution of the spirit of the world toward a final creative union with the absolute through the attractive cosmic force of love energy. Now, the Jesuit priest returned to scientific research. He taught as an associate professor of geology at the Institut Catholique (1920-1923) while resuming work on his doctoral thesis on the mammifers of the Lower Eocene in France and those layers of soil in which they are found. However, his dedicated religious commitment endured throughout this scientific research and beyond.

In 1922, Teilhard received the title of doctor with distinction at the Sorbonne for this thesis on *The Mammals of the Lower Eocene Period in France* (the Jesuit priest was now also a formally recognized professional geopaleontologist).

Père Émile Licent, a fellow Jesuit, had built a museum and laboratory at Tientsin in China. Its purpose was the study of Chinese geology, mineralogy, paleontology, and botany. As director of the project, he invited Teilhard to join the French Paleontological Mission. Thus, on 6 April 1923, the young French Jesuit scientist left Paris for Tientsin. He accompanied Licent on an expedition to inner Mongolia and the Ordos desert. It was during this journey into the interior of that desolate region that on a Sunday and having no means to celebrate Mass on the Feast of the Transfiguration Teilhard finished his religious and mystical poem entitled "The Mass on the World" (1923).[10] For Teilhard, cosmogenesis as evolutionary convergence focused on the earth will lead to the ultimate Christogenetic unification of our humanized and spiritualized planet with the supracosmic and mystical personal God at the Omega Point.

On 13 September 1924, Teilhard left China and returned to France. During the winter months of 1925-1926, he gave four lectures on the theory of evolution and originated his concept of the noosphere, which represents the hominization of this earth considered from the biosocial and psychocultural perspectives. Spiritually, however, the noosphere represents the accumulation of all persons as reflective monads or centers-of-consciousness within an earthbound layer or terrestrial envelope of minds and their thoughts. Through continued evolution, Teilhard held that these sensate units of psychic force are collectively converging and involuting toward a personal God as the Supreme Center of a process reality.

Because of his evolutionary orientation and largely unorthodox interpretation of the Original Sin, the Jesuit superiors now felt compelled to withdraw Teilhard from the Institut Catholique: his lectures were challenging the traditional Catholic view of a more or less static planet with its hierarchy of eternally fixed forms. Moreover, he was confined to scientific research in descriptive geopaleontology, forbidden to publish or teach in the areas of philosophy and theology, and exiled from France back to China in 1926.

On his return to Tientsin, from November 1926 to March 1927, Teilhard wrote his first book entitled *The Divine Milieu;* it is a spiritual essay on life in the light of an inward vision, attempting to reconcile the love of nature with the adoration of God. It advocates the divinization of human activities and passivities while supporting a pervasive application of the theory of evolution. Critical of Christian asceticism, he taught that the goal of human evolution will be reached only through a collective effort of our species. Like some before him (e.g. Bonaventure, Scotus, and Cusa), Teilhard's philosophy of man and nature is not only essentially geocentric and anthropocentric but also, and more importantly, Christocentric. His pantheistic perspective of things is now clearly evident in his writings: God is at once both immanent to and transcendent of the world.

For Teilhard, material evolution is spiritualizing and therefore purifying itself. Cosmic history, which at first reveals itself as a multiplicity of separate evolutions, is actually the unfolding of one single great mystery through the immanent force of Christ as the ultimate source of meaning and purpose to the whole process of universal and directional development. The process philosopher was convinced that through a concerted effort a new and better earth is being slowly engendered. As a result, the evolutionary fulfillment of Christogenesis at the point of Parousia will bring about the synthetic and mystical union of a united humanity with a personal God, jointly forming the Pleroma. It is evident that this particular position of Teilhard's rests upon profound faith rather than certain knowledge. In any event, he was refused the official Church permission for the publication of this, his first theologically somewhat unorthodox book.

Nevertheless, Teilhard continued to see Christ in planetary matter and God as the beginning and end of cosmic evolution.

He was never to abandon this highly personal position, but he would later attempt to give it scientific support and theological justification in his second and major work, entitled *The Phenomenon of Man*.

In 1929, Teilhard was appointed scientific advisor to the National Chinese Geological Survey as well as advisor and collaborator of the Cenozoic Research Laboratory established by Dr. Davidson Black, then director of the Peking Union Medical School.[11] He also participated in both the Central Mongolian Expedition (1930) and the Yellow Expedition into Central Asia (1931-1932). During his twenty years in China, Teilhard managed to take part in geopaleontological field trips to India, Burma, Java, Ethiopia, and the United States. As a natural scientist, he was primarily a geologist, secondarily a paleontologist specializing in mammals, and only in the third place a paleoanthropologist.

It is ironic that Teilhard was relocated in China because of his evolutionary convictions. Only twenty miles southwest of Peking is Chou-Kou-Tien, an area that was soon to become and remain one of the world's most significant human paleontological sites. In 1928, fossil remains of the jawbone of a hominid had been found, and in the following year, the largest part of a fossil cranium of a hominid was also unearthed. The jawbone and cranium from Chou-Kou-Tien were held to belong to Peking Man (*Sinanthropus pekinensis*), estimated to be at least 350,000 years old.[12] Teilhard received worldwide recognition for his scientific articles that popularized the series of findings from this site in the Western Hills, although he neither discovered the fossil evidence nor analyzed it. He did, however, contribute to an understanding of and appreciation for the geological and paleontological contexts of these important finds and always remained aware of their broader philosophical and theological implications. Unfortunately, he was not allowed to express any of these metascientific insights in his published writings (although he continued to record his private thoughts and visions).

The discovery of the Sinanthropus material (1928-1934) reinforced Teilhard's evolutionary perspective and gave the theory of human evolution additional scientific support.[13] For this singular man of genius, not only is evolution applicable to general geology and biology but also, and more particularly, to the origin as well as sociocultural history and psychic destiny of our earthly species.

Teilhard's knowledge of the Sinanthropus material was supplemented by scientific field trips to Central Asia with George Barbour (1934), to India and then Burma with Helmut de Terra (1935, 1937-1938), and also twice to Java to investigate the pithecanthropine evidence at the invitations of G.H.R. von Koenigswald (1935, 1938). In 1937, after years of scientific research and rigorous reflection, the ordained geopaleontologist started his preliminary notes for a grand synthesis of humankind within natural history. The work progressed steadily, with the Jesuit naturalist writing one or two paragraphs each day. Finally, after two years from June 1938 to June 1940, he completed his major work, *The Phenomenon of Man*. On 6 August 1944, however, its author learned that ecclesiastical permission to publish his manuscript had been refused.

Teilhard remained in Peking during World War II. In 1946, he left for France and was never to return to China again. Although he suffered his first heart attack in 1947, the following year he journeyed to Rome to obtain official permission for the publication of both *The Divine Milieu* and *The Phenomenon of Man* (which he had slightly altered for that purpose) as well as to succeed the late Abbé Breuil as professor of prehistory at the College de France. Unfortunately, all these requests were rejected.

Discouraged but still optimistic, Teilhard now wrote his third book, entitled *Man's Place in Nature: The Human Zoological Group* (1949). Although it is a somewhat more scientific and lucid restatement of the arguments found in his major work, it was not to be published during his lifetime. Nevertheless, he continued to express his original thoughts and unconventional beliefs in essays privately circulated among close friends as well as through prolific correspondence. Fortunately, at least his contributions to science were publicly recognized and appreciated.

In 1947, Teilhard was elected to the French Academy of Sciences for his outstanding work in geopaleontology throughout his long distinguished career as a natural scientist: he held a position as corresponding member of mineralogy in this prestigious society.

In 1951, Teilhard made his seventh visit to New York City. He now accepted a research post at the Wenner-Gren Foundation for Anthropological Research (Viking Fund). At the invitation of C. Van Riet Lowe, the organization sponsored his two trips to the australopithecine sites in South Africa (1951, 1953). In fact, during his life, he had managed to examine the major australopithecine, pithecanthropine, and neanderthal fossil evidence.

Earlier in the year Teilhard had expressed a wish to die on the Feast of the Resurrection, and, ironically, at the age of 74, he died of a sudden stroke on 10 April 1955 in New York City. With a certain poignant appropriateness, his death actually occurred on Easter Sunday evening. He was buried at Saint Andrew on the Hudson, in the cemetery of the Jesuit novitiate for the New York Province. With wise foresight, the brilliant but controversial thinker had entrusted his unpublished manuscripts to the care of Jeanne Mortier, a loyal and trustworthy friend. By the fall of 1955, the original edition of *The Phenomenon of Man* was published in France in its author's native language (Teilhard's complete bibliography amounts to over five hundred titles).

The Phenomenon of Man is a remarkable synthesis of science, philosophy, religion, and poetic mysticism within a Catholic interpretation of cosmic evolution. In its preface, Teilhard wrote that he is offering a scientific treatise dealing with the whole phenomenon of humankind, i.e. an introduction to an explanation of our species's place in a dynamic universe solely as a phenomenon. Yet, this book abounds in assumptive reasoning and is ultimately grounded in personal faith, process theology, and a pervasive mysticism. Why, then, did Teilhard claim it to be a scientific treatise? First, probably because he wanted thereby to gain the attention of the scientific community, and second, because he actually believed his vision to be scientifically sound.

In the foreword, Teilhard held that either one must adequately see the phenomenon of man in its entirety or face the prospect of probable human extinction. With the employment of seven categories (space, time, quantity, quality, proportion, motion, and organic unity) for the illumination of one's vision, he taught that the result of a scientific investigation of man's place in nature would reveal that our species is still the structural and spiritual center of the physical universe. As such, his frank anthropocentrism is steadfast and unequivocal.

Within an evolutionary framework, Teilhard saw the continuity of three fundamentally unique events: the emergence of prelife, life, and thought. He gave thought a privileged position in the universe, clearly wanting to supplement the previous analytical studies of the external structures of material evolution with a complementary phenomenological investigation of the internal emergence of spirit in the historical convergence of psychic evolution. He claimed neither to have given a final explanation of things nor to have constructed an original metaphysical system. *The Phenomenon of Man* is a largely tentative, personal, suggestive but incomplete essay on humankind within nature founded upon a mystical presupposition: fuller being is closer union with God. In order to understand and appreciate his message adequately, one must consider each of Teilhard's four fundamental and interrelated assumptions that give coherence to his evolutionary scheme of things: (1) spiritual monism, (2) law of complexity/consciousness, (3) critical thresholds, and (4) the Omega Point.[14]

Teilhard's Catholic philosophy of evolution rests upon an ontological monism. As a result of giving a privileged position to mind or consciousness, he held that the universe is ultimately psychic or spiritual in nature and is developing toward greater unity and perfection. His position of objective idealism as evolutionary panpsychic is religious in both inspiration and orientation and is more implicit than explicit in his writings.

Teilhard desired to take account of the meaning and purpose of consciousness in the cosmos. He started from both objective and subjective considerations of man as a biological, social, and spiritual event in nature. He extrapolated the necessary conditions of our universe required to bring the human phylum into existence and guarantee its survival and fulfillment. In an early essay entitled "Cosmic Life" (1916), he presented an embryonic form of his intellectual and spiritual testament.

Although Teilhard had adopted a position of panpsychism, he retained an unclarified distinction between matter and spirit. What is clear, however, is that he held cosmic evolution to be a process of the ever-increasing spiritualization of matter. In "The

Mystical Milieu" (1917), he committed himself to an ontological monism, maintaining that there is but a single substance created to sustain the successive growths of consciousness in the universe.

In "Creative Union" (1917), Teilhard sounded in some ways very much like Leibniz, who was, however, though a teleologist and a spiritualist, neither a finalist nor an evolutionst. In "The Eternal Feminine" (1918), he taught that matter is a tendency or direction: it is the side of spirit that we meet as we fall back. He never doubted the spiritual evolution or ultimate unity of the entire cosmos, claiming that both plurality and matter are merely temporal phenomena of reality. In "The Universal Element" (1919), he wrote, "strictly speaking, there is in the universe only one single individual (a single monad), that of the whole conceived in its organized plurality."

Written during World War I, Teilhard's early essays reflect a religious mind struggling to reconcile the material world of the naturalist with the spiritual orientation of the theologian. In his "Hymn of the Universe" (1916), he even tried to describe his mystical experiences as a soldier-priest.

Unlike Descartes, whose subjective methodology unfortunately reintroduced into the history of philosophy an ontological dualism between mind and matter (in fact, Descartes held that there are three separate substances: thought, extension, and God), Teilhard developed a monistic metaphysics. It is evident that his thoughts were inclined to give a preference to spiritual monism. In "My Universe" (1924), he gave a clear summary of his theistic and spiritualistic conception of the world.

Teilhard maintained that the order of interacting monads is not a continuum but rather manifests significant levels differing in kind. Each successive ascending level of monadic existence is held to be relatively higher and closer to perfection than the preceding lower level, i.e. more conscious and therefore more Godlike (like Leibniz's monads, the fragmentary elements of the Teilhardian universe mirror God imperfectly). Leibniz held that the world is an unbroken series of an infinite number of developing monads, each mirroring the Absolute Monad by degree in such a way that no two monads could ever be psychically identical. In "The Phenomenon of Man" (1928), Teilhard emphasized that sidereal evolution is not only spiritual but also irreversible. Like Leibniz, the Jesuit priest held that God is responsible for the pervasive order in the monadic development of the cosmos.

During his years in China, free from teaching responsibilities and public appearances, Teilhard had time to reflect upon his scientific research and the far-reaching implications of the theory of evolution from a Christian perspective. Although immersed in geological and paleontological investigations, his preoccupation with matter did not divert him from his primary concern for the spiritual fulfillment of the human species. His participation in the Sinanthropus excavations and field trips to other major human fossil sites in India and Java strongly reinforced his view that our species is the most recent product of an organic evolution that had its origin billions of years in the past. Although covering a span of several million years, man's existence on earth represents, nevertheless, a unique cosmic event. Through the emergence of humankind, the spiritual universe has become at least once aware of itself. Becoming less concerned with the origin and emergence of the human phylum, the Jesuit scientist now concentrated on developing a view conducive to a suitable outcome for the future spiritualization and convergence of a collective mankind (in his popularizations of the Sinanthropus finds, Teilhard could only hint at the theological implications he was discerning from the design he believed to be manifested in planetary evolution).

Teilhard did write philosophical and theological essays, hoping that someday his superiors would grant the permission needed to see them in print. There was always present in him the desire to write a book in which he would synthesize science, philosophy, and theology in order to give a proper account of the significance of man's place in the flux of the world.

Teilhard held that the existence of a cosmic development of spirit is the greatest discovery made by modern science. He also argued that the human species is actually the end product of cosmic evolution. If the essence of man is that he is a person, the unfolding universe is therefore ultimately a process of self-personalization. In the continuation of this argument, he also assumed that our planetary evolution will terminate in the formation of a Supreme Person at the Omega Point.

Teilhard's clearest position on the nature of the universe is presented in his major work, *The Phenomenon of Man*. Plurality, unity, and energy are held to be the three basic characteristics of the cosmos, i.e. the boundless world as a whole is represented as manifesting a system, a totum, and a quantum respectively. Matter is ultimately composed of homogeneous units of psychic energy that are dynamic, interdependent, and infinitesimal. These cosmic units of matter/energy reveal themselves in an indefinite development within duration as a converging space-time continuum. The philosopher's atomistic position reminds one of the seeds of Anaxagoras and the atoms of Leucippus, Democritus, and Lucretius as well as the monads of Bruno and Leibniz. It may also be noted that although Teilhard sometimes refers to the universe as being boundless, one may safely assume that he actually held it to be finite (although of enormously indeterminate size).

For Teilhard, energy evolves from a homogeneous base to a future limit, i.e. the Omega Point as the supreme upper pole of the world. Structurally, he held that the evolution of energy represents a cone, pyramid, or spiral. This spiral structure of evolution is absolutely crucial to his cosmological model, as is the finite sphericity of our planet, and both must constantly be kept in mind (contrast with Bergson's view of only diverging evolution).

Teilhard considered the ontological status of a cosmic embryogenesis through which the potentialities of the universe are being actualized within a converging space-time continuum. He held that cosmic evolution is divinely self-enclosed and self-sufficient, there being no addition of external energy to replace the energy irrecoverably lost through entropy as the cost of creative syntheses. In short, cosmogenesis is subject to the two principles of the conservation and dissipation of energy. However, he actually held that the evolution of love energy is in a direction antithetical to cosmic entropy: while the physical energy of the universe is diverging and dispersing to a state of equilibrium, the psychic energy of the cosmos is evolving and converging to a point of dynamic concentration and final fulfillment.[15]

What is the ultimate nature of this cosmic energy? Can

Teilhard's ontological position be fairly described as materialistic, spiritualistic, or perhaps dualistic? At first, he attempted to unite the two positions of materialism and spiritualism by merely making a distinction between determined tangential energy (the without of things as the level-oriented physical or horizontal component of nature) and free radial energy (the within of things as the dynamic or vitalistic vertical component of evolution). These two forms of energy are interrelated so that everything in the universe appears to have both a physical and a psychic aspect or pole.

Teilhard held that while scientists are concerned only with the structure of matter and the Marxists only with historical material-ism as economic determinism, they fail to do justice to the emergence of human consciousness and its spiritual significance. Similarly, he held that the phenomenologists and existentialists do not take the theory of evolution and its far-reaching conse-quences seriously enough, while naturalists do not sufficiently appreciate the significance of Christianity for human survival and fulfillment. Although his own approach was a bold and unique attempt to do justice to both physical and psychic evolution, preference is clearly given to spirit over matter or energy. However, Teilhard did not retain the distinction between radial and tangential energies when he left the phenomenal level of investigation to consider the ultimate ontological status of the universe. In a crucial passage in his principal work, *The Phenomenon of Man,* he not only gives radial energy a privileged position in the cosmos but clearly commits himself to an ontological monism grounded in spirit. For the Jesuit priest, there must be a single energy operating in the world, and it is held to be essentially psychic in nature (in the first English translation of *The Phenomenon of Man,* this one fundamental energy had been mistranslated as being physical rather than psychic in nature. This unfortunate error has been responsible for many past ambiguities concerning Teilhard's ontological position).

This fundamental and crucial aspect of Teilhard's philosophy of evolution has surprisingly been either entirely overlooked or largely misunderstood. He continued to use the term matter in his later writings, while equating progress with the growth of consciousness. However, this ambiguity may be resolved by making the philosophical distinction between appearance and reality, i.e. the distinction between epistemology and ontology. From the epistemological perspective, one may distinguish between matter (i.e. that which is independent of human experience and is referred to as matter) and one's own perception of it. In contrast to this perspective, Teilhard preferred to assume that all reality is ultimately spiritual in nature: instead of spirit being a temporary emanation of matter, matter itself is only a temporary alienation of spirit.

At this point, a crucial distinction is called for between objective and subjective methodologies and the crucial status of the concrete world. If it is to be intelligible at all within his own view of things, Teilhard's position must be acknowledged to support pervasive spiritualism. He held that the beginning of cosmogenesis was prior to and independent of human experience. Philosophically, he even held that consciousness causes the further evolution of itself. Theologically, however, he went so far as to maintain that the immanent nature of God is manifested as radial or spiritual energy in the process of becoming, while the transcendent nature of God is the Supreme Center of Love as Being itself. By equating God's immanent nature with a cosmic interpretation of the Christ, he believed that this divine presence in matter allows God's transcendent nature to make things make themselves. Teleology and vitalism are both important elements in this dynamic philosophy and process theology. In brief, universal evolution is ultimately the cosmic Christ unfolding and fulfilling Himself toward the future Omega Point. Teilhard has been the only evolutionist to equate the Christ with an evolving cosmos (or at least the evolving earth).

Teilhard held that the universe has an internal structure not accessible to objective investigation. Therefore, it is necessary but not sufficient to deal with the external structure and complexity of things. He taught that the internal realm is more significant than the external one, as its consideration provides the means for understanding the purpose and goal of planetary evolution. The emergent evolutionists had seen consciousness only as a recent quality. Unlike Teilhard, they did not trace its origin as consciousness back to the outset of cosmic evolution.

Teilhard held that all the objects of nature have an interior as well as an exterior. Bergson had acknowledged this when he postulated that a stream-of-consciousness causes (and runs through) the evolution of the otherwise inert matter of this universe, therefore formulating an unwarranted vitalistic dualism. Teilhard's phenomenology of cosmic evolution was to account for the development of matter and mind. Yet, to do justice to the evolving interiority of the universe, Teilhard did an injustice to its material foundation.

As a scientist and Jesuit priest, Teilhard sought to reconcile naturalism with theism but could not satisfy himself with mere agnosticism or pantheism. As a devout Christian, he had to reject both atheism and materialism in favor of a spiritualistic metaphysics. Since Christian eschatology was always foremost in his thought, there was no compromising as far as ontology was concerned. To allow for the future convergence of nature with God, his faith took preference over science. Unlike the orthodox conceptions of theism, he established a process and spiritualistic panentheism. For a scientific naturalist, the Kantian separation of the phenomenal realm from the noumenal world is untenable, as it leads to an unnecessary and superfluous ontological dualism. One might look at the universe from the position of perspectivals: there are an infinite number of possible perspectives of the material cosmos, and the human perspective is only one of these.

Teilhard had erroneously overextended the ontological status and importance of mental activity. His spiritualist monism is colored by religious bias, and this universe is, alas, seldom in accordance with one's needs, wishes, and desires.

The Jesuit priest held that from its beginning to its end the cosmos is nothing other than the growth of consciousness. However, the emergence of consciousness on earth as a natural activity is, as a matter of fact, a relatively recent event of cosmic history and biological evolution as we know them. It is anthropocentric in the extreme to claim the fleeting phenomenon of human consciousness to be of primary importance in this material universe.

Ideally, philosophical concepts should supplement empirically established scientific facts, making the kaleidoscopic process of the world rationally intelligible. On the realistic assumption that this world is a natural macrophenomenon, science and reason ought to be sufficient to account for evolutionary creativity. On this assumption, cosmic evolution appears to be an endless series of interrelated objects and events grounded in the underlying and pervasive continuity of matter and energy. A more rigorously scientific philosophy would have acknowledged the primacy of matter or energy and seen the human mind as merely a form of natural activity resulting from biological evolution. Specific questions concerning the structural and functional aspects of matter should be left to be resolved by further scientific inquiry and logical reflection.

As a natural scientist, Teilhard devoted his life to descriptive geopaleontology. He was an expert on Chinese geology and specialized in vertebrate paleontology, concentrating on mammalian evolution. In later years, his interest shifted to primate evolution in general and human paleontology in particular. His work was both analytic and synthetic, for every fact was quickly incorporated into an evolutionary framework. He taught that philosophy and science supplement each other, and he never failed to extend the implications of his vision to the future evolution of our species. Although he had been permitted to publish his strictly scientific articles, he was repeatedly warned not to delve into philosophy or theology. Fortunately, his popularizations of the Sinanthropus material brought him worldwide fame and academic recognition.

Copernicus had held that the geocentric model of the universe could be replaced by a heliocentric theory. Galileo Galilei gave the latter view, while it was a hypothesis, scientific grounding. Giordano Bruno argued further for an eternal and infinite universe with its center everywhere and its circumference nowhere. As a result, man's terrestrial habitat no longer occupied even in human thinking a central position in the cosmos. The earth with its sun was now recognized as merely one heavenly body among innumerable other planets and stars; as anticipated by Cusa, Bruno, and others, it is even possible that sentient beings exist in other solar systems. Consequently, the tempting supposition about man's special place in nature had taken a serious blow from science and reason.

In spite of these impressive developments, however, Teilhard taught that humankind is still the center of the universe, occupying a unique and crowning place in the process of planetary evolution. According to him, man is capable of investigating both the psychic (internal) and physical (external) structures of the cosmos. He taught that since the human species is the latest and highest product of primate evolution, it has to be of great biological significance, and as a self-conscious phenomenon in nature qualitatively above all other biological species, it has to be of singular religious and philosophical importance as well.

To Teilhard, the universe is a divinely created but self-unfolding and progressive process. All things and events are viewed correctly if and only if seen as parts of this universal becoming. The old conception of a geocentric and geostatic cosmology had misrepresented the dynamic history and material unity of the cosmos. Teilhard's new vision called for neologisms, and he wrote of a cosmogenesis that included geogenesis, biogenesis, and noogenesis. The entire structure of the universe, from the stars to human self-awareness, represents more or less a continuous process of progressive development. Evolution is not merely a provisional and useful hypothesis but rather a cosmic principle governing all inorganic, organic, and psychic phenomena. Teilhard even held that he had discovered a law of evolution that gives meaning and purpose to the multiplicity of cosmic objects and events: a universal law, he taught, not only explains the past history of the cosmos but, if extended to its final logical conclusion, can also foretell the condition of the ultimate earth and the end of human development upon it. He referred to this regularity as the law of complexity/consciousness. To be more exact by taking into consideration the centralizing or converging nature of the evolving universe, he referred to it as the accelerating cosmic law of increasing centro-complexity/consciousness.

Theologically, Teilhard sought to reconcile planetary evolution with the existence of a personal God as well as the personal immortality of the human soul, the responsible freedom of the human will, and a divine destiny for humankind as a whole. He taught that theology, philosophy, and science are merely three distinct but not separate levels of intellectual investigation. Each presents one major perspective of a single cosmic process, and each supplements the incompleteness of the other two in giving a comprehensive and intelligible view of man's place in the universe.

Being a mystic, Teilhard experienced the evolving unity of a spiritual universe as a cosmogenesis that is progressively developing toward a Personal Center. Yet, he never neglected the sciences. In fact, it is through his contemplation of evolving nature that he arrived at this mystical vision of a God-surrounded and God-imbued world. At the end of the whole terrestrial process, he taught, panentheism will finally be transcended in a true pantheism in which God shall be all in all in the future at the Omega Point.

Teilhard is concerned with the whole phenomenon of humankind, especially in terms of its unique appearance within the evolving cosmic scheme of things from our planetary perspective. In fact, one of the major difficulties in giving a definitive assessment of the overall value of Teilhard's attempt at an evolutionary synthesis is the regrettable fact that one cannot say without hesitation whether his perspective was consistently merely planetary or all-inclusively cosmic in scope. He, despite some cursory references to other possible processes of evolution elsewhere in the universe, leaves this crucial point unsettled and indeterminant. Every descriptive methodology necessarily implies an ontology (there is no inquiry entirely free from presuppositions conscious or unconscious, stated or implied). Therefore, although he claims to be giving merely a scientific description of the process of evolution, one is not surprised when the Jesuit priest ultimately grounds his philosophical view of man within nature in a process theology.

What Teilhard gave, at least in part, was a phenomenological analysis of planetary evolution. His orientation was objective and subjective in order to do justice to both external (material) and internal (conscious) evolution, respectively. The result is a description of the essential structures, relationships, and essences within the phenomena of global history. Without doing an injustice to established facts, he wanted to reduce cosmic phenomena to their basic formations in order to reveal a fundamental law that would disclose the meaning and purpose of evolution as well as man's place and destiny in the universe. It is unfortunate, however, that his phenomenology of evolution did not manage to bracket out his methodologically dispensable religious presuppositions and pervasive mysticism.

For Teilhard, the entire universe manifests a cosmic design that gives meaning and a single direction to evolution in general, as well as a purpose and ultimate goal to the human phylum in particular. In addition to Pascal's two abysses (the infinitely great and the infinitely small), the Jesuit scientist recognized and emphasized the evolutionary significance of a third and no less perilous abyss in the universe: the abyss of the infinite complexity of elemental arrangements.

Teilhard viewed the universe from two major perspectives: a horizontal or synchronic perspective concerned with the size of things and a vertical or diachronic one concerned with the evolution of things. The former recognizes the tremendous range of size in cosmic objects from the subnuclear elemental parts of atoms in the microcosm to the immense stars and comets of the macrocosm. In this breathtaking range, man has been held to occupy a position midway between the infinitely small and the infinitely great. From such a horizontal viewpoint, man's place in the cosmic scheme of things may seem insignificant and inconsequential. However, the Jesuit scientist argued that this observation is, in fact, erroneous.

Teilhard taught that if one takes a vertical view of the universe (i.e. if one considers the position of things within the historical growth and development of the cosmos), man becomes the most significant object in the totality of terrestrial nature. This follows from observing the degree of interior complexity and the resultant radial energy manifested in the sequential emergence of things in terms of consciousness. According to the process philosopher, there are three discernible stages of planetary evolution thus far: prelife, life, and thought. Within each stage there is a great spectrum of diversity and degrees of consciousness. Yet, as one ascends the atoms of Mendeleev's periodic table and follows through organic and psychosocial emergence, one finds that each successive manifestation of evolution is ever more complex in the internal configuration of elements and each at the same time displays increasing degrees of radial energy. The simplest living things, e.g. bacteria and, under some conditions, viruses, are infinitely more complex in their interior structure than the composition of mere stars and geological formations. Similarly, the metazoans are infinitely more complex than the protozoans, while the vertebrates are even more complex than the invertebrates. As one moves through the fishes, amphibians, reptiles, birds, and mammals toward the appearance of the human animal, one notices that this natural tendency toward ever greater complexification continues at an accelerating rate. In Teilhard's interpretation of this tendency, the degree of consciousness is directly proportional to the degree of interior complexification and concentration of the underlying and conditioning structural elements that determine the position of any object in the evolutionary scale of terrestrial nature. He further argued that man is the most significant creature in the universe because he is the most complex and therefore the most conscious of all (again, Teilhard does not clearly distinguish between the planetary and extraterrestrial cosmic perspectives). This anthropocentrism is unfortunately pervasive and persistent in his thought. He would have been on less controversial ground had he limited this plausible claim to the earth alone instead of extending it to the whole unknown universe.

Teilhard believed that to write the true natural history of the world, one would have to be able to follow it from within. Therefore, to correlate this evolutionary increase in complexity or interiorization of matter with the resultant increase in psychic activity, he held that there must be a certain law on which reality is based (a hierarchical law of increasing complexity in creative unity). He referred to his qualitative curve of the universe as the accelerating cosmic law of increasing centro-complexity/ consciousness: this law provided him with a means for viewing geological, biological, and psychosocial evolution as three distinct stages of one continuous process converging and involuting toward an ultimate spiritual end on this planet.

Teilhard saw the law of complexity/consciousness manifested in the successive developments of planetary evolution: units of psychic energy (the monadic stuff of the universe), elementary corpuscles of the cosmos (electrons, protons, neutrons, etc.), systems of atoms (molecules and megamolecules), and cells (protozoans and metazoans). In organic evolution, the acceleration of this law becomes obvious as one moves up through the fishes, amphibians, reptiles, birds, and mammals to the sub-human primates and finally the appearance of man. The acceleration of the increasing complexification has led to a concentrated manifestation in the differentiation of the tissue in the central nervous system and the expansion of the brain. The French scholar saw the plant and insect kingdoms as unsuccessful deviations from this general tendency throughout organic history on earth. Strangely enough, Teilhard did not see the interdependent plant and insect evolutions as supportive of and supplementary to mankind's physical survival and development or as essential preconditions for its spiritual maturation and fulfillment on earth. He held that the insects (particularly the Hymenoptera and Lepidoptera orders) represent a great proliferation in complexity and diversity but their radial energies have been arrested in their growth and ossified as instinct. Likewise, the psychism in plants is too diffuse to be of any consequence. Whatever the objections to this scheme, the law provided the philosopher with an argument for direction, meaning, and purpose within global evolution. He always held that planetary evolution has a precise orientation and a privileged axis in terms of psychic energy or spirit.

Teilhard emphasized the special importance of the primate order. During the seventy million years of primate evolution, one sees the successive emergence of the prosimians, the monkeys of the New World and Old World, and the lower and higher apes as well as the human animal. In general, the early Miocene dryopithecinae complex of hominoids eventually emerged into two separate directions: the line of pongids and that of the hominids (the latter line led to man).

The Jesuit priest noted that from the evolution of some tree shrewlike form to *Homo sapiens sapiens* the accelerating law of increasing centro-complexity/consciousness concentrated with rapidity on the central nervous system and brain. Although one may be impressed with the size of the dinosaurs that dominated the earth for about 140 million years during the Mesozoic era before the appearance of the primates of the following Cenozoic era, they nevertheless had relatively very small brains.

Teilhard speaks of cephalization and cerebralization: the former refers to the increase in the size of the brain in general, while the latter refers to the increase in the size of the cerebrum in particular. The frontal lobes of the cerebrum are necessary for abstract thinking and, therefore, intelligence. Primate evolution is important because it manifests this rapid development in the central nervous system and brain. As such, man's superior intelligence is due not only to the fact that his brain is three times as large as the average brain of the gorilla but also, even more importantly, to the fact that its neural structure is infinitely more complex. The philosopher maintained that the continued acceleration of the law of centro-complexity/consciousness within the ongoing evolution of the human phylum is directly responsible for the emergence of that central phenomenon referred to as reflective conscious and self-conscious thought. He argued that from a planetary perspective the future outcome of this unique phenomenon could only be a superconscious and supercomplex collectivity of reflective persons converging and involuting around the earth.

In his essay "Science and Christ" (1921), Teilhard insisted on the fundamental groping of evolution through a process of "directed chance" toward ever higher convergent integration within a precise orientation and toward a specific target (i.e. a teleologically preordained terminal synthesis as the unity of humankind with God).[16] Although he was aware of the influences of mutations and the external environment on the process of biological evolution, he proposed that from a cosmic perspective the whole process reveals a trend too obvious to be merely the result of cumulative chance genetic variability and necessary natural selection. By underestimating these two major influences while holding that tangential energy is determined but radial energy is free, he never satisfactorily resolved the problem of freedom versus determinism. In fact, he did claim that through cosmic evolution the universe is growing increasingly free. Presumably, then, there are degrees if not even kinds of emerging freedom.

The cosmic law of centro-complexity/consciousness is obviously as crucial to Teilhard's thought as is his spiritual monism. He willed himself into making a daring leap of faith from the alleged rational order of cosmic evolution to the existence of a Supreme Intelligence, i.e. from the discernment of an apparent design in the universe to the conclusion that there must be an Absolute Cause behind it. He never doubted that evolution revealed such a creative design and its Creative Designer. In his article "A Defence of Orthogenesis in the Matter of Patterns of Speciation" (1955), written only three months before his death, he still clung to the assertion that there is pervasive teleology in paleontology.

Like many others, Teilhard saw psychosocial development as the continuation of biological evolution. The law of complexity/consciousness now also manifests itself in scientific and technological development. He held that, for the most part, man has remained biologically stable for the last fifty thousand years or so. Except for its large brain, the human animal remains, somatically speaking, relatively unspecialized. The innovation of tool making, however, was analogous to a major and most favorable biological mutation in the psychosocial realm conducive to an increasingly successful coping with an ever-wider range of environments. For over three million years the human animal has been using tools and weapons, from the crudest stone and bone tools/weapons of the deepest paleolithic times to the most sophisticated metallic and electronic instruments of the present space age science and technology. The Jesuit scientist correctly saw these tools and weapons as specialized extensions of the human body that enhance adaptation and survival. Through them, man can actively modify and transform the environment to suit his own needs and desires instead of merely passively responding to its challenges and opportunities. As a direct result, during the last fifty thousand years or so, the acceleration of complexity in tools/weapons has been mirrored in and paralleled by the increasing complexity of human social structures, functions, and relationships. Although human social organization has its origins in that of the earlier primates, man's superior intelligence has allowed him to become ever more capable of directing his own further evolution.

Teilhard always remained an optimist. He even held that wars are merely the "growing pains" of a converging human evolution and simply rejected the possibility that all of mankind could be aborted by a global catastrophe. As a result of the continuing increase in human population with the consequent rise in demographic pressure on the closed surface of the earth, accompanied by further acceleration in the advancements of science and technology, the Jesuit priest envisioned a psychically unified planet as the ultimate goal of complexity/consciousness in terms of a global and spiritual synthesis; that is, the eventual formation of a final creative unity of persons as the inevitable result of the evolutionary concentration of a monadic universe that is, in the last analysis, psychic or spiritual in nature. From this perspective, he reinterpreted the Aristotelian "Great Chain of Being" as an evolutionary process and progressive scale of ever-increasing complexity/consciousness ranging from the homogeneous psychic units of the universe to the ultraconscious and ultracomplex human world of the future.

Lastly, the law of involuting complexity/consciousness may be understood as a dialectical process as long as one remains merely concerned with the phenomena of cosmic evolution. In Teilhard, there is a direct relationship between matter and psychic energy in their historical development. Also, one will see that he did believe that throughout evolution increasing changes in quantity do eventually result in changes in quality. Whereas Hegel implied the dialectical process to be eternal, Teilhard viewed it as finalistic because it was conducive to an ultimate eschatological end.

All of the philosophers of evolution have been aware that its process displays a chain of products ranging, by and large, from the simple to the ever more complex. As with the Jesuit scientist but before him, Spencer had also formulated a cosmic law to account for this trend throughout universal history. In his phenomenological analysis of evolution, Teilhard has reduced this process to the fundamental accelerating law of increasing centro-complexity/consciousness. There are, however, serious errors in his impressive interpretation of cosmic evolution; for example, the law itself is not cosmic because Teilhard has chosen a perspective that is essentially planetary in scope. To date, there is no scientific evidence that biological and psychosocial evolutions have been taking place elsewhere in our solar system, let alone throughout the rest of this universe. It is possible, but not probable, that evolution as it has taken place on our earth has been occurring elsewhere on other celestial bodies in the exactly identical manner as here. The astronomical number of variables involved that would have to be duplicated in precisely the same sequence staggers the human imagination. As with the rest of us, the only evolutionary process about which Teilhard was factually knowledgeable is the one occurring on this planet.

It is simply assumptive reasoning to hold dogmatically that evolution must always result in greater complexity and ever-increasing consciousness. As Teilhard knew, various orders of animals have evolved without necessarily displaying a tendency toward reflective self-awareness. As already noted, he has generally ignored the evolution and relative value of plants and insects because of his almost exclusively anthropocentric orientation. For instance, Bergson's philosophy of creative but solely divergent evolution is clearly less dogmatic and substantially less anthropocentric because it recognizes the relative importance of plants and insects as well as the human animal. Man is not, as Teilhard argued, indispensable to the cosmos if one assumes that humankind is merely a relatively recent outcome of a biological evolution explained in terms of random genetic variability and necessary natural selection taking place in a seemingly eternal and infinite universe (in fact, some plants and insects may yet inherit the earth).

Teilhard dealt with scientific descriptions. As such, his alleged *a priori* cosmic law is actually an *a posteriori* planetary generalization at which he arrived through an *ex post facto* review of the available empirical evidence. It is not a universal law but rather a retroactive synthetic generalization limited in application to terrestrial history as we now know it. One recalls Alfred North Whitehead's clarification of the fallacy of misplaced concreteness: a crucial distinction is necessary between human idealizations and physical reality, the latter being both independent of and prior to the human knower. The extreme complexity of nature and the probability that it, whether in part or as a whole, may evolve into a plurality of new structures and functions within limitless space and endless time make the assumption of an *a priori* law of evolution in a process universe very unlikely; that is, there is no empirical guarantee that the end of the earth from a human perspective will result in the formation anywhere of a superconscious and ultracomplex spiritual entity in the future.

Teilhard's preferential judgments are clearly anthropocentric. According to his law, something increases in value as it increases in complexity/consciousness. Man is held to be the most valuable object in the universe because he is the most complex and conscious animal on earth. One may, however, value something for a plurality and variety of reasons: its necessity, simplicity, practicality, beauty, uniqueness, ephemerality, and many others as well.

Evolution may be, but is not necessarily, synonymous with progress. Distinctions between the two are, therefore, called for. The current situation on our earth clearly demonstrates that scientific and technological progress is no guarantee of concomitant ethical and moral developments. The perception of a general tendency in natural history does not constitute proof of a preestablished design in cosmic or planetary evolutions. To infer from the presently available empirical data that evolution is the result of the activity of a Supreme Mind or Personal Center is to make an intellectually unconvincing and religiously perilous leap from science to theology. Both terrestrial and universal evolutions are natural processes and, as such, subject to human investigation and rational verification.

The proper conception of the nature of man and his place in the universe is crucial to the philosophical quest for truth. A sound philosophical anthropology requires a rigorous scientific ontology and cosmology. However, it is unfortunate that many exponents of an allegedly philosophical anthropology have little or no scientific evidence to justify their conceptions of man and nature. The limiting of inquiry to a single method or the disregarding of the established facts of the special sciences when formulating a philosophy of man in nature is very unfortunate indeed. A myopic view usually yields unwarranted assumptions resulting in an untenable overall position concerning the issues under discussion. A sound philosophical anthropology must first consider the established facts of the natural and social sciences if it is to be at all meaningful and true. The evolutionary perspective upon the cosmos is also indispensable to relating mankind to the rest of the universe.

It is to Teilhard's credit, however, that he had organized and mastered such a wide breadth of knowledge from the natural and social sciences. His twenty years of academic and social isolation in China were beneficial not only in providing opportunities for geological and paleontological research but also in giving him ample time in which to do a vast amount of reading, reflection, and writing.

Whatever Teilhard's shortcomings in his interpretation of the facts, his inquiry into the nature of man had the advantage of being, at least to a significant extent, scientific in approach. No other philosopher of evolution in the twentieth century can match his extensive and versatile scientific training, field research, and global experiences. Yet, he unfortunately separated man from the rest of nature in too sharp a manner by asserting that the history of planetary evolution has been a series of critical and progressively arranged thresholds resulting in the periodic formation of new kinds of phenomena. This assertion is also theologically oriented, for Teilhard relied upon it especially as an argument for the personal immortality of the human soul.

One need only walk through a large museum or zoo to become very impressed with the great diversity and complexity of life on the earth. It is even more amazing when one realizes that this represents but an infinitely small part of all the living forms that have resulted from a biological process that, to the best of our present knowledge, has been taking place for over four billion years. At the same time, one also becomes aware of the similarities among this vast array of organisms. Whatever the significance of their undeniable mental differences, there is no scientific evidence justifying an ontological separation between the apes and man. There are no apparent ontological separations in the evolutionary process, merely distinctions due to natural differences. For theological reasons, however, Teilhard understandably desired man to partake in a level of becoming allegedly higher than the rest of the animal kingdom. Once again, a theological presupposition has distorted scientific evidence in order to save the claim to man's central and crowning position in nature.

For Teilhard, a point of critical value is reached at the end of each successive stage of terrestrial evolution. At such a critical threshold (level or phase), there occurs a new and quasispontaneous change of state (aspect, condition, or nature), and this results in a creative movement forward and upward. Although he held that each critical juncture is crossed instantly and only once, he was too scientifically sophisticated to believe in uncaused spontaneity. The result has been a series of successive layers (zones, spheres, envelopes, or orders), every one different from each and all of the previous ones not merely in degree but also in the kind of its psychic development. The crossing of a critical transition regularly results in a qualitative change, while a graded spectrum of quantitative differences is manifested within each level. Every living organism manifests its own particular degree of psychic development according to its position in the evolutionary hierarchy. Man's psychic energy differs in kind from all its earlier manifestations in the biosphere, belonging as it does to a new and higher level of existence (i.e. the sphere of thought).

The total evolution of the earth has thus far generated three unique events. The first critical threshold is said to have ocurred at the granulation of the stuff of the universe, resulting in the sudden formation of elementary corpuscles. Further evolution resulted in the phenomena of crystallization and polymerization.

The second critical threshold distinguishes between the inorganic level, or prebiosphere, and the formation of a living level, or biosphere: a qualitative leap from units of preliving atoms to units of living cells had taken place. Teilhard claimed that organic evolution is qualitatively different from and higher than inorganic evolution. Organic evolution is not only homogeneous and coherent but also manifests the emergence of new kinds of phenomena, which he refers to as elementary movements of life: reproduction, multiplication or duplication, renovation or diversification or ramification, conjugation, association or aggregation, and controlled additivity. In such an organic evolution, the phenomenon of controlled additivity acts as a vertical component and is, therefore, responsible for biological evolution in a predetermined direction. He saw life disclosing groping profusion, constructive ingenuity, and indifference, yet structurally forming a global unity. He saw the solidarity of organic evolution as a single gigantic organism developing toward an ultimate planetary goal of spiritual unity.

The crossing of the latest critical threshold did result in the qualitative distinction between mere life and reflective thought. One has here the emergence, vitalization, and hominization of substance as radial energy. Atoms, cells, and persons represent the building blocks of these three successive stages of planetary evolution respectively.

For Teilhard, the emergence of man as a reflective animal represents a unique event not only on the planetary scale but, strangely enough, on the cosmic one as well. He taught that man constitutes the axis of evolution pointing the way to the final unification of our terrestrial world in an all-embracing thought at the Omega Point.

Crucial to Teilhard's philosophy of evolution is also the finite spherical surface of the earth. This factor prevents the indefinite dispersion of each distinct evolutionary layer. Prelife, life, and thought are able to converge or involute about our finite planet. As a result of the law of succession (which refers to the necessary qualitative leaps in evolution when energy has reached a certain critical point of internal concentration) and this finite sphericity of the earth, Teilhard abstracts superimposed but coextensive terrestrial circles. Each successive envelope differs in kind from the previous one. From the perspective of a converging evolution, one has the successive formations of the geosphere (includes the barysphere, lithosphere, hydrosphere, atmosphere, and stratosphere), biosphere (includes the plant, insect, and animal worlds), and noosphere: inorganic, organic, and reflective envelopes respectively. Each sphere represents a distinct stage of planetary emergence and history, for the axis of geogenesis was extended through biogenesis to noogenesis. Hence, there have been sudden leaps that represent gaps and discontinuities in the otherwise continuous, converging, and involuting evolution of our planet.

As a direct result of evolution having crossed a third critical threshold, Teilhard stressed, man differs in kind, not merely in degree, from all other and earlier animals (let alone plants and rocks). With the emergence of humankind, planetary evolution took a qualitative leap forward and upward. Because of the finite surface of the earth, man has been evolving collectively and thus forming, convergingly, a pervasive thinking envelope around his planet earth. Teilhard calls this envelope the noosphere.

In order to give a proper evaluation of man's position in the cosmos and a realistic anticipation of his planetary destiny, Teilhard held that it is necessary to consider the "within" as well as the "without" of things. Looked at from without, the human animal has acquired an erect position and a larger, more complex brain (particularly a structurally sophisticated and intricate cerebrum allowing for symbolizing and the subsequent production of artifacts). Looked at from within, which for Teilhard revealed more the accurate measure of qualitative development, our species represents a unique phenomenon in the universe. For the first and only time in biological evolution (and at one single stroke), the evolution of radial energy had taken a dramatic and irreversible leap forward and upward: this resulted in a chasm between all nonhuman life and human thought. Man not only knows, but he knows that he knows (at best, all other animals can only know). In other words, at a certain critical level of biological evolution and sociocultural development, consciousness ceased to be merely an adaptive device for the perceptual handling of other objects and became the conceptual instrument for a systematic and rigorous reflection about itself.

For Teilhard, the human animal is unique in the universe because he is a *person* (i.e. a reflective and self-reflective center-of-consciousness or intelligent sociocultural being). As a result, the Jesuit priest claimed that man still occupies a special place in nature: from without the human zoological group represents just one more qualitatively distinct biological species, but from within our phylum now represents the advancing spiritualization of the cosmos. He saw planetary evolution as a preparation for the emergence of man and held that the ultimate self-realization and global fulfillment of the human species would be accomplished at the Omega Point.

Teilhard also argues that the immortal human soul is a natural product of biological evolution: an inference from his assumption that the radial energy of man differs in kind from the radial energies of all other animals. Since for Teilhard there was no First Couple but a plurality of such couples to start the human race, the doctrine of the Original Sin takes on a cosmic dimension. Everything in the universe is imperfect proportional to its spatio-temporal distance from God. Through evolution, however, the universe is perfecting itself.

Teilhard argued that since mankind represents a qualitative leap in biological evolution it therefore represents a difference in kind (not merely in degree) from all other animals. As a direct result, the human mind (i.e. human radial energy) is held to differ in kind from the degrees of radial energies manifested within the ascending hierarchy of biological complexification. What is the significance of this radical change of state in the evolution of radial energy? The answer is a theological assumption: personal immorality of the human soul.

Teilhard held that each species in the animal kingdom manifests its own particular degree of radial energy as a direct result of its degree of centro-complexification. The more advanced a species is in its complexification, the higher its position in the hierarchy of nature and the greater its degree of consciousness. At the death of an animal, its radial energy is transformed back into and reabsorbed by tangential energy. However, Teilhard argues that this is not the case at the death of a human being. Since human life resulted from the crossing of the last (most recent) critical threshold, human radial energy is unique among the psychic energies manifested by all other animals in the biosphere. Human radial energy does not change its nature as a result of the physical transformation of the human body at death but rather endures forever as a disembodied center-of-consciousness.

It should be noted that Teilhard used a philosophical assumption to verify a theological speculation (both positions are scientifically unwarranted). It is necessary to point out the unique nature of man as the symboling animal, particularly if one disregards the distinction between radial and tangential energies (Teilhard, however, ultimately argued for a position of spiritual monism). If there are no critical thresholds or distinctions between the two postulated energies in the universe, either all souls are immortal or no souls are immortal. From Teilhard's position, the intensity and centration of the psychic centers in all animals prior to man have not evolved to the point where nonhuman centers are able to maintain their identity without a physical substructure: hence, the alleged uniqueness of man in the Teilhardian view.

One may or may not hold that the difference in degree between the mental faculties of the great apes and man is so essential that the distinction may be said to amount to a difference in kind.

Such a distinction between the levels of mere consciousness and reflective self-awareness does not, however, warrant Teilhard's assumption that there are quantum jumps or discrete discontinuities in the otherwise progressive continuity of psychic evolution. (There is no sudden leap from the consciousness of the world to additional self-awareness during the development of the human mind but rather a gradual increase in the awareness of one's self as distinct from one's environment. Briefly, one may speak of the gradual emergence of personality.) Even if a qualitative distinction is made between the psychic manifestations of the apes and man, it does not necessarily lead to a belief in the personal immortality of the human soul. Teilhard's philosophy of man, as with his view of the cosmos, is religiously oriented. If he had not had to reconcile his evolutionary perspective with his own fideistic predilections and the dogmas of the Catholic Church, his philosophical anthropology would certainly have been significantly different.

With the rise of biology in the last century, some philosophers started to return to an organismic view of the universe. The theory of evolution substantiates such a turn away from the crude mechanistic modes of an earlier materialism.

For centuries, it has been taught that man's superior mental powers are due to his participation in a rational world of existence somehow bridging the gap between the state of Being and the process of Becoming. The Medieval Church had once forbidden inquiry into the nature of man, holding that the human body is sacred and therefore not subject to scientific investigation. It also taught that man is endowed with a simple, nonmaterial, immortal soul (this position inferred an ontological dualism, which only resulted in relational problems). Only within the last century have biology, anthropology, and psychology become serious areas of study. The results have been most productive and innovative: all scientific evidence points to the position that the human being is a relatively recent product of biological and sociocultural developments rooted in the general evolution of matter, totally within the all-encompassing system of the physical universe.

It may be unwarrantedly egocentric and anthropocentric for man to place himself on an ontological plane higher than the rest of the biological kingdom. Similarly, the world is independent of and prior to human experience. Such a perspective, grounded in the special sciences and given coherence by a typically disciplined theory of evolution, seems to be incompatable with any form of idealism if mental activity is only a function of material neurological structures.

No single characteristic separates man from the great apes, but a uniquely human orchestration of several characteristics clearly distinguishes man even from his closest primate relatives and makes him increasingly capable of choosing the direction of his own future evolution.

Teilhard's mature thought was less interested in the geological and paleontological bodies of evidence in support of evolution and much more concerned with the future survival, direction, and goal of the human species. He had always maintained that naturalism is necessary but not sufficient to provide an adequate interpretation of planetary evolution. Neodarwinism, or the modern synthetic theory of evolution (which adds the understanding of population genetics to the time-tested principles of Darwin's original teaching), is in Teilhard's view incomplete, for it neglects the religious dimension of man's existence in general and his psychosocial development in particular. This is why the

Jesuit scientist resorted to theological beliefs and philosophical concepts that seemed to supplement the purely scientific evidence.

If planetary evolution is spiritual, as claimed by Teilhard, then it expectedly requires a spiritual end. He stressed that for survival a collective mankind must supplement science with a belief in progress and unity under the guidance of a personal God.

Within this organismic and spiritual framework, Teilhard focused upon the human phylum. In man, consciousness has been raised to the second power, i.e. awareness has involuted back upon itself so that it has become, as actually shown, an object of its own contemplation and reflection. The material human brain folding or convoluting back upon itself has resulted in human thought having the capacity for abstract self-reflection. Another planetary transmutation had, so to speak, taken place: this new transformation resulted in the eventual formation of a single and unbroken global tissue, a quasimembrane consisting of a complex network of reflective consciousness and all its contents (i.e. the noosphere).

For thousands of years, the human species has remained a single biological unit: it has not undergone further adaptive radiation resulting in actual speciation as has occurred over long periods of time in many plants and other animals. Due to this biological uniqueness and the global finitude of our planet, mankind has begun to converge upon itself. Like geogenesis and biogenesis, but even more so, noogenesis is a single and unified process (conscious reflection or personalization and planetary reflexion or planetization are, according to Teilhard, the two major aspects of the phenomenon of man).

In a general sense, hominization is the progressive phyletic spiritualization of the human layer. The superhominization of this layer is occurring as a result of the noosphere closing in upon itself under the structural pressure of the finite sphericity of the earth. Thoughts are increasingly encircling and enveloping our planet, forming a psychic membrane, or the noosphere.

For Teilhard, no evolutionary future awaits man except in association with all other men. He saw cultural evolution as the extension of biological evolution, holding that the axis of biogenesis culminated in a new movement toward neolife or superlife (i.e. the personalization of evolution). Psychosocial evolution manifested itself in two phases: the socialization of expansion or divergence represented by the early stages of socio-cultural evolution followed by the socialization of compression or convergence started during the Neolithic Metamorphosis (i.e. during the last ten thousand years). As a result of biosocial convergence, the modern earth is experiencing both global migration and consequently global acculturation. Always at least partially optimistic, Teilhard held that although the human layer will reach a physical end it will not experience a spiritual death. Since biosocial evolution and involution are both inevitable, he envisioned an ultimately personalized earth. In short, there are two major stages of human evolution: (1) socialization of divergence followed by (2) socialization of convergence. Psychological evolution as well as intellectual and moral convergence will culminate in the planetary collectivity of mankind.

Teilhard offers a quasi-Lamarckian interpretation of cultural evolution, emphasizing the cumulation of knowledge from generation to generation. As a crucial part of noogenesis, technology is accelerating and thereby complements the bio-psychic and symbolic convergence of the human species (along with certain predominantly progressive, optimistic, and eschato-logically-oriented philosophies of history). As a result of this biotechnological accumulation (a phenomenon aided by the already mentioned roundness of the planet), our philosopher was aware that it would be feasible for man, through euthenics and eugenics, to both control and improve upon his own further evolution on earth. In brief, man himself will more and more direct the very process of superhominization to its ultimate end.

There are striking similarities but major differences between aspects of Tielhardism and Marxism. One way or another, all of them originated from a preponderantly naturalist attitude, which emphasizes the humanistic values of a collective mankind and expresses a deep concern for the present human condition as well as its survival and ultimate fulfillment. Teilhard saw the resolution of human problems through an increase in love (the amorization of the human species). Of course, the Marxists as passionate atheists do not accept the priest's theological orientation. Yet, it is ironic that certain Marxists have been more sympathetic and receptive to the humanist core of Teilhard's thought than are some members of his own Jesuit order. In a manner reminiscent of their treatment of Hegel, what some Marxists did was to turn Teilhard's metaphysical system upside-down. For obvious reasons, Christianity played a very important role in Teilhard's evolutionary philosophy. He saw the "Christian phenomenon" itself as a progressive continuation of psychosocial evolution. Through the influence of Christianity, noogenesis is extending itself into a Christogenesis (cosmic evolution or cosmogenesis is said to be essentially Christocentric).[17] For Teilhard, Christianity is playing an important role in uniting the human species through its emphasis on love and compassion. Within the noosphere, evolution continues to progress foward and upward, although the movement may be almost imperceptible.

Teilhard envisioned a "Human Front," with faith in futurism, universalism, and personalism.[18] He taught that only Christianity is capable of guiding spiritual evolution to its completion: only it is capable of saving the person (in contrast to the mere individual) within a collective by acknowledging the primacy of spirit. From without, human beings are merely individuals of their species. From within, however, they are souls endowed with personalities. An individual is an agent of psychocultural interaction endowed with both rights and responsibilities, while a personality is the soul's inner structure, i.e. the psychological and moral organization of a person's feelings, thoughts, values, and attitudes. Although a human being may submerge his physical individuality within a collective, he enhances his personality as a result of emotional, intellectual, and spiritual interaction and sharing with others.

Christianity replaces pessimism and the passivity of isolated individuals with optimism and the activity of collective persons. The end of Christogenesis is the formation of a collective superhumanity manifesting hyperreflection. Christ is no longer merely a historical figure; He takes on a cosmic and dynamic dimension. Teilhard's mystical vision is bold enough to identify a dynamically conceived Christ with terrestrial (if not even cosmic) evolution! Through Christ, in fact, evolution is made both whole and holy. The consummation of the world will be brought about through love-energy, the highest form of radial energy.

Teilhard taught that the anguish of modern man (who fears the possibility of collective self-destruction and total death) is caused by an acute awareness of his seemingly diminutive stature when contrasted with the enormity of space, time, and number. The scientific discoveries of the past centuries, especially the theory of evolution, have jarred and jolted the human mind from its geocentric and geostatic torpor. The meaning and purpose of human existence are being questioned and debated, and competing forms of essentialism and existentialism divide the intellectual circles. For Teilhard, the answer is still found in Christianity. He taught that only love is capable of fulfilling and completing noogenesis, which as one has seen is nothing more or less than a Christogenesis. The survival of mankind requires an increase in scientific knowledge coupled with a parallel increase in love and compassion. He taught that the cosmic energy of love, which is everywhere in the universe in some extended form, is capable of preserving and perfecting that which it unites. Christianity represents a phylum of love, and Christogenesis is the increase in radial energy (now in the form of love-energy) and a concomitant decrease in tangential energy. Through love-energy, cosmic evolution represents a divinizing convergence toward God-Omega at the Omega Point.

For Teilhard, human survival and fulfillment demand a superior form of existence.[19] He held that the ultimate source and object of love is above and ahead of the evolutionary process. It is not something but rather a supreme Someone: the ultimate object of love can only be a personal God as the Great Presence guaranteeing the stability of process reality.

It is very important to note Teilhard's distinction between God-Omega and the Omega Point. God-Omega is the transcendent focus and end of the self-personalizing universe (i.e. God-Omega is the Personal Center of universal convergence), while the Omega Point is the final event of planetary evolution. God-Omega is the transcendent Ultimate Monad, and the evolutionary process is sustained by this Prime Mover Ahead as the ultimate Gatherer and Consolidator of Souls. Teilhard held that autonomy, actuality, irreversibility, and transcendence are the four attributes of God-Omega. Being already in existence as Provider and

Revealer, God-Omega is directing, unifying, and purifying the personalizing and converging cosmogenesis. As we have seen, through love this cosmogenesis is self-extended into a Christo-genesis that will be fulfilled at the Omega Point.

In Teilhard's view, God-Omega does not directly create matter but rather causes it to create itself by being immanent in planetary evolution as Christ (the vital principle or force) and transcendent as the Final Cause. Teilhard's conception of an Omega Point is surely his most original but most vulnerable assumption.

By way of a grand metaphor, the Mystical Body of Christ is literally the collective of all monadic souls that have been emerging throughout the human phase of terrestrial evolution. As such, this planet is acquiring the equivalent of both a brain and a heart. Extended to its logical conclusion, Teilhard taught that the earth will acquire a semblance of a single mind and a single heart that will become united in only one center of thought and feeling. Remembering that he envisions planetary evolution as a spiraling pyramid, or as a cone, the process is now narrowing and converging upon its apex, which is the Omega Point. The future union of a superconscious humankind on a global scale is to be followed by its fusion with the personal God-Omega (this final creative synthesis as the theosphere is the fulcrum of human destiny and the human apogee of evolution).

Teilhard assumed that the physical death of a person merely liberates an immortal center-of-consciousness. For thousands of years since the first reflective beings died this side of the last critical threshold between life and thought, all such liberated reflective monads have been forming a planetary layer of thought. Thus, hominization extended into a superhominization will be finalized in a global arrangement or megasynthesis of persons forming a unified and harmonized superconsciousness around the finite spherical geometry of the earth.

For Teilhard, the Omega Point is an inevitable extraplanetary event, i.e. the final event of human evolution, which will transcend space and time. His principle of emergence is extended to its very limit in the formation of a conscious pole as the result of a supercentration of persons into an absolutely original Person or single Supreme Center. In turn, this monadified aggregate of persons (this spiritual macromonad so to speak) converges upon and unites with God-Omega.

Teilhard held that everything that rises must converge and that closer union is a fuller being. The Omega Point is the final, creative, and clearly differentiated metaunion of persons as a result of a converging cosmogenesis. With the attainment of the Omega Point, Teilhard's law of the recurrence of creative unions is manifested for the last time. He taught that evolution unites like to like. With the Omega Point, one has the ultimate synthesis in which the universal and personal fuse as the result of the interior totalization of the spirit of the earth. Hence, Teilhard has given us a simple argument in his own version of a dynamic natural theology: if cosmic evolution is spiritualistic, directional, and converging, it can terminate only in a Personal Center of ultimate complexity and consciousness.

The Omega Point represents his most daring assertion. By repeatedly resorting to theologically oriented philosophical assumptions and mystical conceptions, he has not limited himself to a strictly naturalist phenomenology of planetary evolution. Far from being a naive and shallow optimist guided by wishful thinking, Teilhard realized and admitted that there may be an evolution of evil (there seemed to be an excess of evil in the world as it is). He made the common distinction between physical and moral evil, yet he claimed that there is only one ultimate form of it. Evil is matter, determinism, multiplicity, and death. It is that which retards and prevents evolution toward creative union. However, growth itself requires and necessitates evil. The evolution of the universe represents an infinite number of steps from absolute evil (i.e. total nothingness or total plurality) to the Supreme Good, which is total being in total unity. Evolution is the progressive movement from imperfect plurality to perfect oneness. Evil is a necessary element and the price of cosmic evolution. It is statistically inevitable as a by-product of evolution proceeding by trial and error even in directed chance.

If evil continues to increase, an alternative possibility that Teilhard realistically acknowledged, the end of the world process will represent a bifurcation of the noosphere. An internal schism within this layer would result in a polarity between the love of the physical and the love of the spiritual. As a result, only that segment of the noosphere that has synthesized itself across space and time and beyond all evil will be united to God-Omega at the Omega Point.

From Teilhard's theological perspective, God's love vitalizes the cosmos from within and divinely attracts directional evolution from above. The evolving universe is supposed to have emanated from God at the Alpha Point, and humankind will presumably return to God at the Omega Point.

For a mind that thought in terms of billions of years and held that man's psychosocial progress would perhaps continue to accelerate for millions of years to come, it is astonishing that Teilhard never allowed for the possibility of human divergence on a solar or cosmic scale. Actually, his whole system of converging evolution would collapse if one day man should extend his evolution beyond the confines of the spherical surface of this planet.

Teilhard speculated that perhaps Omega Points have been, are being, or will be reached on other planets elsewhere in the universe. Even so, he held that communications with other worlds and their Omega Points are unlikely. In any event, the only Omega Point that mattered to him was our own.

Teilhard held that a theological goal is necessary in order to motivate into action an otherwise apparently meaningless process. But his own personal needs and desires are not necessarily those of the rest of the human zoological group. A naturalist may accept a form of mysticism if it is simply an awareness of the spatiotemporal unity of the seemingly eternal and infinite universe. Certain forms of mysticism may lead a legitimate naturalist to monism, while spiritual mysticism tends to advocate an unwarranted spiritual monism or ontological dualism.

Teilhard the man and the thinker will be a subject debated decades if not even centuries from now.[20] Had he done nothing else, he would be remembered for having mightily contributed not only to the advancement of science but also to the transformation of religion. We are, of course, all deeply indebted to Teilhard's moral fiber and intellectual integrity. He was truly a great scholar as well as a noble human being who, despite inevitable human shortcomings, had the courage to attempt a synthesis and thereby saw above and beyond the horizons of most of his contemporaries in science, philosophy, and theology.

As a natural scientist, process philosopher, and deeply religious man, Teilhard created a lofty vision that is very likely to be even more fully appreciated a hundred years from now. Time will surely put him in his proper place of honor in the annals of human thought. His greatest accomplishment, for which he will be remembered by posterity, is neither his science alone nor his theologically inspired philosophy alone but rather the supreme artistry and integrity of the magnificient tapestry of thought into which he managed to weave them. It will continue to elicit intellectual admiration, moral respect, and aesthetic delight long after our world has forgotten the profound personal drama of Pierre Teilhard de Chardin.

NOTES AND SELECTED REFERENCES

[1]For the evaluations of Sir Julian Huxley, George Gaylord Simpson, and P. B. Medawar refer to Philip Appleman, ed., *Darwin* (New York: W. W. Norton, 2nd ed., 1979, pp. 342-362).

[2]Cf. Claude Cuénot et al., *Evolution, Marxism and Christianity: Studies in the Teilhardian Synthesis* (London: Garnstone Press, 1967).

[3]Cf. Pierre Teilhard de Chardin's *The Divine Milieu* (New York: Harper Torchbooks, rev. ed.,1968), *The Phenomenon of Man* (New York: Harper Torchbooks, 2nd ed., 1965) and *Man's Place in Nature: The Human Zoological Group* (New York: Harper & Row, 1966).

[4]Cf. Pierre Teilhard de Chardin's *Letters From Egypt, 1905-1908* (New York: Herder and Herder, 1965), *Letters From Hastings, 1908-1912* (New York: Herder and Herder, 1968), *Letters From Paris, 1912-1914* (New York: Herder and Herder, 1967), *The Making of a Mind: Letters From a Soldier-Priest, 1914-1919* (New York: Harper & Row, 1965), *Letters to Léontine Zanta* (New York: Harper & Row, 1969), *Letters From a Traveller* (New York: Harper Torchbooks, 1968), *Letters to Two Friends, 1926-1952* (New York: The New American Library, 1968), and *Letters From My Friend Teilhard de Chardin, 1948-1955* (New York: Paulist Press, 1976).

[5]Cf. George B. Barbour, *In the Field with Teilhard de Chardin* (New York: Herder and Herder, 1965), Nicolas Corte, *Pierre Teilhard de Chardin: His Life and Spirit* (New York: Macmillan, 1961), Claude Cuénot, *Teilhard de Chardin: A Biographical Study* (Baltimore: Helicon, 1965), Henri de Lubac, *Teilhard de Chardin: The Man and His Meaning,* (New York: Hawthorn Books, 1965), Helmut de Terra, *Memories of Teilhard de Chardin* (New York: Harper & Row, 1964), Robert T. Francoeur, ed., *The World of Teilhard* (Baltimore: Helicon, 1961), Paul Grenet, *Teilhard de Chardin: The Man and his Theories* (London: Souvenir Press, 1965), Mary Lukas and Ellen Lukas, *Teilhard* (New York: McGraw-Hill, 1981), Jeanne Mortier and Marie-Louise Auboux, *Teilhard de Chardin: Album* (New York: Harper & Row, 1966), Michael H.

Murray, *The Thought of Teilhard de Chardin: An Introduction* (New York: The Seabury Press, 1966), Olivier Rabut, *Teilhard de Chardin* (New York: Sheed and Ward, 1961), Charles E. Raven, *Teilhard de Chardin: Scientist and Seer* (New York: Harper & Row, 1962), René Hague, *Teilhard de Chardin: A Guide to his Thought* (London: Collins, 1967), Robert Speaight, *Teilhard de Chardin: A Biography* (London: Collins, 1967), Claude Tresmontant, *Pierre Teilhard de Chardin: His Thought* (Baltimore: Helicon, 1959), and N. M. Wilders, *An Introduction to Teilhard de Chardin* (New York: Harper & Row, 1968).

[6]Cf. Ronald Millar, *The Piltdown Men* (Canada: Paladin 1974), K. P. Oakley, "Further Evidence on Piltdown" in Robert F. Heizer, ed., *Man's Discovery of His Past: Literary Landmarks in Archaeology* (Englewood Cliffs, New Jersey: Prentice-Hall, 1962, pp. 37-40) as well as "Further Contributions to the Solution of the Piltdown Problem" in *Bulletin of the British Museum (Natural History), Geology*, 2 (3): 244-248, 253, 256-257, J. S. Weiner, *The Piltdown Forgery* (New York: Oxford University Press, 1955), and J. S. Weiner et al., "The Solution of the Piltdown Problem" in Robert F. Heizer, ed., *Man's Discovery of His Past: Literary Landmarks in Archaeology* (Englewood Cliffs, New Jersey: Prentice-Hall, 1962, pp. 30-37) as well as "The Solution of the Piltdown Problem" in *Bulletin of the British Museum (Natural History), Geology*, 2 (3): 141-146.

[7]Cf. Stephen Jay Gould's "Piltdown Revisited" in *Natural History*, 88 (3): 86-87, 89-90, 94-97, "The Piltdown Conspiracy" in *Natural History*, 89 (8): 8, 10-11, 14, 16, 18, 20, 22, 25-26, 28, "Vision with a Vengeance" in *Natural History*, 89 (9): pp. 18-20, and "Piltdown in Letters" in *Natural History*, 90 (6): 12, 14, 16, 18, 20-28, 30. Also refer to Stephen Jay Gould's "Piltdown Revisited" in *The Panda's Thumb: More Reflections in Natural History* (New York: W. W. Norton, 1980, pp. 108-124).

[8]Cf. Alan Houghton Broderick, *Father of Prehistory: The Abbé Henri Breuil, His Life and Times* (New York: William Morrow, 1963, esp. 191-204).

[9]Cf. Pierre Teilhard de Chardin, *Writings in Time of War* (New York: Harper & Row, 1968).

[10]Cf. Pierre Teilhard de Chardin, *Hymn of the Universe* (New York: Harper & Row, 1965, esp. pp. 19-37). Also refer to Thomas M. King, *Teilhard's Mysticism of Knowing* (New York: The Seabury Press, 1981).

[11]Cf. Dora Hood, *Davidson Black: A Biography* (Toronto: University of Toronto Press, 1964, esp. pp. 29, 62, 72, 74, 83, 100, 105, 112, 123, 125, 127).

[12]Cf. Christopher G. Janus with William Brashler, *The Search for Peking Man* (New York: Macmillan, 1975, esp. pp. 30-31, 55, 208), and Harry L. Shapiro, *Peking Man* (New York: Simon and Schuster, 1974, esp. pp. 39-41, 53, 79, 81).

[13]Cf. Pierre Teilhard de Chardin's *The Appearance of Man* (New York: Harper & Row, 1965) and *The Vision of the Past* (New York: Harper & Row, 1966). Also refer to Pierre Teilhard de Chardin's "The Idea of Fossil Man" in *Anthropology Today: Selections*, edited by Sol Tax (Chicago: Phoenix Books, 1962, pp. 31-38) and Sol Tax et al., eds., *An Appraisal of Anthropology Today*

(Chicago: Midway Reprint, 1976, esp. pp. 13, 44, 150, 152, 338-339).

[14]Cf. H. James Birx, *Pierre Teilhard de Chardin's Philosophy of Evolution* (Springfield, Illinois: Charles C Thomas, 1972).

[15]Cf. Pierre Teilhard de Chardin, *Activation of Energy* (New York: Harcourt Brace Jovanovich, 1971) and *Human Energy* (New York: Harcourt Brace Jovanovich, 1969).

[16]Cf. Pierre Teilhard de Chardin, *Science and Christ* (New York: Harper & Row, 1968) for similar views.

[17]Cf. Pierre Teilhard de Chardin, *Christianity and Evolution* (New York: Harcourt Brace Jovanovich, 1969) and *The Heart of Matter* (New York: Harcourt Brace Jovanovich, 1978).

[18]Cf. Pierre Teilhard de Chardin, "The Antiquity and World Expansion of Human Culture" in *Man's Role in Changing the Face of the Earth*, edited by William L. Thomas, Jr. (Chicago: The University of Chicago Press, 1962, pp. 103-112). Also refer to Pierre Teilhard de Chardin, *Building the Earth* (Wilkes-Barre, Pa.: Dimension Books, 1965).

[19]Cf. Pierre Teilhard de Chardin, *The Future of Man* (New York: Harper & Row, 1964) and *Toward the Future* (New York: Harcourt Brace Jovanovich, 1975).

[20]Cf. H. James Birx, "Pierre Teilhard de Chardin: A Remembrance" in *The Science Teacher*, 48 (9): 19, and "Teilhard and Evolution: Critical Reflections" in the *Humboldt Journal of Social Relations*, 9 (1): 151-167. Also refer to Stephen Jay Gould, *Hen's Teeth and Horse's Toes: Further Reflections in Natural History* (New York: W.W. Norton, 1983, esp. pp. 201-240, 245-250), Thomas M. King, S.J., and James F. Salmon, S.J., eds., *Teilhard and the Unity of Knowledge: The Georgetown University Centennial Symposium* (Ramsey, New York: Paulist Press, 1983), and Mary Lukas and Ellen Lukas, "The Haunting" in *Antiquity*, 57(219):7-11.

Chapter 18

FARBER

This year, which marks the centennial of Charles Darwin's death, is an appropriate time to recall the distinguished naturalist philosopher Marvin Farber (1901-1980) and his pivotal scholarly writings that owe so much to the theory of evolution in both its terrestrial and cosmological aspects.[1] Concerning evolution, perhaps no other single theory in the annals of natural science so clearly points out that the human animal with its mental activity and resultant sociocultural environs is but a relatively recent (geologically speaking) product of universal history. Humankind is a fragile species on a cosmic speck we call the planet earth, which appears to have no meaning or purpose other than those judgments and values that the human animal creates in order to adapt, survive, and thrive in an otherwise material universe that seems to be totally indifferent to the ephemeral existence of our zoological group.

Marvin Farber never underestimated the predictive, explanatory, and exploratory powers of the theory of evolution. Like Ludwig Feuerbach, whom he greatly admired, Farber saw all religions as being essentially grounded in the psychosocial wants, needs, and desires of our very vulnerable species within the processes of nature itself.

Similar to Karl Marx, whose thoughts and writings had a pervasive and lasting influence on his own philosophy of mankind within society and nature, Farber saw the human predicament with all of its problems and aspirations as being historically and to a significant extent economically conditioned:

one cannot remove a scientist, philosopher, or theologian from the influences of a particular sociocultural milieu.

In the history of western philosophy, there has been a perennial dialogue between the naturalists and the subjectivists concerning a sound interpretation and proper evaluation of man's place in the cosmos. This ongoing exchange of viewpoints is directly related to basic ontological and epistemological questions about the very nature of reality and even the possibility of knowing it. Both idealists and materialists in the rich history of western philosophy have, in fact, contributed to our understanding of and appreciation for human experience and the position our species occupies in the universe. However, their philosophical commitments pertaining to the relationship between natural existence as the object to be known and the human being as the knower are, in most cases, radically different. These two schools of thought, often diametrically opposed, are represented by Edmund Husserl (1859-1938) as a philosophical idealist giving preference to a rigorously critical subjectivist methodology (which metaphysically aspires to support a transcendental idealism) and Marvin Farber as a philosophical materialist drawing from the advances in the special sciences within a naturalist framework that recognizes the need for rigorous reflection.

In recent American and world philosophy, Farber clearly represented a scientific naturalist and rational humanist viewpoint that was ontologically grounded in an uncompromising and unapologetic materialism. To a significant degree, this pervasive materialism was the direct result of his steadfast commitment to the far-reaching theoretical implications and, at times, sobering if not devastating physical consequences of the theory of evolution and sociocultural development (as well as his serious adoption of a truly cosmic perspective). There can be no doubt that he championed the process philosophers and scientific evolutionists, unequivocally maintaining a materialist explanation of the cosmic process of universal evolution in sharp contrast to the various idealist interpretations of dynamic reality also present in the serious philosophical literature.

Farber spoke of the philosophical quest and saw as one of its major functions the critical evaluation and synthesis of the natural and social sciences. His lectures stressed the perennial "themes of inquiry" in the history of philosophy (giving special attention to the major thinkers of the recent past since Kant and Hegel). He maintained that the fundamental questions in the great tradition of reflective thinking are always open to critical analysis and rigorous reevaluation in light of the advancing special sciences and a plurality of cooperative logical and methodological procedures.

Farber especially respected anthropology as an academic discipline and thought it to be a highly valuable preparation for philosophy. In that sense, at least, he was also a major philosophical anthropologist. Clearly, there is a qualified optimism in his healthy-minded and sound view of the physical universe and the tenuous place of human existence within it. He gave preference to the empirical facts and natural relationships of the material world rather than to religious beliefs and personal opinions. His rational orientation acknowledged that reality is increasingly knowable to the human intellect. He rejected completely the Kantian dichotomy between the knowable and unknowable realms of reality, allowing only for the difference between the already known and the as yet unknown.

Marvin Farber's own emerging and maturing materialist viewpoint had been particularly influenced by the works of three American philosophers: Ralph Barton Perry's *General Theory of Value* (1926), a major contribution to axiology as a whole and especially to a naturalist ethics, Alfred North Whitehead's *Symbolism: Its Meaning and Effect* (1927), which he admired, and Clarence Irving Lewis's *Mind and the World-Order* (1929), which outlines a theory of knowledge in terms of a common ground where philosophy and mathematics meet.[2] Similarly, as with his critically revered teacher Edmund Husserl, logic played a prominent role in Farber's evaluation of all philosophical arguments and their implications (also, he had been greatly impressed with the works of Sheffer and Zermelo).

In his writings and lectures, Farber frequently contrasted the naturalist viewpoint with the phenomenological attitude or subjectivist framework, seeing the latter (when isolated from the former) as an unfortunate position taken by some phenomenologists and all idealists who neither recognized the value of the special sciences and the theory of evolution nor resisted a personal inclination to adopt a mentalist or spiritualist epistemology and ontology (often for religious reasons). In his own philosophy, there are no purely subjectivist or idealist elements at all, for he always claimed that scientific evidence and logic should be sufficient to substantiate the recent emergence of human thought as a vitally important adaptive device.

In his classic volume, *The Foundation of Phenomenology* (1943), Farber pointed out that Husserl's subjectivist philosophy neglected both developmental psychology and scientific evolution, not only as aids to a rigorous methodology but even in the great phenomenologist's attempt to ontologically constitute the realms of natural and social existence.[3] Although an admittedly rigorous device for describing mental activity (especially intentionality and symbolic creativity), Farber is quick to point out that the phenomenological method of subjective inquiry alone certainly cannot pretend to give a universally convincing and scientifically sound metaphysics in any meaningful sense of the term.

Farber's *Naturalism and Subjectivism* (1959) is a major statement clarifying the crucial distinction between ontology and epistemology.[4] This volume appeared in the centennial year that celebrated not only the publication of both Darwin's *On the Origin of Species* and Marx's *Critique of Political Economy* but also the births of Samuel Alexander, Henri Bergson, John Dewey, and Edmund Husserl (not to mention the first performance of Wagner's *Tristan und Isolde*). In it, Farber wrote that the process of human experience occupies only an infinitesimal part of the material cosmos and it occurs only when there are sentient beings in action. He held that the documented scientific evidence for the fact of biological evolution is firm, sufficient, and indisputable. One could add that the validity of the theory of evolution is implicitly, if not always explicitly, a basic fact within his own materialist philosophy of process nature.

As a result of incorporating the evolutionary perspective and its obvious implications, Farber's declaredly pervasive materialism taught the all-important "principle of independence": the natural world is prior to and independent of the human knower and its sociocultural environs. As such, the whole range of human experience is seen to be a relatively recent emergence in the vast organic history of our planet earth. He also wisely and astutely appealed to the awesome cosmic perspective for a sound interpretation and proper evaluation of humankind's fleeting place in the flux of the physical universe.

Marvin Farber was particularly influenced by the writings of Herbert Spencer and Ernst Haeckel. He singled out Spencer's *First Principles* (1862) and especially Haeckel's *The Riddle of the Universe* (1900) as having had, each in its own way, a lasting impact on the ultimate orientation of his own thought. Puzzlingly enough, throughout Farber's copious writings there are only occasional, fragmentary, and cursory references to the seminal works of Charles Robert Darwin and Thomas Henry Huxley. However, as a materialist, Farber did have a high regard not only for fellow naturalists among the Presocratics and Greeks (e.g. Aristotle) but also for the Renaissance philosophers of nature (notably Giordano Bruno).[5]

Farber adopted Spencer's cosmic perspective, but he warned about the fallacy of metabasis or illicit transference. This fallacy of reasoning consists in the unwarranted transfer and application of explanatory principles valid in a particular field of inquiry to other areas of investigation and realms of reality to which they cannot validly be extended and applied. A glaring example of this is Spencer's own overextension of the concept of the "survival of the fittest" from biology to sociology with devastating ethical consequences, i.e. from raw nature itself to the human sociocultural realm. The distinction between these two levels of evolution was first recognized by Thomas Henry Huxley: the first of these two levels of existence is guided primarily by necessity in the form of biologically inherited instincts, while the second is guided increasingly by the emerging freedom of choice and social responsibility as the consequences of ever more intellectually directed and pliable human activity. Each field of science has its own peculiar problems, methods, and principles of explanation. Spencer's concept of the "survival of the fittest" may be, to some extent, a useful descriptive generalization applicable to purely biological evolution, but it is neither logically nor ethically applicable when extended to human sociocultural development. As with Darwin, Farber was opposed to Social Darwinism (which actually would be more aptly described as Social Spencerism).

Farber followed in Haeckel's footsteps by embracing both a cosmic and an evolutionary overview grounded in a monistic interpretation of reality. By adopting such a perspective, he rejected the basic thesis of philosophical idealism. For Farber, the existence of the material world is not contingent upon human or divine experiences. He stressed and required the crucial distinction between a sound ontology and a critical epistemology.

Marvin Farber recognized the value of the phenomenological method solely as a rigorous form of subjective inquiry. Yet unlike Husserl, he never accepted the implied idealist metaphysics of a myopic application of this subjectivist methodology taken out of its naturalist context. Instead, Farber's own philosophical commitment was to a broad naturalism and a realistic humanism rooted in the special sciences, logic, and a keen sense of compassion and justice.[6] His writings are lucid, intellectually honest, and scrupulously accurate. They stress mankind's natural, social, and cultural conditions within a historic overview. His philosophy acknowledges the importance of human values, methodological and logical procedures, and rational metaphysical speculations opposed to (and exclusive of) every sense of otherworldliness in precisely that order. For Farber, naturalism is able to incorporate all the rich findings of subjectivism without adopting the latter's idealist ontology and/or cosmology. His own naturalist phenomenology advocates enhancing human freedom, happiness, and longevity through the wise and cautious use of the special sciences within the rational guidelines of a materialist philosophy and a humanist morality.

Farber rightly pointed out that philosophy is a human activity that requires presuppositions.[7] He drew attention to the basic fact that the human animal as knower and doer can never really get outside of its natural, sociocultural, and mental environs. Those who do not suffer from what he refers to as the error of "illicit ignorance" realize that methodological and logical pluralism is required for adequately handling the broad spectrum and diversity of human problems. Therefore, he repeatedly emphasizes the need for a multiplicity of cooperative and complementary methods and logical procedures.

There is always a need to clarify the proper place of the human species within the material cosmos and to assess the status of mental activity in the context of biological evolution. As a frankly pervasive materialist, Farber would sometimes zestfully refer to the human animal as a "bag of bones" to draw attention to the relatively insignificant place that mankind occupies within the sidereal depths of the universe: a cosmos that for all practical purposes is apparently eternal in time, infinite in space, and endlessly changing.

Farber's devotion to philosophy recognized the value of such diverse figures as Kant, Hegel, Feuerbach, and Nietzsche. With icy logic and subtle humor, he fought against obfuscation and ignorance in the recent philosophical literature: he was particularly dissatisfied with Max Scheler's attempted idealist evolutionism, Martin Heidegger's morass of contestable philosophical statements, and Maurice Merleau-Ponty's confusion regarding the crucial distinction between epistemology and cosmology. For Farber, scientific progress destroys any ascientific synthesis (e.g. philosophical anthropology cannot ignore the findings of Galileo and Darwin). The phenomenologist and existentialist must admit that human experience is not necessary for the objective reality of the material universe. Clearly, the needed view is that of the independence of the natural world from human consciousness. As such, Farber found value in the materialist writings of Marx, Engels, and even Lenin.[8]

Marvin Farber summarized Edmund Husserl as a formidable honorable foe of scientific naturalism who, as a result, was not in step with those ideas that constituted the Darwinian evolutionary movement in modern science.[9] However, Farber warned that the emergence of the theory of evolution in the special sciences and philosophy in the last century and, finally, process theology in this century did not bring to an end the old type of idealist and fideist worldviews (but such residually subjectivist and theist perspectives did cease to be as important as they once were). In Farber's learned and wise opinion, materialist naturalism, supported by an impressive array of empirical evidence from evolutionary biology to scientific cosmology, is a major blow to all forms of metaphysical idealism and lingering supernaturalism.

Unlike William James and John Dewey, Farber never limited phenomenological inquiry to merely an analysis of immediate human experience.[10] Instead, he soundly recognized the historical nature of all human experience. He often stressed that no closure should be placed on either the direction or the future reach of human inquiry as long as it is ethically defensible. In short, he claimed phenomenology to be a meaningful but artificial mental construct of definite but limited usefulness to human activity.

According to the Farberian worldview, philosophical activity has four major functions: a clarification of the perennial problems (questions) and basic ideas concerning material reality, a recurring attempt at synthesizing the findings of the special sciences, a rigorous analysis of all methods and human experiences, and the ongoing critical examination of values within the context of sociocultural development as well as natural history and the cosmic perspective.

As did Augustine and Kant, Husserl abandoned his interest in astronomy for an excessively subjectivist approach to things (this unfortunate shift from cosmology to egology is an anathema to all serious naturalists, whether essentially scientific or philosophical in orientation). Farber acknowledged and incorporated the findings of scientific cosmology, evolutionary biology, comparative anthropology, and descriptive psychology. His philosophy of man within nature is free from geocentrism, zoocentrism, anthropocentrism, ethnocentrism, and egocentrism. Until the end of his life, he followed the developments in the special sciences with keen interest and assessed the recent findings in comparative primate behavior studies as a striking confirmation of the theory of evolution. Farber both espoused and built upon the values of the Age of Enlightenment. As did Dewey, he held human concepts and ideas to be symbolic instruments of adaptive value (in a moment of wit, Farber called a belief the principle of sufficient wishing). Yet, in establishing values within a cosmic flux in which time is brutally real, one must never underestimate the significance of man's fallability and finitude.[11]

Marvin Farber may be regarded as the philosophical founder of materialist phenomenology. He was a dedicated teacher, uncompromising scholar, internationally respected and admired author, and loyal friend to all students and colleagues committed to the pursuit of truth, excellence, and integrity. As a profound philosopher and an original eclectic outside the conventional schools of his day, he belongs nevertheless to the great tradition of serious reflection. Although devoted to social justice and human liberty (and always scrupulously tolerant of ideological viewpoints significantly different from his own), Farber never condoned nonsense, cruelty, fanaticism, or totalitarianism of any sort. However, he neither gave a fixed sociocultural program nor developed a comprehensive philosophical system. As such, he anticipated objections wherever logically foreseeable and did his best to prepare adequate responses.

Farber acknowledged that the arts enrich and ennoble the human condition. As was Einstein, he was an accomplished violinist. He was especially fond of great music (e.g. the works of Bach, Beethoven, Wagner, and Strauss) and literature (e.g. the writings of Goethe, Stendhal,Gogol, and Dostoevski). Indeed, it is refreshing to see such an acute intellect enjoy the aesthetic dimension of human existence.

This aesthetic sensitivity was undoubtedly one of the reasons for the great attraction Marvin Farber felt for Friedrich Nietzsche, a basically kindred spirit, whose iconoclastic sarcasm and dazzling eloquence he not only relished but at times even followed in his own outpourings of original wit.

Marvin Farber always encouraged intellectual development, and is an example of the human intellect at its luminous best. In the history of human thought, he has left indelible and admirable marks.[12]

NOTES AND SELECTED REFERENCES

[1]I first presented some of the ideas expressed in this chapter in the closing paper delivered at the Marvin Farber Memorial Symposium held at the State University of New York at Buffalo (March 12, 1982) as part of the international conference on philosophy and science in phenomenological perspective dedicated to the memory of Marvin Farber (1901-1980).

[2]Cf. Ralph Barton Perry, *General Theory of Value: Its Meaning and Basic Principles Construed in Terms of Interest* (Cambridge, Massachusetts: Har-

vard University Press, 1967), Alfred North Whitehead, *Symbolism: Its Meaning and Effect* (New York: Capricorn Books, 1959), and Clarence Irving Lewis, *Mind and the World-Order: Outline of a Theory of Knowledge* (New York: Dover, 1956).

[3]Cf. Marvin Farber, *The Foundation of Phenomenology: Edmund Husserl and the Quest for a Rigorous Science of Philosophy* (Albany, New York: State University of New York Press, 3rd ed., 1968). One of Marvin Farber's outstanding achievements was his forty-year dedication as the editor of *Philosophy and Phenomenological Research* from 1940 to 1980, an international journal which included scholarly articles and critical reviews from the whole range of philosophical viewpoints.

[4]Cf. Marvin Farber, *Naturalism and Subjectivism* (Albany, New York: State University of New York Press, 1968).

[5]Cf. Ksenija Atanasijević, *The Metaphysical and Geometrical Doctrine of Bruno* (St. Louis, Missouri: Warren H. Green, 1972) translated into English from the French original by Dr. George V. Tomashevich, and Antoinette Mann Paterson, *The Infinite Worlds of Giordano Bruno* (Springfield, Illinois: Charles C Thomas, 1970).

[6]Cf. Marvin Farber's *The Aims of Phenomenology: The Motives, Methods, and Impact of Husserl's Thought* (New York: Harper Torchbooks, 1966), *Phenomenology and Existence: Toward a Philosophy within Nature* (New York: Harper Torchbooks, 1967), and *Basic Issues of Philosophy: Experience, Reality, and Human Values* (New York: Harper Torchbooks, 1968).

[7]Cf. Marvin Farber, "The Ideal of a Presuppositionless Philosophy" in *Phenomenology: The Philosophy of Edmund Husserl and Its Interpretation* (Garden City, New York: Anchor Books, 1967, pp. 35-57) edited by Joseph J. Kockelmans.

[8]Cf. Isaiah Berlin, *Karl Marx: His Life and Environment* (Oxford: Oxford University Press, 4th ed., 1978), Auguste Cornu, *The Origins of Marxian Thought* (Springfield, Illinois: Charles C Thomas, 1957), and V.I. Lenin, *Materialism and Empirio-Criticism: Critical Comments on a Reactionary Philosophy* (New York: International Publishers, 1927).

[9]Cf. Marvin Farber, ed., *Philosophical Essays in Memory of Edmund Husserl* (New York: Greenwood Press, 1968).

[10]Cf. D.C. Mathur, *Naturalistic Philosophies of Experience: Studies in James, Dewey and Farber Against the Background of Husserl's Phenomenology* (St. Louis, Missouri: Warren H. Green, 1971).

[11]Cf. Marvin Farber, "Humanistic Ethics and the Conflict of Interests" in *Moral Problems in Contemporary Society: Essays in Humanistic Ethics* (Englewood Cliffs, New Jersey: Prentice-Hall, 1969, pp. 255-267) edited by Paul Kurtz.

[12]Cf. Dale Riepe, ed., *Phenomenology and Natural Existence: Essays in Honor of Marvin Farber* (Albany, New York: State University of New York Press, 1973) and, more recently, Kah Kyung Cho and Lynn E. Rose, "Marvin Farber (1901-1980)" in *Philosophy and Phenomenological Research*, XLII (1): 1-4.

Chapter 19

THE CREATION/EVOLUTION CONTROVERSY

Evolution is not only a theory but also a fact.[1] Today, most scientists and natural philosophers accept both the process and the doctrine of evolution as an established datum of material reality. The basic empirical evidence now supporting an evolutionary worldview is overwhelming, scientifically convincing, and beyond any serious rational dispute as to its fundamental finds and essential conclusions. Of course, what remains open to free and critical inquiry as well as ongoing expert dialogue are the details of particular patterns, mechanisms, and interpretations of this pervasive natural phenomenon. Geological facts reveal the age of our planet to be approximately five billion years, recently discovered paleontological remains in Australia and Canada indicate that primitive forms of life on earth first appeared over four billion years ago, and human fossils as well as man-made artifacts found in central East Africa place the emergence of incipient humankind back at least three million years. Yet, there are those who, on emotional grounds, still unwarrantedly reject all this evidence with its far-reaching implications and obvious consequences.

In 1857, only three years before the appearance of Darwin's *On the Origin of Species*, the English fundamentalist Philip Gosse made an earnest attempt to reconcile a literal acceptance of the biblical story of creation with the contrary evidence from the developing natural sciences.[2] Despite his commitment to the Holy Scriptures, he could not close his eyes to the compelling facts supporting the conceptual framework of terrestrial evolution. Therefore, his own vision of earth history, which gave priority to the Bible, led him to the following conclusion: this planet is at the same time as old as demonstrated by science and as young as claimed by the Holy Scriptures. How is this possible? It is, if one assumes with Gosse that God created our earth with a built-in past, with all the stratigraphic sequences of rocks as well as plant and animal fossils in their respective places. This evolution without evolution accepts the consequences of the process while at the same time rejecting the process itself. Poor Gosse was heartbroken and could not understand why his valiant and Sysiphian effort at reconciling science and religion was rejected with equal force by both scientists and religionists. The law of parsimony should have suggested to him that there was no reason on earth or in the heavens to appeal to such a fantastically improbable possibility that today flies in the face of all empirical evidence and common sense (let alone its unacceptability to creationists, including the fundamentalists).

Early in the nineteenth century, after studying medicine and then theology, the brilliantly developing scientist Charles Robert Darwin (1809-1882) accepted the tenuous and somewhat ambiguous position as a naturalist and traveling companion on the survey ship H.M.S. *Beagle*, under Captain Robert FitzRoy. At the time, Darwin was especially devoted to historical geology; his main hobby was collecting beetles, and he also enjoyed the sport of hunting. The eccentric aristocrat FitzRoy personally wanted the curious and open-minded Darwin to amass empirical evidence for the purpose of factually documenting a literal interpretation of divine creation as recorded in the story of *Genesis* in the Old Testament of the Holy Bible. As a result, the creation/evolution controversy had its origin in the last century, especially in the private but serious disagreement that emerged between fundamentalist creationism as defended by the captain and scientific evolution ("descent with modification" or the mutability of species) as strongly suggested by the free-thinking naturalist. In microcosm, this personal dispute actually anticipated the contemporary international debate of the macrocosm.

During the five-year voyage of the *Beagle* (1831-1836), Darwin's own critical observations of surface life forms and rock formations, particularly his discovery of giant mammal fossils in Argentina, resulted in the awakening in him of a sober doubt concerning the then prevalent teaching of special creations and the alleged immutability of plant and animal species. Soon enough, a personal conflict ensued between the fundamentalist creationist FitzRoy and the would-be clergyman turned scientist Darwin; as such, one may safely argue that the creation/evolution controversy had its actual inception aboard the *Beagle* as it circumnavigated the globe. The captain dogmatically clung to the conventional belief in the eternal fixity of flora and fauna types, while his scientific friend slowly but steadily advanced toward a dynamic interpretation of life on this planet as a process of incessant transformation and continuous variation as a result of natural causes or forces.

Again, FitzRoy had his own religious motivation for having a naturalist as his traveling companion aboard H.M.S. *Beagle*. It was the captain's personal desire to have a scientist collect sufficient empirical evidence in order to factually demonstrate as best as possible the truth of holding to a literal interpretation of the creation story of *Genesis* as presented in the Old Testament of the Holy Bible. Interestingly enough, this task was given to the would-be minister but soon to be scientific father of the evolution theory, Charles Darwin (this is only one of several ironic turns of events during Darwin's life in particular and the historical development of the theory of evolution in general).

During this famous trip, the relationship between Darwin and FitzRoy became seriously estranged. The more empirical evidence amassed by the young naturalist Darwin and the more he critically reflected on his own experiences within nature (as well as the far-reaching consequences of the data), the more he became certain that a series of divine special creations or the vitalist interpretation of life is not a true explanation for the origin and history of living things on this planet. In fact, the evidence suggested an incredibly different story. Darwin's scientific attitude was clearly challenging the then held age of our earth, interpretation of the geological history of this planet, and religious accounts given to explain the origin and development of all life (including even the assumed special place of humankind within the universe).

Darwin was curious, open-minded, and always alert to the overwhelming facts and wide-ranging experiences during this trip: reason was his teacher and nature was the classroom. However, FitzRoy was generally indifferent to the emerging natural sciences of his day and remained both closed-minded to the implications of a rigorous study of nature as well as rigidly dogmatic in his own religious commitment to understanding and appreciating humankind in the cosmos: faith was his only guide and the Holy Bible was the only authority.

Darwin took time, change, and biological evolution (as the gradual "descent with modification") seriously. The private controversy between Darwin and FitzRoy represented the ongoing and unfortunate struggle between science and religion, respectively. Darwin represented the emerging process vision of man, life, and earth history grounded in science and reason. On the other hand, FitzRoy represented the traditional static Judeo-Christian view of man's place within the cosmic scheme of things.

At the end of the voyage of the *Beagle*, Darwin was convinced that evolution had taken place and would continue to bring about new living forms on our planet. FitzRoy felt responsible for Darwin's evolutionary interpretation of things and developed a deeply rooted guilt about his having accepted the young naturalist as a member aboard his ship. Darwin lived to see his theory of evolution accepted by the scientific community (although modifications are required as our empirical understanding and appreciation of the origin of life and biological history increase thanks to continued scientific inquiry and rational deliberation). However, FitzRoy eventually committed suicide. One may suspect that his guilt in having aided the scientific founder of evolution played a key role in this unfortunate decision. Perhaps symbolically, Darwin's triumph argued in favor of a naturalist overview of process reality while FitzRoy's death demonstrated that the old fixed interpretation of things grounded merely in faith had been superseded by science and reason.

Influenced by the Hutton/Lyell geological theory of uniformitarianism (which argued for the slow evolution of the crust of the earth), a fortunate five-week visit to the unique Galapagos Islands in the Pacific Ocean, and Thomas Robert Malthus's essay on population, Darwin became definitively convinced of the scientific truth that biological evolution, primarily by means of natural selection, could explain the creative history of all life on our planet far better than any alternative hypothesis or theory, let alone a resort to a religious myth. After about two decades of hesitation and indecision, Darwin was finally prevailed upon by sincere friends to publish at last his long-maturing major work, *On the Origin of Species* (November 24, 1859).[3] It remains a comprehensive and intelligible argument for the truth of biological evolution in terms of science and reason, although expectedly modified and complemented by discoveries made since Darwin's time. The venerable Englishman, however, never debated or defended his own theory of evolution, with its awesome theoretical implications and disquieting cosmic consequences, in public or in print.

On 30 June 1860, less than a year after the publication of Darwin's *Origin*, organized science and organized religion openly collided in a now amusing public debate under the auspices of the British Association for the Advancement of Science held in Oxford's University Museum Library. The clash between these two institutions occurred in the persons of the eminent vertebrate zoologist and natural philosopher Thomas Henry Huxley and Anglican Bishop Samuel Wilberforce, or "Soapy Sam," respectively.[4] (Recall the furious controversy between the ovists and the spermatists in the seventeenth century and the debate between the Vulcanists and the Neptunists in the eighteenth century.)

Wilberforce unconvincingly and impudently railed against the Darwinian worldview, which he was simply not competent to understand or qualified to discuss. The theory of evolution, however, was successfully and eloquently defended by Huxley. In his attempt to refute the emerging theory of biological evolution, Wilberforce as a Prince of the Church of England revealed his complete ignorance of the contemporary natural sciences and consequent incompetence to speak on this grave scientific and philosophical issue at all. Although the Bishop's delivery was elegant, his emotionally charged and not very Christian invective got him into unexpected trouble. This incident can best be summarized as follows: concluding his believed arguments against Darwinism (suggested by Owen) the eloquent Wilberforce sarcastically inquired whether the learned scientist Huxley considered himself descended from the monkey through his grandfather's or his grandmother's side? With exemplary dignity and self-control, the great evolutionist replied: "If the question is put to me, 'Would I rather have a miserable ape for a grandfather, or a man highly endowed by nature and possessed of great means and influence, and yet who employs these faculties and that influence for the mere purpose of introducing ridicule into a grave scientific discussion—I unhesitatingly affirm my preference for the ape." (Although Darwin was not present, his theory obviously won its first major public victory.)

As a semi-invalid and quasirecluse at Down House in Kent, England, the shy and gentle Charles Darwin now rigorously reexamined his earlier traditionalist religious beliefs. As an admitted agnostic, although never an intolerant or militant atheist, his pervasive naturalism and materialism were distressing and unacceptable to his affectionate and loyal wife, Emma Wedgwood Darwin, who deplored the probability that her beloved husband would be wrongly remembered as a blasphemous infidel rather than the kindhearted man and towering genius of science that he actually was.[5] It is to Darwin's lasting credit that he kept his own personal religious opinions discreetly to himself because he viewed them as irrelevant to the public import of his scientific legacy. Darwin died on 19 April 1882 and is buried in Westminster Abbey with the other greats of Britain's artistic, political, and intellectual history. However, as can be seen even today, the creation/evolution controversey did not die with him.[6]

In 1925, the contrived and infamous John Scopes "Monkey Trial" of Dayton, Tennessee, reawakened the old arguments between fundamentalist creationism and scientific evolutionism.[7] The gifted orator William Jennings Bryan of "Cross of Gold" fame defended a literal interpretation of *Genesis* with his usual fire-and-brimstone speeches, while the brilliant criminal lawyer and atheistically inclined agnostic Clarence Darrow argued for the freedom and dignity of human thought and unimpeded scientific research. Although Mr. Scopes was found technically guilty of breaking an antievolutionist law of Tennessee, the originally state-confined issue brought worldwide attention to the growing empirical evidence for and logical implications of the theory of evolution.

At present, Carl Sagan gives astronomical evidence to support even the nonbiological concepts of cosmic evolution,[8] while Stephen Jay Gould presents the synthetic biological theory of terrestrial evolution in modern science chiefly in terms of the accumulating geological and paleontological evidences.[9] Yet astonishing as it may seem, the struggle between scientific evolution and fundamentalist creationism continues in some quarters unabated and even essentially unmodified more than a century later.[10] Although Darwin never claimed that humankind descended from the now living higher or later apes (let alone from monkeys) but rather that man and they have once shared a very distant common ancestry, his position nevertheless continues to be grossly misprepresented either through ignorance or malice or as a result of both. The field of paleontology and comparative studies of primate behavior combined with other special sciences (such as comparative anatomy and physiology, embryology, immunology, biochemistry, and genetics) clearly demonstrate that man is, in fact, closer to his evolutionary cousins the three great apes (the orangutan, chimpanzee, and gorilla) than even Huxley, Haeckel, or Darwin could have imagined in the nineteenth century.

In making a necessary distinction between the religious account of the origin of life in general (and man in particular) and the theory of evolution as the only plausible empirical alternative, one must keep in mind the essential and crucial difference between the metaphoric imagery of religious myth and poetry on the one hand and the literal truth of scientific literature and reason on the other. Those who fail to make this quintessential distinction are insensitive to both the Bible as great literature and to science as our most powerful tool and reliable method of empirical/rational inquiry.[11] Imaginatively interpreted, even the obviously legendary and didactic parts of the Holy Scriptures (such as the story of Creation and of the Great Flood) contain a great deal of symbolic wisdom and perhaps morally useful instruction. But taken literally and imposed as would-be substitutes for critical scientific inquiry and its results, these dogmatic religious traditions may become deadly impediments to the further advancement of scientific and philosophical learning.

What is science? It is an ongoing search for truth based on an empirical method of formulating and testing hypotheses and theories validated wherever possible by experience, experimentation, critical reasoning, and other rigorous means of verification. As such, science seeks systematically related bodies of factual knowledge and causal relationships that lead to predictions in terms of degrees of statistical probability if not even empirical laws or principles.

Unlike religion, advancing science is primarily rational, self-critical, self-correcting, self-revising, and self-improving (it is essentially and irreconcilably antidogmatic, selectively cumulative, and always open-ended). Of course, one must distinguish between science and scientists (just as there is a difference between religion and its practitioners). Many religionists delight in speaking of evolution as if it were merely a theory, confusing this concept with that of a hypothesis. Far from being a mere hypothesis, the theory of evolution is a tightly reasoned explanation of natural phenomena now sufficiently supported by overwhelming empirical evidence, past and present human experience, and rigorous logic with no resort or appeal to an alleged supernatural influence however personally imaged and theologically defended.

In western history, other theories at first vehemently rejected were (after a reasonably short period) ultimately accepted by both the scientific community and the general public. One cannot help but recall the tragic cases of Copernicus, Bruno, and Galileo as well as other pioneers of critical thought and modern science. The profoundly significant facts and insights of these and other scientists and natural philosophers challenged the emotionally charged and well-entrenched opinions of their day so strongly that the immediate reaction of the religious and political establishments threatened their very lives. Yet, not too long thereafter, these remarkably insightful paradigms and resulting scientific predictions became accepted knowledge, illuminated and confirmed by the ever-growing empirical evidence as well as the use of systems of mathematics and technological advances (especially in the reach and precision of recent optical instruments, from the electronic microscope to the giant reflecting telescope). This acceptance was further enhanced by the socioeconomic shift in transalpine Europe from a feudal nobility to an emergent bourgeois merchant class.

Why, then, is the Darwinian framework still being greeted with so much hostility and unyielding opposition? Could this resistance be motivated by unconscious guilt, triggered in some believers by the mere thought of a serious alternative to what they have heard in childhood as the truth taught by parents and teachers concerning the creation of the world?[12]

It was one thing for Copernicus, Bruno, and Galileo to argue against geocentrism (thus removing the earth from its wrongly alleged central position in the universe), and quite another for Darwin and his followers (including Freud and his school) to demonstrate that the anthropocentric position is a conceited illusion. The glaring truth remains that the human being is a long-emerging product of primate evolution in particular, rooted in the history of life in general, and totally within the natural world. In short, our planet does not occupy a central position in the cosmos, and humankind is not a privileged species in the universe.

Concerning the age of the earth itself, recent astrophysical and astrochemical evidences clearly demonstrate that this universe (including our planet) is billions of years old. Obviously, these facts cannot be harmonized with a literal interpretation of *Genesis* as recorded in the Old Testament of the Holy Bible.

It is bewildering that any enlightened thinker should think of this ongoing dispute as a conflict between creationism and evolutionism. Is not evolution as both process and theory self-evidently and richly creative? And is not creation a process of change and development? As such, creative evolution and evolutive creation (properly interpreted) are only different conceptions and terms for the complementary scientific and philosophical aspects of the same cosmic process and planetary history.

As for God, some scientific evolutionists even continue to believe in His existence; for example, God may be held to be ontologically and logically identified with nature itself as the all-encompassing totality and unity of infinite Being as eternal Becoming or, as the recent example of Carl Sagan, one may simply state that evolution is a fact (not a mere theory), irrespective of one's being a self-professed atheist. Finally, as in the case of Teilhard de Chardin, one may make a heroic but unsuccessful attempt at reconciling one's acceptance of the scientific facts of evolution with one's personal religious beliefs.[13]

To be sure, interpretations of the theory of evolution occasionally differ in details and emphasis as well as mechanisms and means of illustration. However, this does not invalidate either the basic suppositions or the empirical evidence accumulated in its support. To turn away from science is to doom oneself to needless self-isolation and eventual self-defeat. Regardless of all our wishful thinking, reality is continuously reasserting itself in utter indifference to humankind and its works (including its illusions). The theory of evolution remains an amply demonstrated fact that human biases, prejudices, and emotional insecurities can never disprove: just as the earth moves, so do species change.

Once again the scientific theory of biological evolution has come under scathing criticisms by the special creationists, particularly the fundamentalist movement. Even after the Darwin-FitzRoy conflict (1831-1836), Huxley-Wilberforce debate (1860), and the John Scopes "Monkey Trial" (1925), the evolution framework continues to be challenged or rejected by orthodox religions and static philosophies. Given the strong differences today in commitments to religious faith or natural science, the ongoing controversy between the Bible and evolution is perhaps unavoidable. A critical examination of the arguments given to refute the claims of the evolution viewpoint is required. Such a serious investigation shows the fundamentalist creationist position to be untenable as a scientific theory or rational philosophy.[14]

Proponents claim that the fundamentalist creationist theory is scientific, but they actually disregard any facts that contradict their religious viewpoint. More alarming, they are even insisting that special creationism be taught on an equal basis with the evolution theory in science courses in public schools (especially in biology classes). Some creationists have introduced bills in at least fifteen states to force school board committees to choose textbooks that include the biblical story of creation along with the scientific theory of evolution. This is obviously a glaring violation of the separation of church and state as well as the freedom of thought. This movement is primarily the result of efforts from the Institute for Creation Research in San Diego, California. It is made up of chemists and engineers, but few if any biologists. Seven staff scientists all have doctoral degrees and spend most of their time promoting fundamentalist creationism on college and university campuses and writing books advocating their position (the texts are published by the Creation-Life Publishers, also in San Diego).

What is the case for evolutionism?

Influenced by Lyell and Malthus, both Darwin and Wallace argued for biological evolution by means of natural selection, or the survival of the fittest. In *On the Origin of Species* (1859) and *The Descent of Man* (1871), Darwin presented a strictly mechanist explanation for organic evolution, which remains the essential foundation of the living sciences. As a science, modern evolution biology is a complex system of ideas that explains similarities and differences among organisms within space and throughout earth history. The evolutionist holds that new species of plants and animals emerged from earlier, different forms over vast periods of time. As a result of selective forces (especially the major explanatory principle of natural selection) and the chance appearance and accumulation of favorable mutations, whether slight variations or major modifications, some individuals in a population have an adaptive and survival advantage and therefore a reproductive advantage over others. These individuals are favored in the struggle to exist in a changing environment.

Appealing to biblical chronology, the Ussher and Lightfoot calculations together held that God created the world in 4004 BC on October 23 at 9:00 AM.[15] Many fundamentalists still believe this account of creation to be true. They claim the earth is less than 6,000 years old and that every kind of plant and animal is eternally fixed in an ordered nature (although they do acknowledge varieties within these static types). These fundamentalists now refer to their religious view as "scientific creationism" while rejecting the tenets of organic evolution. Despite the separation of church and state, they demand that biology textbooks give equal attention to both the story of divine creation as presented in

Genesis and the modern synthetic theory of biological evolution grounded for the most part in natural selection and genetic variability.

Nevertheless, the evidence to support organic evolution is sufficient to convince any open-minded intelligent thinker. Admittedly incomplete at this time, the fossil record is the single most important body of evidence to support the fact of organic evolution. Continuous discoveries in paleontology are filling up the gaps in the fossil record. They support the evolution model rather than the alleged worldwide Noachian Deluge account as presented in *Genesis* (although rare, intermediate or transitional forms do exist in the known paleontological evidence).

Both taxonomy and comparative studies, from biochemistry and embryology to anatomy and physiology, support the implications of the scientific theory of biological evolution. Likewise, genetic research demonstrates the historical continuity and essential unity of all life on this planet (especially in terms of the DNA molecule). Like does not beget like, and kind does not beget kind. All populations are variable and subject to the forces of evolution. In fact, no two plants or animals even in the same population are ever absolutely identical. The biological evolutionist argues that over long periods of time the accumulation of genetic changes in a population interacting with a changing environment may result in a variety becoming a new species in its own right.

Biogeography demonstrates the explanatory principle of natural selection. The distribution of organisms and their physical/behavioral adaptations to different environments clearly support the survival of the fittest. There is a direct and ongoing relationship between living things and their habitats. This is especially illustrated among the plant and animal forms throughout the Galapagos Islands (e.g. the differing species of finches, iguanas, and tortoises inhabiting their own unique environments within the archipelago).

Finally, the story of evolution does not violate the second law of thermodynamics or entropy. The earth is not a closed system, for energy from the sun is always available for the creative process of organic evolution to occur.

Taken together, all of this evidence provides sufficient ground for accepting the scientific theory of organic evolution. However, fundamentalists believe in the literal truth of the biblical account of creation. They hold to a scriptural explanation for the origin of living forms within a fixed and ordered view of nature. Yet, the Bible is not a valid scientific document. Special creationism is theology and not an empirico-logical explanation for the origin and history of life on this planet. As a religious view, it appeals to the supernatural and the authority of the Bible (not to mention that it is biased in excluding all other creation stories except the Judeo-Christian account).

Fundamentalist creationism is neither falsifiable nor verifiable in principle. It ignores the established facts of the evolutionary sciences (e.g. the evidence from historical geology, comparative paleontology, and prehistoric archaeology as well as the recent advances in genetic research and the use of radiometric dating techniques). Creationism consistently misrepresents the evolution theory and ultimately breaks down under rigorous scientific and logical scrutiny. Special creationism is irrational in principle and, therefore, not admissible as a scientific doctrine. To suggest that a scientific investigation of and rational explanation for the rock and fossil and artifact records support a literal interpretation of *Genesis* is clearly ludicrous: such a view ignores both facts and logic in favor of biblical authority and religious assumptions.

In sharp contrast, the evolution model has not been refuted by either empirical tests or the principle of falsifiability. Evolution remains a meaningful theory in its explanation of evidence and prediction of events in modern biology, including physical anthropology. Most western religions do, in fact, accept the evolution theory as the creative process throughout natural history. God remains as the First Cause of the universe, if not also the Creator of only the common source or first forms of all life on our earth.

Of course, there is a crucial distinction between the scientific fact of evolution and the various interpretations and perspectives of this natural process in the world literature. Bold attempts to reconcile scientific evolution with religious beliefs have failed, being poor compromises grounded in obscurity. The natural and supernatural are not compatible in terms of facts and logic. Such undertakings by Wallace, Bergson, and Teilhard de Chardin resulted in giving preference to process spiritualism (*The World of Life*, 1910), an intuitive metaphysics (*Creative Evolution*, 1907), and theistic mysticism (*The Phenomenon of Man*, 1940), respectively.

Does God design harmful mutations, allow the spread of fatal diseases, destroy countless plant and animal species from time to time, and deliberately confuse believers with the age and sequence of rocks and fossils and artifacts? If one believes in a personal Supreme Being, could Darwin and Mendel have been divinely inspired to understand organic evolution in order to appreciate His creative powers? Darwin's importance in intellectual history is enormous, and his theory remains as powerful as ever. It explains a great deal. No alternative to organic evolution gives an equally comprehensive and satisfactory explanation of the facts and relationships in modern astronomy, biochemistry, geopaleontology, anthropology, and even psychology. The modern synthetic theory of biological evolution is grounded primarily in the explanatory concept of natural selection (Darwin) and genetic variability (Mendel). Organic evolution is profound genetic change in populations throughout the history of life, resulting at times in the appearance of new species. It does not necessarily result in progress, and such progress is not excluded. Evolution supports neither the necessity of increasing complexity or perfection nor a preestablished meaning, purpose, direction, end, or goal within natural history. The guiding principle is simply survival by adaptation to the changing environment and subsequent reproduction.

Although Darwinian natural selection still remains an important aspect of modern biology, science has not as yet exhausted an understanding of the evolution mechanisms or forces (especially in the areas of genetics and group behavior). There is even a possibility that some Lamarckian element plays a minor role in biological inheritance.[27] However, present uncertainty or incompleteness in science does not warrant turning to the fundamentalist view. There may be a psychological need to believe in the supernatural, which is subject to scientific investigation.

One might argue that the biblical creation story may be taught in a history class, but to compel its teaching in a biology course is clearly not appropriate (just as astrology and faith healing are no substitute for astronomy and medicine, respectively). As dogmatic religion, the fundamentalist view is a danger to science and its teaching in the classroom, especially if it discourages free empirical inquiry in the special sciences and philosophical reflection.

Special creationism ignores the overwhelming scientific evidence that now supports the fact of evolution. Therefore, it actually threatens the advancement of free thought and the special sciences. Science, in principle, does not acknowledge the supernatural and requires free inquiry into natural history and human existence. Scientific inquiry is a liberation from religious dogmatism, blind faith, vacuous myths, and human emotions. In fact, all scientific explanations as such are naturalistic and rational in principle.

Evolution theory is a natural process: a scientific framework supported by facts, experience, and reason. It is open to modifications and interpretations in light of new empirical evidence, rigorous reflection, and logical procedure. Unlike the special creationist or fundamentalist, the evolutionist as naturalist and humanist accepts the far-reaching implications of the evolutionary sciences. The planet earth is not the center of reality, and man does not occupy a privileged position within the natural history of the material universe.

Although questions remain to be answered at this time, evolutionists continue to make progress through the ongoing and self-correcting method of scientific investigation. What is clearly needed is more science, i.e. the continued free inquiry into the natural origin and historical development of life on our earth and its probable existence elsewhere in the cosmos.[17]

In view of what we do know of this singularly great man of geopaleontology and biology (including anthropology and psychology), particularly in light of his own scholarly writings as well as copious correspondence and various private documents preserved by his family and friends, it is safe to say that Charles Darwin would be both shocked and dismayed by the recent resurgence of the original controversy between dogmatic literal creationists and open-minded scientific evolutionists.

It is unfortunate that Darwin's thoughts have been grossly distorted in order to belittle his majestic theory with its profound insights and pervasive implications. He neither claimed that our species descended from the now-living higher apes (let alone lower monkeys) nor advocated the so-called Social Darwinism, which actually should be described as Social Spencerism. Both of these allegations are merely caricatured misrepresentations of his true positions. Briefly, he taught that human beings clearly share with the apes and monkeys only a distant common anthropoid ancestry in the biological history of the primate order.[18] Similarly, in sociopolitical matters, Darwin was neither an extreme rightist conservative nor an extreme leftist radical. Instead, he was always a reasonable, compassionate, and moderate liberal.

In one form or another, this counterproductive and regrettable controversy, fomented again by the fundamentalist creationists, will no doubt continue to plague both the serious scientists and the enlightened religionists. It is to be hoped that this unfortunate and sterile quarrel eventually will evolve toward a convergence of viewpoints sufficiently close to allow everyone concerned to exercise and practice mutually respectful tolerance conducive to increasing factual knowledge, philosophical understanding, and religious wisdom. On the difficult road to this lofty goal, Darwin's exemplary life of intellectual honesty and quiet diligence should serve as an effective guide and a lasting model.

NOTES AND SELECTED REFERENCES

[1]Some of the ideas expressed in this chapter were first presented in my article "The Creation/Evolution Controversy," which appeared in the premier issue of *Free Inquiry*, ed. by Dr. Paul Kurtz, 1(1):24-26, and in my special feature "Creationism Debate Would Shock Darwin" to *The Buffalo News* (Viewpoints, p. F-6, Sunday, 18 April 1982). Adapted with permission.

[2]Cf. Philip Henry Gosse, *Creation (Omphalos): An Attempt to Untie the Geological Knot* (London: John Van Voorst, 1857).

[3]Cf. Charles Darwin, *The Origin of Species by Means of Natural Selection* (New York: Pelican Classics, 1981).

[4]Cf. Gertrude Himmelfarb, *Darwin and the Darwinian Revolution* (New York: W. W. Norton, 1968, pp. 272-277, esp. 287-294, 359-360).

[5]Cf. Paul H. Barrett, ed., *Metaphysics, Materialism, and the Evolution of Mind: Early Writings of Charles Darwin* (Chicago: Phoenix Book, 1980), and Neal C. Gillespie, *Charles Darwin and the Problem of Creation* (Chicago: Phoenix Book, 1979). Also refer to Philip Appleman, ed., *Darwin* (New York: W. W. Norton, 2nd ed., 1979, pp. 295-386), John N. Deely and Raymond J. Nogar, *The Problem of Evolution: A Study of the Philosophical Repercussions of Evolutionary Science* (New York: Appleton-Century-Crofts, 1973, esp. pp. 251-338), and Eugenia Shanklin, "Darwin vs. Religion" in *Science Digest*, 90(4):64-69, 116.

[6]Cf. Roger Bingham, "On the Life of Mr. Darwin" in *Science 82*, 3 (3):34-39. Of special relevance, refer to the excellent articles as well as features and reports in the issues of the *Creation/Evolution* journal ably edited by Frederick Edwords. This is the only scholarly publication in the world devoted exclusively to the ongoing controversy between fundamentalist creationism and scientific evolutionism. Write: *Creation/Evolution*, Box 5, Amherst Branch, Buffalo, New York 14226.

[7]Cf. Ray Ginger, *Six Days or Forever? Tennessee v. John Scopes* (New York: Signet Books, 1960).

[8]Cf. William J. Harnack, "Carl Sagan: Cosmic Evolution vs. The Creationist Myth" in *The Humanist*, 41(4):5-11.

[9]Cf. Stephen Jay Gould and Niles Eldredge, "Punctuated Equilibria: The Tempo and Mode of Evolution Reconsidered" in *Paleobiology*, 3(2):115-151, and the articles in the special evolution issue of *Scientific American*, 239(3). Also refer to Stephen Jay Gould, "The Importance of Trifles" (a tribute to Charles Darwin) in *Natural History*, 91(4): 16, 18, 20-23, and "In Praise of Charles Darwin" (an essay) in *Discover*, 3(2):20-25, as well as "Evolution as Fact and Theory" in *Ibid.*, 2(5):34-37.

[10]Cf. Isaac Asimov and Duane Gish, "The Genesis War" in *Science Digest*, 89(9):82-87; Stephen G. Bush, "Creationism/Evolution: The Case AGAINST 'Equal Time'" in *The Science Teacher*, 48(4):29-33; Laurie R. Godfrey, "The Flood of Antievolutionsim: Where is the Science in 'Scientific Creationsim'?" in *Natural History*, 90(6):4, 6, 9-10; James Gorman, "Creationists vs. Evolution" in *Discover*, 2(5):32-33, and "Judgment Day For Creationism" in *Ibid.*, 3(2):14-18; Allen Hammond and Lynn Margulis, "Creationism as Science: Farewell to Newton, Einstein, Darwin . . . " in *Science 81*, (10):55-57; David H. Milne, "How to Debate with Creationists — and 'Win'" in *The American Biology Teacher*, 43(5):235-245, 266; Norman D. Newell, *Creation and Evolution: Myth or Reality?* (New York: Columbia University Press, 1982); John Skow, "Creationism as Social Movement: The Genesis of Equal Time" in *Science 81*, 2(10):54, 5760; Linda D. Wolfe and J.Patrick Gray, "Creationism and Popular Sociobiology as Myths" in *The Humanist*, 41(4):43-48, 50.

[11]Cf. George V. Tomashevich, "Reflections on Science and Religion" in *Free Inquiry*, 1(3):34-36.

[12]Cf. "Evolution vs. Creationism in the Public Schools" and "The Resurgence of Fundamentalism: A Symposium" in *The Humanist*, 37(1):4-24 and 36-43,

respectively. Also refer to "Creationism: The AAUP Resolution — A Discussion" in *ACADEME*, 68(2):6-36.

[13] Cf. Pierre Teilhard de Chardin, *The Phenomenon of Man* (New York: Harper Torchbooks, 2nd ed., 1965).

[14] Cf. See the articles in the special issue of *Free Inquiry*, ed. by Dr. Paul Kurtz, devoted to "Science, the Bible, and Darwin" (an international symposium), 2(3): Summer 1982. Also refer to Niles Eldredge, *The Monkey Business: A Scientist Looks at Creationism* (New York: Washington Square Press, 1982); Francis Hitching, "Was Darwin Wrong?" in *Life*, 5(4):48-52; Michael Ruse, *Darwinism Defended: A Guide to the Evolution Controversies* (Reading, Massachusetts: Addison-Wesley, 1982) and "Darwin's Theory: An Exercise in Science" in *New Scientist*, 25 June 1981, pp. 828-830; Mark Ridley, "Who Doubts Evolution?" in *Ibid.*, pp. 830-832.

[15] Cf. Ronald Lane Resse, Steven M. Everett, and Edwin D. Craun, "The Chronology of Archbishop James Ussher" in *Sky and Telescope*, 62(5):404-405.

[16] Cf. Stephen Jay Gould, "Shades of Lamarck" in *Natural History*, 88(8):22, 24, 26, 28, and Roger Lewin, "Lamarck Will Not Lie Down" in *Science*, 213(4505):316-321.

[17] Cf. H. James Birx and Gary R. Clark, "The Cosmic Quest" in *Cosmic Search*, 4(1):29, 38.

[18] Cf. John Gribbin, "The 1% Advantage: Human vs. Gorilla" in *Science Digest*, 90(8):72-77.

Chapter 20

HUMANKIND WITHIN THIS UNIVERSE

Hegel once wrote that no single individual could ever experience everything. Nevertheless, it is intriguing to speculate on the existence of a hypothetical ageless and indifferent extraterrestrial observer on the moon watching the vast sweep of earth history. What incredible events and relationships would this imaginary spectator of the cosmos view during his five-billion-year witness to the process of planetary evolution?

Early in this century, meteorologist Alfred Wegener proposed the then controversial theory of continental drift to explain the geological history of the surface of our planet (a dynamic viewpoint that subsequently gave rise to the science of plate tectonics). He argued that the seemingly rigid and fixed continents on the crust of the earth are, in fact, undergoing imperceptible movements, while the web of flora and fauna that covers most of them has changed across enormous stretches of planetary time. Briefly, from the original land mass Pangaea surrounded by the ocean Panthalassa, natural forces have slowly but continuously shifted this single structure apart so that the surface of our earth now consists of several distinct continents: the evolving biosphere is superimposed upon the moving geosphere.

However, where does one draw the line between the inorganic and the organic realms of material reality?

From the naturalist speculations of Leonardo da Vinci and Alexander von Humboldt to the recent writings of Lewis Thomas and Gerrit Verschuur, there have been those visionaries who conceived of our planet (if not the universe) as a living thing. Da

Vinci and Humboldt saw the earth as analogous to an organism, whereas Thomas sees this planet as most like a single cell.[1] Going beyond what researchers call the Gaia hypothesis, the notion that our earth's biosphere plus its atmosphere equals a living entity, Verschuur boldly views the entire cosmos as an organic system.[2] As such one may imagine that there are microorganic, cellular, planetary, stellar, and galactic levels of metabolic processes and evolutions (each with its own time scale from seconds and minutes through centuries and millennia to billions and billions of years, respectively).

From sidereal galaxies to human ethics, the universal thread of evolution pervades the macrocosm and the microcosm (linking the atoms of stars to the cells of life). One may argue that there is in this grand view that sweeps across space and time a mystical dimension to the human awareness of the totality of things. Among some evolutionists, from Wallace in the last century to Teilhard de Chardin in our own, the unity of reality implies a oneness that transcends matter and energy (also refer to the "evolutionary vision" of Erich Jantsch). Three visionary evolutionists who have offered views of cosmic mysticism are Renan, Unamuno, and Soleri.

In the middle of the last century, Joseph Ernest Renan (1823-1892) abandoned the priesthood and subsequently left the Roman Catholic Church. His ideas were greatly influenced by a philological study of the Holy Bible, the advances of the natural sciences (particularly chemistry), and German thought (especially the dynamic worldview of Hegel). He sought to expand scientific rationalism and to reconcile it with idealist philosophy within a process interpretation of human history as endless progress. Reminiscent of Spinoza and Comte, Renan was interested in the historical development of languages and religions. He taught that the further development of the human mind is the key to understanding and appreciating the directional evolution of our concrete universe. He also claimed that the so-called mysteries of our cosmos would inevitably yield before the advancements of knowledge in both the natural and social sciences as well as the use of reason.

In his volume *Philosophical Dialogues and Fragments* (1871),

Renan speculated on the higher organisms of the centuries to come.[3] Although his interests ranged from archaeology to philosophy, this unorthodox thinker voiced his anxieties about the future of science and humanity. Paleontology reveals the remote history of our species and, within a naturalist framework, suggests the ongoing direction of organic evolution (characterized by ever-increasing consciousness): the immediate goal of the earth is the development and fulfillment of mind as the realization of God in the fullness of planetary time.

As did Nietzsche, Renan envisioned an immense advancement of the human consciousness resulting in the emergence of an elite group of superior individuals devoted to the irrefutable evidence of science and the power of reason. Through continued progress in biology and selective breeding, he saw the improvement of the human intellect leading to this superior race of beings above and beyond the natural capabilities of reality. He imagined that these divine beings would be as far beyond man as man today is above the lower animals.

Although Renan anticipated the persecution of scientists, particularly physiologists and chemists, he foresaw the inevitable triumph of the mind and the supremacy of science. Within an organismic view of dynamic nature, he suggested that the whole universe as an emerging cosmic zygote might become a single being that enjoys perceptions and sensations analogous to those enjoyed by finite man on earth. In fact, everything that has ever existed in this world would eventually culminate in this distinct center of ultimate consciousness as the last goal or final term of the God-developing cosmic evolution. As the total of the universe, this celestial being of trillions of beings will emerge out of living multiplicity into organic unity (just as a terrestrial animal does emerge as an organism of billions of cells from the countless atoms making up its genetic information); the French visionary and futurist believed that all existence will participate in this perfect unity of infinite diversity.

In brief, Renan conceived of a future age in which all the matter in the infinity of space that is now present in countless star-systems or galaxies would become organized into a single being: a living universe that thinks and enjoys not as trillions of individuals but rather as a distinct organism pervading reality. Therefore, for Renan, God is that vast consciousness yet to arrive in which everything will be reflected and echoed back throughout all eternity.

In Spain at the beginning of this century, the Basque philosopher and poet Don Miguel de Unamuno y Jugo (1864-1936) gave a passionate view of life that is more emotive than intellectual: he thought with his heart. Like Kant, Unamuno never traveled out of his own country (Salamanca was his Königsberg). He was a concrete man of flesh and blood whose eyes penetrated into nature, life, and the existential condition of the human situation. He emphasized the supreme value of the individual and was committed to belief and action. His intense mind embraced the spiritual origin and development of humankind and also gazed into the vast mystery of the changing universe itself: he was preoccupied with the destiny of the whole human species and obsessed with the endless longing for personal immortality (one is reminded of Nietzsche's ardent wish to live forever).

Unamuno's masterpiece is *The Tragic Sense of Life* (1912),[4] a volume rich in ideas whose essential theme is the quest for eternal life despite the excruciating conflict between faith and reason: above life's doubt and agony and despair, there is the religious hope for the personal immortality of the individual soul beyond the evidence of science and the argument of logic. The author longed to overcome human finality through the passionate hope of eternal life. He gravitated to feelings, the will, and the imagination rather than to science, logic, and mathematics. His faith in God and the Universe may be summed up in Tertullian's phrase, *credo quia absurdum* (I believe because it is absurd).

Unamuno rejected those evolutionists who, as did Darwin and Haeckel, understood organic history merely in terms of science and reason. His own interpretation of consciousness in the cosmos gives preference to feelings and beliefs; his vision is far closer to the process systems of Bergson, Whitehead, and Teilhard de Chardin than to any of the proponents of a mechanist/ materialist explanation for the emergence of life and thought on our planet. His view owes more to the ideas in process philosophy (Leibniz, Hegel, Schopenhauer, and Nietzsche) than to the facts of the evolutionary sciences.

Since everything that exists seems to be doomed to nothingness, Unamuno has universal pity for all finite things in the universe (from lowly amoebas through human beings to distant galaxies). Organic evolution is simply the perpetual struggle of beings to actualize their consciousness through suffering (despite death) in an endless aspiration to preserve their existence while at the same time striving to become a part of something above and beyond their own limits: this applies to the whole range of natural objects from stars and rocks to plants and animals. An objective idealist, the Spanish thinker saw transitory matter as unconsciousness evolving into spirit as consciousness. In this metaphysics of existence, one has a brilliantly intuitive but philosophically incomplete conception of man within nature.

Unamuno was deeply concerned with the "why" of man's origin and the "wherefore" of his destiny. He argued that the Darwinian zoology sees hydrocephalous and erect-standing man as merely a "diseased" species of orangutan, chimpanzee, or gorilla (a curious and intelligent species with an opposable thumb and articulate speech): the great apes and their kind must look upon man as a feeble and infirm animal, whose strange custom it is to store up and guard his dead. In fact, Unamuno suggested that perhaps feeling rather than reason differentiates man from all the other animals (for the merely exclusively rational man is an aberration and nothing but an aberration).

Must the end of all personal consciousness (the future of the human spirit on earth) be the absolute nothingness from whence it sprang? Will the entire world be extinguished, with only naked silence and profound stillness remaining? Unamuno maintained that man philosophizes in order to live and yearns to reach all space, all time, and all being: "Either all or nothing!"

Unamuno introduced a mystical vision of cosmic evolution grounded in vitalism, pervasive consciousness, teleological progress, and a profound faith in a personal God as the beginning and the end of the spiritual Universe. From his panpsychic and organismic view of all things and through an extraordinary use of analogy, he saw everything from minute living creatures through cells and organs to organisms as ever-higher and more complex distinct units of consciously communicating arrangements of elements (fancying that cells may actually express an awareness that they form a part of an advanced organism endowed with a collective personal consciousness). Through extensive extrapolation, he entertained the possibility that human beings are more or less similar to globules in the blood of a Supreme Being, whose own personal consciousness is the collective consciousness of the entire universe. Just as all the cells of a human animal actively interrelate to produce the consciousness of a personal being with an immortal soul, perhaps the immense Milky Way Galaxy (a cosmic ring or stellar group of which our solar system is merely an insignificant part) is but a sidereal fragment of the Body of God as the ultimate collective of all consciousnesses throughout reality that is preserved forever in His Supreme Consciousness. The whole cosmos mirrors the human organism, and love leads one to personalize the all of which everything is a part: it is universal love that sustains and fulfills the spirit of the world. In short, there is a dialectic between a living and personal God and all the immortal souls of humanity. The end of the Universe is that Supreme Person who, at least in His Memory, gives everlasting life to all that has ever existed.

Unamuno's evolutionist cosmology, imagination-bound and human-centered, is an eschatology rooted in the persistence of consciousness. Reminiscent of Whitehead and Teilhard de Chardin, he embraced a form of panentheism in which every spiritual thing and God are continuously interrelating in a divine Society as the collective unity of imaginations through which God and the Cosmos eventually merge into a single Will of the Universe.

Unamuno suggested that in this vast universe other worlds may be inhabited by living organisms (even consciousnesses akin to our own) and that there may also be a plurality of universes. Reminiscent of Galileo, the unorthodox Unamuno spent the last year of his life under house arrest. Yet, in his incredible speculation on the nature of things, Miguel de Unamuno offered a mystical worldview that attempts to explicate the evolution and goal of the consciousness of this dynamic universe in terms of passion and intuition.

Paolo Soleri, futurist architect of Arcosanti in Arizona and author of *Fragments* (1981), suggests that planetary evolution from matter to spirit has a mystical goal.[6] In fusing architecture with ecology into his humanistic theory of arcology, Soleri envisions a future humanity living in self-sufficient cities of dense population and intellectual creativity (there will be a harmony between man and nature). He was influenced by Marx, Freud, and Einstein as well as Frank Lloyd Wright and, especially, Teilhard de Chardin. His own writings do not represent a rational system of ideas but rather a personal kaleidoscope of speculations and intuitions. This aesthetic view of things is a cosmological scheme clearly grounded in theology and metaphysics rather than science and logic.

For Soleri, reality is time or becoming as creative change from matter to spirit. As such, his evolutionary theology argues for the following eschatological hypothesis: the entelechy of our dynamic universe as a cosmogenesis is love and grace. Obviously, his views on morality reflect a Christian bias.

Despite a galactic perspective, Soleri focuses on earth history and human evolution: our species has evolved from "ape" through man to city-man with the resultant urban effect imperative for human survival and creative fulfillment on this planet. His process worldview is essentially an eschatological model: cosmic evolution moves away from the material Big Bang or Alpha Being through ever-increasing monadic consciousness (the complexity-miniaturization paradigm) toward the spiritual God or Omega Seed as both the divine goal of space-time/mass-energy and the transcendent end of reality as the implosive resurrection of all past beings into the total perfection and fulfillment of ultimate oneness in pure duration.

An evolutionist must seriously ask, Will trilobites and eurypterids of the Cambrian period, *Ichthyostega* and *Seymouria* of the Paleozoic era, Mesozoic dinosaurs and *Archaeopteryx* of the Jurassic, and Tertiary mammals with *Gigantopithecus* of the Quaternary (along with all of our fossil hominid ancestors) appear once again all together in this world to satisfy man's acute longing for personal immortality and cosmic justice in the face of evil, suffering, entropy, and certain death?

Soleri's cosmic optimism, evolutionary framework, and concern for humankind are commendable. Yet, this poetic vision is admittedly one of hopeful anticipation for universal equity. Scientific naturalists and rational humanists will dismiss most of this cosmic vision as being merely cryptic mysticism. The distinction between wishful thinking and physical reality is always crucial.

Early in this century, theoreticians in astronomy such as Slipher and De Sitter argued for the model of an expanding universe (not to forget the relevant contributions of Eddington, Hubble, Jeans, Shapley, and Struve). Of particular importance was the speculative book, *The Primeval Atom: An Essay on Cosmology* (1931, 1950), by Georges Lemaître, as well as the scientific writings of George Gamow, e.g. *The Creation of the Universe* (1952).[6] The theory of an evolving and expanding universe challenged the steady-state interpretation of the cosmos as represented in the works of Einstein, Bondi, Gold, and Hoyle.

Somewhat reminiscent of earlier cosmologies (notably the views of Buffon, Kant, Laplace, and Swedenborg), Lemaître offered a fascinating cosmogonic hypothesis for the origin of this material universe: a synthesis of facts and ideas that attempted to account for the birth, recession, and disintegration of galaxies in noneuclidean space-time as a result of the violent explosion of a single, unique, and unstable primeval atom. Today, the big bang theory refers to this initial cosmic particle as a singularity of unimaginable energy that existed seemingly outside of space and time at least twenty billion years ago.

The Belgian priest and scientist wrote, "The evolution of the world can be compared to a display of fireworks that has just ended: some few red wisps, ashes and smoke. Standing on a well-chilled cinder, we see the slow fading of the suns, and we try to recall the vanished brilliance of the origin of the worlds. . . . Moreover, we can reassure ourselves by stating that space is still extending and that, even if the world must finish in this manner, we are living in a period that is closer to the beginning than to the end of the world."[7]

The recent writings of astronomer Carl Sagan, theoretical physicist Gerard K. O'Neill, and geopaleontologist Stephen Jay Gould bring new ideas and perspectives to the current thought on the theory of evolution. Each has expanded the modern conception of developing humankind within this material universe.

In four major works, renowned astronomer Carl Sagan has explored the most recent discoveries in the natural sciences from process cosmology to biological evolution. He is especially dedicated to the scientific search for life, intelligence, and civilization on other clement worlds among the billions and billions of galaxies in this unfolding universe.

Stretching the human mind from a planet-bound framework to the universe-expanding view of things, Sagan's *The Cosmic Connection: An Extraterrestrial Perspective* (1973) focuses on space exploration and current findings in our solar system, presenting rational speculations on the nature of this galaxy and even beyond the Milky Way (including such awesome projects as astroengineering and the intellectual unification of the material universe).[8]

In *Broca's Brain: Reflections on the Romance of Science* (1974), Sagan critically investigates the marvels of the human mind and the wonders of outer space.[9] His bold praise of science and technology within the spectrum of naturalism and humanisn is both optimistic and refreshing.

In *The Dragons of Eden: Speculation on the Evolution of Human Intelligence* (1977), this astronomer traces the biological development of the human brain from early forms of life to the extraordinary capabilities of modern man, who is aware of both his lowly origin and cosmic destiny.[10] Claiming that the entire evolutionary record of our planet illustrates, in general (as is especially documented in the paleontológical record), a progressive tendency toward intelligence, he envisions the future convergence of machine intelligence with human intelligence in the continued search for life and thought and civilization elsewhere among the countless stars.

In *Cosmos* (1980), Carl Sagan has written a magnificent book that is fresh, intellectually honest, and particularly relevant to the human quest for the scientific understanding and rational appreciation of our universe.[11] With stunning illustrations and lucid text, it is a comprehensive and exciting investigation of material reality from the perspectives of modern science and space technology as well as logical speculation. With *Cosmos*, Sagan brings the starry heavens above and earth history below into the mental reach of both the general reader and the specialist. His own inquiry into humankind within nature is free from superstition, pessimism, metaphysical illusions, and theological obfuscation. He advocates a cosmic perspective, defends the evolutionary framework, and upholds open expression and free inquiry within the rational guidelines of ongoing science as well as humane and responsible behavior.

This impressive volume explores everything from the space-time continuum, other galaxies, and black holes to the DNA molecule and organic development on earth in terms of genetic variability and natural selection. Topics include cosmic evolution, the marvel of the human brain, and unmanned spacecraft missions of discovery throughout our solar system (especially the Mariner, Viking, and Voyager expeditions) involving the mechanical explorations of Venus, Mars, and Jupiter. Special attention is given to the search for extraterrestrial life, intelligence, and civilization.

Cosmos highlights the major contributions of Copernicus, Galileo, Kepler, Newton, Huygens, and Einstein. The important advances in human thought from Aristarchus through Bruno to Darwin are also included. Emphasis is placed on humankind's growing awareness of cosmic unity within seemingly eternal time, infinite space, and endless change.

In the footsteps of the great Italian genius Giordano Bruno, who pioneered the cosmic perspective by placing astronomy definitively on correct scientific foundations, and in the honored tradition of the gifted French astronomer Camille Flammarion, who popularized science without vulgarizing it, Sagan makes a major contribution to the revitalization and dissemination of public knowledge about and interest in the further investigation of our universe as the most worthwhile endeavor for the entire human species. Rich in ideas with a provocative message and challenging vision, Sagan's latest book is that rare volume in the current flood of space literature that is very likely to remain valuable for many years to come.

With his major book, *The High Frontier* (1977), physicist/inventor Gerard K. O'Neill became the first thinker seriously to present a realistic vision of future space colonies.[12] Overcoming planetary chauvinism, he argues for the humanization of our solar system and then deep space: material resources are obtained from mining the moon and asteroids, special modules are built for farming/research/industry, unmanned scientific probes search other galaxies, and eventually living habitats will be designed for interstellar voyages. As a result of continued advances in science and technology, this futurist visionary imagines a whole universe of floating worlds. No longer confined to the earth, the human species will be able to drift throughout the endless abyss of deep space.

In *2081: A Hopeful View of the Human Future* (1981), O'Neill takes a close look at the next century.[13] He refers to some early futurists (Verne, Kipling, Wells, and Haldane), pointing out that they grossly underestimated the speed of advances in both science and technology. The author himself is especially indebted to Konstantin Tsiolkowski and Arthur C. Clarke.[14] He tells us that his own predictions about what our planet will be like one hundred years from now are based on informed conjectures.

O'Neill explores the growth and interaction of five key technological forces as "drivers of change" that will determine the course of the next century: computers, automation, space colonies, abundant energy, and communications; there may even be radical scientific breakthroughs. He foresees enclosed cities, electronic mail, and fast underground magnetic flight in vacuum (floater vehicles). There will be remarkable uses of household robots, greenhouse agriculture, genetic engineering, and solar as well as matter/antimatter energy. Laser, solid-state, holographic, and liquid-crystal technologies will greatly alter the human life-style of 2081.

O'Neill anticipates that individuals in the future will enjoy greater leisure time for sports, hobbies, and travel. Antiaging drugs will prolong life, perhaps resulting in practical human immortality (recall the prediction of the enlightened French philosopher Condorcet in the eighteenth century). Yet, in sharp contrast to Sagan, O'Neill does not take seriously the probability of human contact with extraterrestrial life or sentient beings with superior civilizations. Nevertheless, he maintains that the only limitations to human progress may be the physical limits of the material universe itself, e.g. the number of atoms and the cosmic speed of light.

O'Neill values peace, change, beauty, privacy, and especially freedom. He writes that the twin mysteries of this next century are the meaning of consciousness and the uniqueness of life; our paramount responsibility is to keep the freedom of choice alive. Interestingly enough, there are no references to religion.

In frequent lectures and numerous publications, geopaleontologist and science historian Stephen Jay Gould has defended the scientific theory of organic evolution against empirical misrepresentation, conceptual rigidity, and religious fundamentalism. His major work *Ontogeny and Phylogeny* (1977)[15] critically explores the origin and history of Haeckel's idea of recapitulation, which is a superficial answer to the biological question, What is the relationship between individual development (ontogeny) and the evolution of species and lineages (phylogeny)? Gould demonstrates that Neodarwinism renders the idea of recapitulation untenable. He then focuses on the crucial theme of heterochrony (changes in developmental timing, producing parallels between ontogeny and phylogeny) and its significance in an understanding of gene regulation, the key to any rapprochement between molecular biology and the theory of organic evolution in terms of immediate ecological advantages for slow or rapid maturation. Likewise, neoteny is claimed to be the most important determinant of human evolution, e.g. the large brain of our species is the result of the prolonged retention of rapid fetal growth and development.

In *Ever Since Darwin: Reflections in Natural History* (1977), Gould both critically penetrates and champions the Neodarwinian theory of biological evolution.[16] His prologue points out that Darwin's theory of random variation and natural selection argues for a philosophy that is purposeless, nonprogressive, and materialistic (obviously a severe challenge to a rigid set of entrenched Judeo-Christian attitudes that one is not likely to abandon easily). Thirty-three essays cover a variety of fascinating topics, including Paley's natural theology, Cuvier's geological catastrophism, Wegener's continental drift, Haeckel's comparative embryology, Wilson's sociobiology, and the recent discoveries in primatology and human evolution. Special attention is given to the evolution of the human brain/intelligence and the current racist literature.

The Eldredge/Gould hypothesis of punctuated equilibria favors the role played by major mutations or saltations in the "sudden" origin of new species in the biosphere rather than the slow process of accumulating slight variations, as was Darwin's own explanation to account for descent with modification throughout organic history. However, it must be noted that the views of Darwin (gradualism) and Eldredge/Gould (episodism) are not necessarily mutually exclusive naturalist interpretations to account for the appearance of new species in biological evolution on earth.

The present creation/evolution controversy clearly demonstrates the ongoing conflict between dogmatic religious faith and free scientific inquiry. Gould writes, "The Western world has yet to make its peace with Darwin and the implications of evolutionary theory."

In *The Panda's Thumb: More Reflections in Natural History* (1980), Gould further investigates the theory of evolution. [17] He examines a wide range of evidence and ideas from Darwin's speciated finches and warm-blooded dinosaurs to the explanatory mechanism of natural selection and the current issues in racism, with sections devoted to early life and human history (distinguishing between biological evolution and sociocultural development).

In *The Voyage of the Beagle* (1839), Charles Darwin wrote, "If the misery of our poor be caused not by the laws of nature, but by our institutions, great is our sin."[18] Early anthropologists were eager to extend both taxonomy and the theory of evolution to account for a range of biological variations throughout our species (especially among isolated populations). Craniometry as the science of human measurement in the last century was more or less replaced by psychological/intelligence-testing in this century. In fact, the misuse of anthropometry and these examinations has resulted in tragic misery for countless poor, outcast, and uneducated individuals.

In *The Mismeasure of Man* (1981), Gould gives a brilliant and significant refutation of so-called scientific racism by showing it to be only an ideology grounded in two major fallacies of abstraction: the reification of human intelligence as a single, innate, and fixed quantity called IQ (an entity assumed to be measurable, heritable, and located somewhere at a specific physical area within the brain) and the naive quantification of human intelligence as one number for each individual that unfortunately allowed for classifying and ranking peoples in a unilinear and hierarchical series of worthiness from the supposedly inferior to the allegedly superior.[19]

Would-be scientific racism is a subjective theory of rigid limits supporting "biology is destiny" and maintaining that those differences among populations, classes, and sexes arise primarily from biological determinism as a result of genetic inheritance. An objective analysis of this theme and some of its inferences discloses that certain scientists have taken uncritically *a priori* sociocultural biases, prejudices, and bigotries mixed with their own creative interpretations founded on personal preconceptions (these scientists consciously or unconsciously projected and perpetrated their unwarrantedly chosen views).

In historical perspective, Gould rigorously examines the arguments for scientific racism and even reanalyzes the supporting data. From Morton to Jensen, he critically explores the racist literature (especially the American hereditarianism of Goddard, Terman, and Yerkes). This documented survey considers the findings and opinions of Broca, Galton, Binet, Burt, Spearman, and Thrustone. It also exposes the erroneous but influential ideas of "criminal anthropology"(Lombroso) and "ontogeny recapitulates phylogeny" (Haeckel) as well as the deplorable misapplication of neoteny and eugenics.

Gould's investigation of the "allure of numbers" reveals the use and abuse of correlation coefficients and factor analyses, not to mention some deliberately fraudulent studies of questionable empirical evidence. The distinction between merely mathematical generalization for simplification and true causal entities in the complexity of material reality is crucial and always necessary.

The so-called scientific racism is humanly vicious, methodologically fallacious, intellectually unsound, and socioculturally destructive. It is ludicrous to reify and rank human intelligence, which is a complex phenomenon produced by a combination of known and unknown aspects of both nurture and nature. Concerning the various causes of what is called human intelligence, one must guard against extreme hereditarian as well as extreme environmentalist views. In demonstrating its fatal flaws, Gould has clearly shown the racist/sexist self-serving conceptual framework to be scientifically vacuous and logically invalid.

Growing ethnic prejudices throughout the world demand a rigorous adherence to science and reason for a proper understanding of and appreciation for the place of our species within process nature. Gould's *The Mismeasure of Man* is a convincing, sorely needed, and therefore unusually important volume in the recent evolutionary literature.

Evolutionists are concerned with the origin of species and the end of species. One may ask, Why have the trilobites and dinosaurs vanished from the face of the earth? Both terrestrial and celestial influences, from volcanic eruptions to cosmic radiation, have probably played a role in bringing about the total extinction of most plant and animal species that have emerged on earth.

In their timely volume *Extinction: The Causes and Consequences of the Disappearance of Species* (1981), biologist Paul Ehrlich and researcher Anne Ehrlich explore the critical and fundamentally human problem of endangered species threatened with extinction.[20] They present a grim picture of the future of this planet if the alarming acceleration of vanishing plants and animals continues as a direct result of the unwise exploitation of our finite earth. In fact, species are now disappearing at such a rapid rate that natural processes cannot replace them with new forms of life as has been the case throughout countless millions of years of organic history on this planet. Briefly, on a global scale, technology is both impoverishing nature and putting an end to biological evolution as we know it: there is the wanton destruction of habitats and wildlife (especially in the tropical rain forests of our earth's biota).

The Ehrlichs claim that the capacity of this planet to support human society is an inescapable function of biological enrichment. As such, they argue that the imminent loss of genetic variability and species diversity will become, inevitably, a serious threat to the survival of our own zoological group. It is alarming that at this time among the primates the lemurs and three pongids (orangutan, chimpanzee, and gorilla) are considered endangered species. In short, a promising and successful future for humankind on earth and perhaps even elsewhere requires an immediate check to the soaring extinction of life on this planet.

The volume *Biotic Crises in Ecological and Evolutionary Time* (1981), edited by Matthew H. Nitecki, is a collection of those papers presented at the proceedings of the third annual Spring Systematics Symposium at the Field Museum of Natural History in Chicago on 10 May 1980.[21] These previously unpublished articles explore the causes/effects of major but infrequent and unpredictable biological or physical "crises" that have changed or eliminated species during long (evolutionary) or short (ecological) space-time frameworks. Topics treated range from astrophysics and plate tectonics to microfossils and island biogeography. It is argued that earth's floras and faunas have been suddenly altered by rare cataclysmic events that directly influence both ecology and evolution: meteorite collisions, spontaneous reversals of our magnetic field, floods, earthquakes, glaciers, volcanisms, diseases, introduced species, and even human activities (e.g. urbanization and plundering agriculture). This fascinating book presents various theories that should stimulate new research projects in systematic biology. Although one may not agree with all the insights of this at present controversial paradigm, this symposium exemplifies the need for free inquiry in the modern scientific community.

In *Conservation and Evolution* (1981), authors O.H. Frankel and Michaele Soulé offer a comprehensive and detailed investigation into the processes of extinction from the evolutionary viewpoint.[22] Their central theme is that genetic diversity (allelic variation) is necessary for the maintenance of wild biota as well as cultivated plants and domesticated animals, if not even the human zoological group itself; it stresses man's negative interferences upon the biosphere and his assumed responsibility for its protection from further harmful and irresponsible exploitation. They argue persuasively that incisive human efforts are now required to guarantee the continuing existence of threatened biological species. The most feasible methods of preservation and conservation are critically examined: nature reserves, genetic resources, and zoological and botanical gardens. The authors' extensive research and sincere concern are commendable. However, in light of probable future advances in biotechnology like genetic engineering breakthroughs and extraterrestrial organic farming (remembering that in planetary evolution the extinction of biological forms has been the rule rather than the exception), the authors' scientific and aesthetic defenses of enhancing the genetic diversity of currently endangered species will not necessarily seem all that convincing to some readers. Earth's ecosystem as we now know it may not always be essential for the survival of the human animal. One quickly enters the realm of arbitrary preferential judgments.

Should humankind not take serious measures to protect life on this planet (including our own species), the diversified and adaptable insects along with the plants may yet inherit our earth.

Science is a voyage of discovery toward new horizons. In *Our Cosmic Universe* (1980), astronomer and electrical engineer Dr. John Kraus introduces the observable universe in seven steps, taking one from the earth and its rings to quasar OQ172 and beyond to a present celestial horizon or observable limit at a cosmic radius of 15 billion light-years![23] Without modern scientific instruments, from the radio telescope to the electron microscope, the very great and the very small would have remained unknown to the human mind. A speculative person may even ponder the possibility of a superior method superseding both science and reason.

At the midpoint of earth history, man emerged as the only terrestrial creature aware of the very natural process that had brought his species into existence. He uses the most sophisticated modern scientific techniques to examine (and determine the age of) rocks, fossils, and artifacts from the remotest times of planetary evolution. Interestingly enough, the very important potassium/argon radiometric dating technique has a halflife of 4.56 billion years, about the age of the earth itself.

Literally, we owe our birth and death to the stars. Today, matter exists as a result of spontaneous symmetry breaking (should proton decay be an aspect of the cosmos, it is likely that all matter will eventually disappear from the universe). As such, one may anticipate the end of the stuff of reality at a future point in time!

Like an organism, our dynamic universe of quarks and galaxies had a birth about twenty billion years ago and presumably will have an end. Following the primordial big bang, this expanding while crystallizing cosmos has evolved and diversified from maximum homogeneity (a symmetry of energy) to ever-increasing heterogeneity (the asymmetry of matter).[24] The early stages of universal history are still represented by such cosmic fossils as quasars and radiation, with the latter seemingly uniform in its distribution throughout space. Recently, radio and optical astronomers discovered quasar PKS 2000-330, claiming it to be as distant as 18 billion light-years from earth (making it the oldest known object in this universe).

This grandiose perspective of an evolving, if not even perhaps an oscillating, universe has now replaced the previous steady-state or equilibrium model of our cosmos. It seems that through new noneuclidean geometries the human mind can erect the edifice of the material world. The new physics may discover a pervasive unity in the laws and four forces governing the fate of our dynamic universe on both the smallest and largest scales, and these laws themselves will, no doubt, evolve throughout time. Such a unity will shed light on the dawn and destiny of cosmic history.

The entire universe is our ultimate laboratory. The exploration of outer space is an emancipation from the bonds of our earth and, at the same time, an escape into limitless reality: a journey homeward to the galaxies. Presently, the new challenge is to overcome a cosmocentric view of reality. One may imagine that there are an infinite number of distinct or separate universes, each with its own vastly different paradigm of laws and forces.

There is no need to vainly assume that consciousness has emerged only once in the incomprehensible vastness of this material universe. Even if life elsewhere in the cosmos may be rare (and rarer still other intelligences and civilizations), we need to overcome our provincial view of planetary biology. The ongoing search for extraplanetary life and extraterrestrial intelligence is one of those new areas of empirical inquiry where science and philosophy converge. Unfortunately, however, at this time scientists and philosophers are limited to the study of a single biology and therefore a single process of evolution.

Codiscoverer with James D. Watson of the double helix structure for a working model of the DNA molecule, the code of life and language of heredity, Nobel laureate and biologist Francis H.C. Crick has written an informative and provocative work, *Life Itself: Its Origin and Nature* (1981), that boldly presents an unorthodox speculation to account for the first appearance of organic objects on this planet several billion years ago.[25] This book offers the biocosmic hypothesis of directed panspermia, an intriguing idea first developed by Crick and Leslie E. Orgel in a joint paper published earlier in the space journal *Icarus* (1973), edited by astronomer and scientific humanist Carl Sagan.

From Anaxagoras of antiquity to S.A. Arrhenius in the last century and J.B.S. Haldane in 1954, some thinkers have maintained the existence of cosmic seeds or spores that originated elsewhere in outer space but then (drifting to earth) started life as we know it on our own planet. Crick explores nature from the submicroscopic world of atoms and molecules to the vast panorama of this galaxy and the universe (i.e. from the primeval big bang to human consciousness of today). He seriously offers a variant of the panspermia hypothesis: the evolution of life on this planet began only after an unmanned alien rocket carrying microorganisms (bacteria) from another world in this Milky Way Galaxy was deliberately sent into deep space billions of years ago by the intelligent beings of an advanced (but perhaps doomed) civilization and subsequently landed in or near the life-sustaining waters of our earth. Crick's directed panspermia hypothesis assumes that there have been intelligent beings in our galaxy and that the astonishing biochemical unity of all complex life on earth, from amoebas and cilicates to plants and animals, is due to a common source such as simple bacteria of celestial origin.

Crick emphasizes the awesome age (perhaps twenty billion years), unimaginable size, and essential emptiness of the material universe. He presents those steady physical conditions necessary for primitive living things as we now know them to survive and thrive on a planet: free energy from sunlight, liquid water on its surface, a gaseous atmosphere (made up of simple compounds of hydrogen, nitrogen, oxygen, phosphorus, sulphur and especially carbon), and a suitable gravity and temperature. As it is, the complex biosphere in which we live is a frail veneer of matter on the surface of a rather small planet of a rather average star.

Crick's plausible notion does not actually account for (but merely assumes) the prebiotic origin of life from nonlife somewhere in our galaxy. Nevertheless, it is very unlikely that such an aimless life-carrying space ship would ever land on earth about four billion years ago at just the right time and in a suitable location to favor the survival of its organic visitors. This viewpoint does, however, raise important questions: Did life first appear here on earth or elsewhere in the cosmos? Is the origin of life an extremely rare event or an almost certain occurrence? Did the nucleic acid as the DNA molecule or RNA molecule or else protein (amino acids) emerge first, or did they evolve together? Concerning organic history on earth, of particular significance was the decisive evolutionary transition from primitive prokaryotes in a reducing atmosphere to the more recent eukaryotes in an oxidizing atmosphere. In brief, the directed panspermia hypothesis is neither verifiable nor falsifiable in light of the present extreme paucity of relevant evidence.

Finally, more or less reversing his own hypothesis, Crick envisions man seeding our extraordinary universe with life (bacteria, of course) and warns that the process should proceed slowly and wisely: we should not lightly contaminate our galaxy, for one quickly enters the area of cosmic ethics. Anyway, the cosmos is apparently already filled with organic molecules.

In a recent series of four books, astronomer Fred Hoyle and mathematician Chandra Wickramasinghe have also explored the hypothesis of panspermia: *Lifecloud; The Origin of Life in the Universe* (1978), *Diseases From Space* (1979), *Evolution From Space: A Theory of Cosmic Creationism* (1981), and *Space Travellers: The Bringers of Life* (1981).

Reminiscent of Charles Darwin ages from now, one may imagine a critical observer sailing around the world: not as a naturalist aboard a survey ship floating among the prehistoriclike islands of a volcanic archipelago isolated in a sea of space but rather as a scientist on a cosmic vehicle traveling among distant galaxies dispersed throughout the endless ocean of time. What new ideas and perspectives will this brave explorer add to the understanding of and appreciation for life and thought in this universe?

Perhaps we are alone in our galaxy (or the sole reflective inhabitants of this universe). Nevertheless, exobiology is a challenge to advances in science and technology.

In the future, one may anticipate a comprehensive theory of biological evolution as advanced beyond our present understanding of and appreciation for life on earth as Darwin's scientific framework of organic history was a major improvement over Aristotle's philosophical interpretation of biology in terms of eternal fixity. In fact, one may even envision a future synthesis of the ideas of Darwin and Einstein into a metabiophysics as a grand unified theory of matter, energy, life, and thought within the relativity framework of an evolving universe.

In the last century, the scientific debate in America over evolutionism was fought between the paleoichthyologist Louis Agassiz who upheld the fixity of species and the botanist Asa Gray who defended the Darwinian theory of organic evolution.

Early in this century, there was an ongoing debate between the selectionists, who gave preference to Darwin's major explanatory principle for organic evolution, and the mutationists, who emphasized the role that genetic variability plays throughout biological history. The outcome was Neodarwinism, or the synthetic theory of biological evolution, which acknowledges the importance played by both natural selection and the inheritance of mutations (slight variations or major modifications) in transforming species throughout organic history.

Today, there is a controversy between the gradualists as followers of Darwin and the saltationists (e.g. Eldredge and Gould, who have introduced their punctuated equilibria hypothesis into modern paleontology). Perhaps one may speak of punctuated gradualism, to avoid the error of taking an extreme position as to the rate of evolutionary change in populations undergoing speciation.

The scientific theory of biological evolution with its unavoidable implications and startling consequences for all areas of investigation (including theology) has had an enormous impact on ways of thinking and believing: it shattered old views of the universe, earth history, organic development, and the place of humankind within nature. However, the persistence of fundamentalist creationism into the last quarter of the twentieth century is a glaring indication of the irrational rigidity of some toward evolution (especially among the unenlightened and arrogant). Likewise, for some, Christian theism has yet to be successfully reconciled with Neodarwinism. In fact there is an alarming gulf widening between the scientific experts and a large portion of the general public, which could hardly be more strikingly illustrated than by the ongoing creation/evolution controversy. To help clarify this issue, it needs to be pointed out that there is a crucial distinction between scientific statements and religious statements, and attention must always be paid to the established facts of the special sciences.[26] Let us face reality: fundamentalist creationism is not a valid or sound alternative to scientific evolutionism as an empirically documented explanation for organic development on our earth. As an essentially sociopolitical movement, fundamentalist creationism represents not only a historical curiosity but also a clear threat to science, education, and free inquiry. Fundamentalist creationists are a nuisance with their residue of an outmoded theology and static worldview. In sharp contrast, the evolutionary theory is grounded in the tremendous success of the special sciences and rational philosophy.

From Xenophanes and Bruno to Spinoza and Einstein, an impressive list of natural philosophers has extended the physical principle of cosmic unity to embrace the totality of existence and, as a result, had the courage to proclaim God and the World to be one. Among the evolutionists, notably Haeckel and Alexander, pantheism was a very attractive compromise between traditional theism and rigorous atheism (Whitehead and Teilhard de Chardin opted for panentheism). Unlike the bitter warfare between science and religion over the theory of evolution in the nineteenth century, an ironic twist is that in this century the modern evolutionary perspective is providing a rich source of inspiration for process theology and religious mysticism rooted in a cosmic empathy for the creative unity of all reality.

Humankind is now experiencing the emergence of the one-world concept, both on the planetary and cosmic levels of inquiry. In ages to come, the interpretation of God may very well be the result of a philosophical convergence of a scientific thought-system and a religious belief-system into a truly pantheistic materialism as the intense awareness of and deep appreciation for cosmic unity: such a concept of God includes all times, all spaces, and all universes above and beyond the fleeting needs and desires of our emerging species in this dynamic reality (of course, without human beings in the cosmos even this position is meaningless).

Our planet once belonged to the invertebrates, then to the fishes, and later to the reptiles. The earth now belongs to us.

The human animal emerged in central East Africa, where hominid fossils have been found from Laetoli and Olduvai Gorge to Lake Turkana and the Afar Triangle. Nearly four million year old hominid footprints attest to the emergence and survival of our remotest ancestors, who represent the dawn of mankind on the earth. Very likely, both aggressive and cooperative behavior patterns have played a role in the evolutionary success of our species. In the richness of life on our planet, about four billion human beings now inhabit this globe (one for each year of organic evolution throughout earth history).

Man is related not to the angels and devils but rather to the apes and monkeys. He is a newcomer to the universe, who has left the garden of blissful ignorance for the tools of science, technology, and reason (logic and mathematics). Humankind is entirely within the purview of science and therefore accessible to empirical inquiry. Erect man is the symboling animal; the uniqueness of our zoological group is that it is the only evaluating species on earth.

It may be said that the modern electronic computer mirrors the power of the evolved human brain. Further advances in science and technology may result in superior artificial intelligence, which could radically transform our species through a biotechnological revolution. Whatever the destiny of our species, human values will always remain within the cosmic scheme of things, and there is always the need for free and responsible inquiry in an open and humane society.

Concerning evolutionary science and its ethical implications for human survival and the future of our species, of particular importance are three controversial works by biologist Garrett Hardin: *Nature and Man's Fate* (1959), *Exploring New Ethics for Survival: The Voyage of the Spaceship Beagle* (1972), and *Naked Emperors: Essays of a Taboo-Stalker* (1982).[27]

There is always the probability that before our earth freezes over and the sun burns out humankind will have discovered the essential laws of this material universe, enabling it to migrate to some other distant but favorable star-system in this cosmos. If practical natural immortality is ever achieved (a possibility that, in fact, has been one of our species's goals), all that we human beings would then wish for is that those who had come before us could have also enjoyed this conquest.

Concerning the destiny of humankind thousands or millions or even billions of years from now, how may a pervasive materialist and cosmic evolutionist supporting scientific naturalism and rational humanism envision the shape of things to come? The possibilities seem utterly endless. One may anticipate the continuation of our species in this universe as a result of mind manipulating matter and energy so that life can adapt and survive beyond the physical limitations of space and time as we know them today: the mystery of the future, with its inevitable challenges and rewards, should not deter the cosmic quest (in fact, there could be no end to progress, even though ultimate reality may always remain essentially an enigma to human curiosity).

The discreet and prudent Charles Darwin was a true naturalist whose soundly scientific and rigorously reasoned theory of evolution has enormously altered forever the traditional views of life on earth. In his 1871 volume, *The Descent of Man*, written during several years at his Down House in Bromley (Kent, England), the great thinker wrote, "We must, however, acknowledge, as it seems to me, that man with all his noble qualities, with sympathy which feels for the most debased, with benevolence which extends not only to other men but to the humblest living creature, with his god-like intellect which has penetrated into the movements and constitution of the solar system—with all these exalted powers—Man still bears in his bodily frame the indelible stamp of his lowly origin."[28]

As evolutionists, let us venture into the flux of the universe with wide-open eyes and enlightened courage: the future belongs to those who are prepared for it. The ultimate human goal is the understanding of and appreciation for ourselves as sparks of consciousness flickering but for a brief moment in the eternality and infinity of cosmic darkness and emptiness. In the last analysis, recalling the thoughts of Thales in antiquity, evolving man is essentially thinking stardust in living water.

NOTES AND SELECTED REFERENCES

[1]Cf. Lewis Thomas, *The Lives of a Cell: Notes of a Biology Watcher* (New York: Bantam Books, 1975, p. 4) and *The Medusa and the Snail: More Notes of a Biology Watcher* (New York: Bantam Books, 1980).

[2]Cf. Gerrit L. Verschuur, "Is the Cosmos Alive?" in *Science Digest*, 89(6):31.

[3]Cf. Joseph Ernest Renan, "The Higher Organisms of the Centuries to Come" in *French Utopias: An Anthology of Ideal Societies* (New York: The Free Press, 1966, pp. 381-391), ed. by Frank E. Manuel and Fritzie P. Manuel.

[4]Cf. Miguel de Unamuno, *The Tragic Sense of Life* (New York: Dover, 1954, esp. pp. 19-20, 72, 147-150, 183, 240, 254).

[5]Cf. Paolo Soleri, *Fragments* (New York: Harper & Row, 1981) and *The Bridge Between Matter and Spirit is Matter Becoming Spirit: The Arcology of Paolo Soleri* (Garden City, New York: Anchor Books, 1973).

[6]Cf. Georges Lemaître, *The Primeval Atom: An Essay on Comology* (New York: D. Van Nostrand, 1950) and George Gamow, *The Creation of the Universe* (New York: The Viking Press, Compass Books, 1956).

[7]Cf. Georges Lemaître, *The Primeval Atom: An Essay on Cosmology* (New York: D. Van Nostrand, 1950, p. 78).

[8]Cf. Carl Sagan, *The Cosmic Connection: An Extraterrestrial Perspective* (Garden City, New York: Anchor Books, 1980).

[9]Cf. Carl Sagan, *Broca's Brain: Reflections on the Romance of Science* (New York: Ballantine Books, 1980).

[10]Cf. Carl Sagan, *The Dragons of Eden: Speculations on the Evolution of Human Intelligence* (New York: Ballantine Books, 1978).

[11]Cf. Carl Sagan, *Cosmos* (New York: Random House, 1980). Also refer to William J. Harnack, "Carl Sagan: Cosmic Evolution vs. the Creationist Myth" in *The Humanist*, 41(4):5-11. On 18 April 1981, the American Humanist Association named Carl Sagan 1981 Humanist of the Year at their Fortieth Annual Conference held in San Diego, California.

[12]Cf. Gerard K. O'Neill, *The High Frontier* (New York: Bantam Books, 1978).

[13]Cf. Gerard K. O'Neill, *2081: A Hopeful View of the Human Future* (New York: Simon and Schuster, 1981).

[14]Cf. Konstantin Tsiolkovsky, *Beyond the Planet Earth* (New York: Pergamon Press, 1960) and Arthur C. Clarke, *2001: A Space Odyssey* (New York:

Signet Books, 1968) as well as *The Lost Worlds of 2001* (New York: Signet Books, 1972) and *2010: Odyssey Two* (New York: Ballantine, 1982).

[15]Cf. Stephen Jay Gould, *Ontogeny and Phylogeny* (Cambridge, Massachusetts: The Belknap Press of Harvard University Press, 1977).

[16]Cf. Stephen Jay Gould, *Ever Since Darwin: Reflections in Natural History* (New York: W.W. Norton, 1977).

[17]Cf. Stephen Jay Gould, *The Panda's Thumb: More Reflections in Natural History* (New York: W.W. Norton, 1980).

[18]Cf. Charles Darwin, *The Voyage of the Beagle* (Garden City, New York: Anchor Books, 1962, pp. 23, 497).

[19]Cf. Stephen Jay Gould, *The Mismeasure of Man* (New York: W.W. Norton, 1981). In 1982, *Discover*, the newsmagazine of science, chose the eminent paleontologist and evolutionary theorist as its scientist of the year: James Gorman, "Scientist of the Year: Stephen Jay Gould" in *Discover*, 3(1): 56-58, 60, 62-63. As a scientific evolutionist, Gould is a leader in the fight against fundamentalist creationism. Adapted with permission from a review of *The Mismeasure of Man* by the author, see "Human Evolution: New Data, New Thoughts" in *BioScience*, 32(5): 255. Copyright© by the American Institute of Biological Sciences.

[20]Cf. Paul Ehrlich and Anne Ehrlich, *Extinction: The Causes and Consequences of the Disappearance of Species* (New York: Random House, 1981). Also refer to Harold T.P. Hayes, "Animal Kingdom . . . or Animal Farm?" in *Life*, 5(11):116-118, 120-122, 124, 127-130.

[21]Cf. Matthew H. Nitecki, ed., *Biotic Crises in Ecological and Evolutionary Time* (New York: Academic Press, 1981). Adapted with permission from a review by the author in "Natural and Man-Made Crises and Evolution" in *BioScience*, 32(2):149. Copyright © by the American Institute of Biological Sciences.

[22]Cf. O. H. Frankel and Michael Soulé, *Conservation and Evolution* (New York: Cambridge University Press, 1981). Adapted with permission from a review by the author in "Natural and Man-Made Crises and Evolution" in *BioScience*, 32(2):149. Copyright © by American Institute of Biological Sciences.

[23]Cf. John Kraus, *Our Cosmic Universe* (Powell, Ohio: Cygnus-Quasar Books, 1980, esp. pp. 10-19, the universe in seven steps.)

[24]Cf. Joseph Silk, *The Big Bang: The Creation and Evolution of the Universe* (San Francisco: W.H. Freeman, 1980), Steven Weinberg, *The First Three Minutes: A Modern View of the Origin of the Universe* (New York: Bantam Books, 1979), and Walter Sullivan, *Black Holes: The Edge of Space, The End of Time* (New York: Warner Books, 1980).

[25]Cf. Francis Crick, *Life Itself: Its Origin and Nature* (New York: Simon and Schuster, 1981). Also refer to Francis Crick, "Seeding the Universe" in *Science Digest*, 89(10):82-84, 115-117.

[26]For a recent opinion on this controversial issue see Richard S. Laub, "God and Science" in *The Explorer*, 24(3): 12-15.

[27]Cf. Garrett Hardin, *Nature and Man's Fate* (New York: Mentor Books, 1961),

Exploring New Ethics for Survival: The Voyage of the Spaceship Beagle (Baltimore, Maryland: Pelican Books, 1973), and *Naked Emperors: Essays of a Taboo-Stalker* (Los Altos: William Kaufmann, 1982). The second book contains the original essay "The Tragedy of the Commons" as Appendix B, pp. 250-264. Also refer to Richard D. Alexander, *Darwinism and Human Affairs* (Seattle: University of Washington Press, 1982, esp. pp. 219-278).

²⁸Cf. Charles Darwin, *The Origin of Species* and *The Descent of Man* (New York: The Modern Library, 1936, p. 920).

LIFE IN THE COSMOS

F ollowing the publication of Darwin's *Origin* (1859) and *Descent* (1871), the scientific theory of biological evolution slowly but steadily became the major explanatory instrument of the modern astronomical and biosocial sciences. As a result, today, the theory and the fact of evolution represent the principal conceptual thread that runs through the entirety of existence, from sidereal galaxies to human ethics.

The survival and fulfillment of humankind seem to require its eventual venturing to the stars as a continuing outgrowth of the as yet unfinished terrestrial evolution. Through modern science and technology, our species soon should be able to adapt, survive, and thrive elsewhere on other worlds. Yet, there are apparently insurmountable obstacles to overcome. A favorable convergence of the space sciences with genetic engineering should remove many of these physical and psychosocial barriers challenging the fragile and vulnerable human animal as it responds to the cosmic call and vital imperative: explore and colonize the universe or perish prematurely on planet earth!

Our dynamic cosmos is about twenty billion years old, and our planet earth is but a tiny fragment within probably eternal time, infinite space, and endless change. With the relatively recent emergence of our species, the evolving universe has at least once

---

Some of the ideas in this Epilogue were first expressed in the feature "The Cosmic Quest" by H. James Birx and Gary R. Clark in *Cosmic Search*, edited by Dr. John Kraus, 4(1): 29, 38. Adapted with permission.

become aware of itself through many ephemeral personal centers of partial and imperfect consciousness. Yet human curiosity need not remain forever earthbound, confined to our solar system, or even limited to this galaxy.

From Thales to Sagan, cosmologists have marveled at the origin, scope, and destiny of this universe. Other stars, planets, and celestial phenomena are subject to critical scientific inquiry and ongoing rational speculation. Therefore, the challenge of understanding the emergence and nature of life may be endless, in the realm of thought and of the evolving cosmic continuum.

In the last century, naturalists took time, change, and development seriously. As the pioneering father of the scientific theory of biological evolution, Darwin explored the historical continuity and essential unity of all life on earth. He was aware of the dynamic and creative interaction between organisms and their environments. However, he was, as shown, unfortunately unaware of Mendel's discovery of the basic principles and algebraic patterns of heredity. In all cells, chromosomes contain the hereditary units of life. For the most part, behavioral traits and physical features are potentially determined by these particles called alleles. These genetic units (both singly or in intricately combined association and interaction) transfer biological information from generation to generation and, through chemical changes (slight alterations or major mutations) in time, may bring about new varieties and eventually perhaps even new species.

Mutational changes in the genetic makeup or genotype are likely to be externally expressed as physical and/or behavioral changes in an organism's phenotype: slight variations or major modifications in the organism's appearance may (at any one time) be of temporarily positive, neutral, or negative value. Until the advent of man with culture, life was subject to the two major forces of random biological evolution: chance genetic variability and necessary natural selection.

Increasing human interference with hitherto predominantly natural selection through culture (especially science and technology) has already resulted in cultivated plants, domesticated animals, and the growing civilization of our zoological group.

Those early experiments with garden peas, evening primroses, and fruit flies may now be extended to alter higher or later forms of life (including human beings in accordance with ethically, aesthetically, and rationally guided human planning and human design for the future). Such directed evolution potentially holds both great promises and cataclysmic perils. Certainly, the sobering complexity of the enormous risks and vast opportunities involved cause one to pause and reflect upon the possibly ominous intellectual implications and staggering moral responsibilities of all human programming of life. Even minor errors could have major consequences (even in the best of intentions).

In the middle of this century, Watson and Crick discovered the chemical, structural, and functional characteristics of the DNA molecule. This material is the essential code of all life on earth. Altering the DNA molecule, either by accident or deliberately, brings about changes in living forms that may enhance or endanger their survival and reproductive potential. The human organism is also determined by its genetic makeup in interaction with its natural environment as well as its psychosocial and cultural setting (i.e. the physical and symbolic aspects of reality).

Mankind with its culture is totally within the process of nature. The existence of our species is incontrovertibly dependent upon and historically tied to those mutable genes within and the elusive stars beyond.

For about four million years, the emerging human animal has been more or less successfully adapting to various habitats on this planet. Yet, cradle earth is only the birthplace of our zoological group. Our species already has the science, technology, and mathematical skill to leave the physical confines and material limitations of this exquisite world. No longer earthbound, man is becoming a citizen of the universe. He may now daringly soar starward, exploring and enjoying the cosmic spectacle. However, to reach distant suns and prosper on other planets, our form of life will have to adapt to deep space, unfamiliar worlds, and strange environments.

Man's curiosity and imagination are insatiable: long before Sagan and O'Neill, Tesla envisioned engineering other worlds for the sake of peopling the galaxies. The exploration, colonization, and utilization of this universe may require even a partial reengineering of the human animal in order to overcome some of the major obstacles.

Beyond certain arbitrary legal restrictions on our own planet, genetic research in outer space could result in knowledge that will help solve those problems man will face as he journeys among the stars. In short, the universe becomes a cosmic laboratory and a school for accelerated moral/behavioral evolution.

Man is only beginning to understand and appreciate all the wondrous potential within the living cell and the rich resources throughout the sidereal depths of our universe. Through genetic engineering, the blind necessities of nature become the deliberate choices of humanity. To succeed in the cosmic quest, the human organism may need to be partially redesigned to overcome the physical challenges of differing gravities, energies, and as yet unimaginable other forces and impediments (including the relativity of time, space, and change).

The current military and political issues of space exploration and genetic engineering should not lastingly blind us to the overall advantages of ongoing research in these two areas so vital to the qualitative continuation and possible improvement of our species and other forms of life.

From worm through ape to man, the evolution of life has had to struggle against the forces of death and entropy. Time and mortality are the ravages of life as we know it. However, through genetic engineering, the biological sciences and medical technologies of the future may give our species such a dramatic extension of average life expectancy as would amount to what essentially could be considered practical immortality (approximating in the physical sense our deep wish for eternity). Along with retarding and eventually halting the aging process, man's senses may be intensified and his mind expanded to overcome those disturbing periods of social isolation and the possibility of cosmic boredom. Also, one should be able to deal successfully with those as yet unknown illnesses and diseases of outer space.

Gene splicing (recombinant DNA/RNA) and cloning should allow humanity to seed the universe with new forms of plants and animals as well as his own kind. Our species would thereby spread the sparks of life and consciousness throughout the dark abysses of endless space and eternal time. Human hope with visionary science may elevate and transform us not only physically and intellectually but, more importantly, perhaps even in the ethical and moral respects. There could be an eventual self-transformation of man as the cosmic primate into a higher being worthier of the challenges and rewards of its sidereal destiny.

The advancement of human knowledge and understanding has always required courageous and bold individuals working in an open and pluralistic society, encouraging critical inquiry within the rational guidelines of both ethically and legally responsible freedom of choice and behavior.

In the obvious and drastic tampering with the stuff of life, one should not (on any *a priori* grounds) fear the unusual and the unknown. On the contrary, one should approach them with open-minded courage and proper respect. Of.course, the awesome and disquieting prospects of a convergence of modern technology with genetic engineering demand considerable caution, benevolence, and wisdom. The cosmic quest should actually be enhanced through the possibilities of genetic engineering. In fact, as a result, our embryonic species may yet enjoy its finest hours amidst the intergalactic lights of this and perhaps other universes.

Still, with all its frightening aspects and magnificent splendors, this material cosmos is totally indifferent to the human animal. But the more we realize that the universe is indifferent to us, the more concerned we should become with one another's lonely needs and cosmic vulnerabilities. Scattered throughout the dynamic vastness of space, we must continue to respect the freedom and dignity of human thought and feeling as well as value the adaptive plasticity and emerging creativity of life. In a word, let us humanize the cosmic quest!

Although full of ice and snow, not all of reality is frozen: nature is flux, change, and evolution. Yet, pervasive nothingness or a frigid eternity may well be the inevitable outcome of cosmic history. Nevertheless, humanity with its delicate genes and mighty space vehicles will probably have made at least a valiant attempt to fathom out the magnificent tragicomedy and perhaps ultimate absurdity of it all. One may hope that this ennobling effort will, however temporarily and fleetingly, increase the sum of human happiness and wisdom. For what benefit would it be if mankind were to gain the entire universe at the price of losing, in the process, its humanity! The course our species takes is up to all of us, and so is the responsibility for the consequences.

Our earth is nearly five billion years old, and life as we know it emerged on this planet approximately four billion years ago. The physico-chemical transition from geogenesis to biogenesis was a major advance in the ongoing flux of material reality. With the recent origin of the human species, the evolution process becomes aware of itself at least once in organic history. Yet, is there life or intelligence on other clement earthlike planets elsewhere in the universe?

No doubt prehistoric men gazed at the night skies, curious about those distant shining objects strewn across the ominous heavens. In their lack of scientific understanding and fear of the unknown, they resorted to magico-religious thoughts and activities as well as legends and myths to explain a world seemingly inhabited by both good and evil spirits.

In antiquity, as we have seen, some natural philosophers not only advocated the empirical/rational investigation of nature but also speculated on living things elsewhere in the cosmos, e.g. Xenophanes, Epicurus, and Lucretius. The following church-dominated Dark Ages and Medieval Period suppressed scientific inquiry. Turning from nature to the supernatural, Augustine wrote that he could no longer dream about the stars. Later, however, Nicholas of Cusa did envision inhabitants in every region of the universe.

During the subsequent Renaissance, Giordano Bruno boldly initiated the modern cosmology: an orientation revitalized by Carl Sagan in this century.

Bruno went far beyond the Aristarchus/Copernicus heliocentric theory by claiming the universe to be eternal in time, infinite in space, and endlessly changing. His own sweeping worldview is grounded in the metaphysical principles of plurality, uniformity, and cosmic unity. He rejected the peripatetic terrestrial/celestial dichotomy in favor of an infinite number of inhabited worlds more or less analogous to our earth: for him, life and intelligence necessarily exist throughout the infinite universe.

Bruno overcame all geocentric, zoocentric, and anthropocentric models of reality. His theme of extraplanetary life or extraterrestrial intelligence was later also supported by Kepler, Wilkins, Huygens, Fontenelle, Kant, Haeckel, Lowell, Flammarion, and Hoyle as well as Carl Sagan and Stephen Jay Gould (in imagining the future of humankind, even evolutionist Richard E. Leakey considers the possibility of contact with interstellar planetary travelers).*

In the last century, the rigorous search for fossils led to discovering those unusual life forms of the remote past. The evolutionary framework held the human animal to be a recent product of organic history. Yet, how accidental or inevitable is biological evolution? Although living forms did emerge at least once in this noisy and violent universe, exobiologists speculate that they will be found elsewhere (perhaps based on different chemistries other than carbon, e.g. silicon or ammonia). Thus, man's intellectual penetration into the infinity of cosmic space was followed by an equally important intellectual penetration into the depths of evolutionary time.

Logic does not dictate that man or the earth must be unique. The birth of stars may inevitably entail the formation of planets engaged in prebiotic chemistry producing organic molecules like nucleotides and/or amino acids (hydrogen, liquid water, and energy are abundant throughout the cosmos). If the basic laws of biology and physics are applicable everywhere in our universe, the emergence of life from inorganic matter seems inevitable given the environmental prerequisites.

To date, the life sciences have restricted their concerns to the earth. However, humankind is no longer confined either intellectually or physically to the finite sphericity of this globe. The new challenge is to overcome the myopic planetary view of things by adopting a properly contextual cosmic perspective of reality.

---

*For recent articles on this subject, refer to "Special Report: Life in Space" in *Discover* (March, 1983), 4(3): 29-31, 34, 36, 38-40, 42, 44-45, 48-49, 52, 56-57, 60-65.

Although there is as yet no irrefutable empirical evidence for astrobiology or sentient life beyond our planet, this alone should not deter us from the cosmic quest: it is too soon to abandon scientific speculation about exobiology. No dogmatic closure should be placed on human wonder or free inquiry in an intellectually open and ethically concerned society.

Fortunately, there are those who are still eager to engage in a serious search for organisms and civilizations elsewhere among innumerable galaxies. Unmanned interplanetary and interstellar vehicle probes as well as radio telescopes penetrate the abysses of outer space in the hope of detecting empirical evidence of extraterrestrial life, including intelligent forms of nonhuman life. Further advances in space technology, incorporating mini-computers and cosmic telescopes, will greatly enhance the human search for protoplasmic existence and intelligence on other celestial bodies throughout this universe.

Apparently the moons and other planets of our own solar system do not support indigenous living things. However, astrobiology maintains that this physical universe may be a zoocosmos teaming with life. Beyond our Milky Way are billions of other galaxies, each with billions of stars if not also countless sun/planet systems. A myriad of other worlds may be swarming with awesome biological forms, disquieting alien beings, and perhaps even organic phenomena beyond man's present imagination and cognition. Thus, although as a form of highly organized matter life may be a relatively rare occurrence in each galaxy, it is statistically speaking still bound to be very abundant in the cosmos as a whole in view of the staggering number of the galaxies.

Special stony carbonaceous chondrite meteorites contain water, minerals, and complex organic molecules such as hydrocarbons or amino acids (the precursors of life on earth). Comets, asteroids, and cosmic gas/dust clouds do harbor these molecules of prebiotic chemistry that anticipate organic phenomena. Physical uniformity or the seemingly pervasive and inexorable laws of nature, statistical probability, empirical inferences, and a temporal framework of billions of years all support both the logical possibility and empirical probability of life-yielding star/planet systems elsewhere in the universe.

Even if the event is centuries from now, the anticipated conclusive empirical evidence for extraplanetary life or human contact with extraterrestrial intelligence will be the most profound discovery in the history of scientific inquiry and a singularly awesome turning point in the life of our species. This event will stimulate a retrospective reevaluation of the nature of existence and at the same time revolutionize our conception of the meaning and destiny of humankind.

Man never has been the ultimate measure of all things. A new and challenging paradigm is emerging: the universe is our ultimate frontier. In the search for life/intelligence beyond our earth, sidereal biophysics is emerging as an intriguing but legitimate area of scientific inquiry. In its continuing exploration of this universe, let us hope that humankind will take (with the rest of its nature) love, compassion, and ethical wisdom to the stars and even beyond. It is this cosmic quest that ennobles and dignifies the human species, and the saga is just beginning!

In the final chapter of *The Voyage of the Beagle* (1839), Darwin, while walking through the dense, wild, and luxuriant splendor of a Brazilian tropical rain forest in early August, wrote the following words: "How great would be the desire in every admirer of nature to behold, if such were possible, the scenery of another planet!" There will probably be those future naturalists who will one day study life forms if not even intelligent beings on other celestial bodies elsewhere in this physical universe. One exciting result of such activity would be the science of comparative evolution. From Thales to Sagan, philosophical speculation about and scientific inquiry into the origin and nature of the universe and life within it have preoccupied the best thoughts of the greatest thinkers. Charles Darwin is clearly one of these. As can be seen from the above quotation, his vision caught a glimpse of what may yet be the future of both the process and theory of evolution. From this awesome perspective, Darwin's intellectual legacy remains as intriguingly relevant as it is rich, cosmic, and indispensable.

BIBLIOGRAPHY

Agassi, Joseph: "Continuity and discontinuity in the history of science" in *Journal of the History of Ideas*, 34 (4): 609-626.

Alexander, S.: *Space, time, and deity* (1920). 2 vols. New York, Dover, 1966.

Andrews, Peter, and Elizabeth Nesbit Evans: "The environment of Ramapithecus in Africa" in *Paleobiology*, 5 (1): 22-30.

Anfinsen, Christian B.: *The molecular basis of evolution*. New York, John Wiley & Sons, 1963.

Appleman, Philip, ed.: *Darwin*. 2nd ed. New York, W. W. Norton, 1969. Reviewed by H. James Birx in *The American Rationalist*, 25(5): 85.

Asimov, Issac: "The 'threat' of creationism" in *The New York Times Magazine*, 14 June 1981, pp. 90, 92, 94, 96, 100-101.

Asimov, Isaac, and Duane Gish: "The Genesis war" in *Science Digest*, 89 (9): 82-87.

Atanasijević, Ksenija: *The metaphysical and geometrical doctrine of Bruno*. Trans. by George V. Tomashevich. St. Louis, Warren H. Green, 1972. Reviewed by H. James Birx in *The American Rationalist*, 25 (6): 103.

Atkins, Hedley: *Down, the home of the Darwins: the story of a house and the people who lived there*. London, Phillimore, 1976.

Aulie, Richard P.: "The doctrine of special creation" in *The American Biology Teacher*. 34 (4): 191-200, and 34 (5): 261-268, 281.

Avers, Charlotte J.: *Evolution*. New York, Harper & Row, 1974.

Axelrod, Robert, and William D. Hamilton: "The evolution of cooperation" in *Science*, 211 (4489): 1390-1396.

Ayala, Francisco J.: "The mechanisms of evolution" in *Scientific American*, 239 (3): 56-69, 242.

Azar, Larry: "Biologists, help!" in *BioScience*, 28 (11): 712-715.

Baker, William: "Herbert Spencer and 'evolution' — a further note" in *Journal of the History of Ideas*, 38 (3): 476.

Bakker, Robert T.: "Dinosaur physiology and the origin of mammals" in *Evolution*, 25 (4): 636-658.

Bakker, Robert T: "Dinosaur renaissance" in *Scientific American*, 232 (4): 58-78.

Balmas, Kenneth S.: "Protein molecules as clues to evolution" in *Cornell Countryman*, 62 (5): 2-3.

Bannister, Robert C.: "'The survival of the fittest' in our doctrine: history of histrionics?" in *Journal of the History of Ideas*, 31 (3): 377-398.

Barkow, Jerome H.: "Culture and sociobiology" in *American Anthropologist*, 80 (1): 5-20.

Barlow, G. W.: "Evolution of behavior" in *Science*, 139 (3557): 851-852.

Barnett, S. A., ed.: *A century of Darwin*. Cambridge, Harvard University Press, 1958.

Barrett, Paul H., ed.: *Metaphysics, materialism, and the evolution of mind: early writings of Charles Darwin*. Chicago: University of Chicago Press, 1980.

Barrett, Paul H., ed.: *The collected papers of Charles Darwin*. Chicago, University of Chicago Press, 1977.

Baumel, Howard B.: "Alfred Wallace: man in a shadow" in *The Science Teacher*, 43 (4): 29-30.

Beddall, Barbara G., ed.: *Wallace and Bates in the tropics: an introduction to the theory of natural selection*. Toronto, Macmillan, 1969.

Benton, Michael J.: "Ectothermy and the success of dinosaurs" in *Evolution*, 33 (3): 982-997.

Bergh, Sidney van den: "Size and age of the universe" in *Science*, 213 (4150): 825-830.

Bergson, Henri: *An introduction to metaphysics*. 2nd rev. ed. New York, Bobbs-Merrill, 1955.

Bergson, Henri: *A study in metaphysics: the creative mind*. Totowa, New Jersey, Littlefield/Adams, 1965.

Bergson, Henri: *Creative evolution* (1907). New York, Modern Library, 1944.

Bergson, Henri: *Duration and simultaneity: with reference to Einstein's theory*. New York, Bobbs-Merrill, 1955.

Bergson, Henri: *Matter and memory*. London. George Allen/Unwin, 1962.

Bergson, Henri: *The world of dreams*. New York, Philosophical Library, 1958.

Bergson, Henri: *Time and free will: an essay on the immediate data of consciousness*. New York, Harper Torchbooks, 1960.

Bergson, Henri: *Two sources of morality and religion*. Garden City, New York, Doubleday, 1935.

Birx, H. James: "Charles Darwin: a centennial tribute" in *Creation/Evolution*, 3 (1): 1-10.

Birx, H. James: "Charles Darwin and fossil man" in *Free Inquiry*, 2 (4): 34-37.

Birx, H. James: "Darwin on animals" in *Zoolog*, 6 (3): 6.

Birx, H. James: "Gigantopithecus" in *Zoolog*, 6 (2): 6.

Birx, H. James: "Knežević and Teilhard de Chardin: two visions of cosmic evolution" in *Serbian Studies*, 1 (4): pp. 53-63.

Birx, H. James: *Man's place in the universe*. Arcade, New York, Tri-County, 1977.

Birx, H. James: "Pierre Teilhard de Chardin: a remembrance" in *The Science Teacher*, 48 (9): 19.

Birx, H. James: *Pierre Teilhard de Chardin's philosophy of evolution.* Springfield, Illinois, Charles C Thomas, 1972.

Birx, H. James: "Teaching evolution: the relevance of philosophical literature" in *BioScience*, 21 (12): 573.

Birx, H. James: "Teaching Pierre Teilhard de Chardin" in *AIBS Education Division News*, 1 (4): 7-8.

Birx, H. James: "Teilhard and evolution: critical reflections" (an invited paper read by the author at the closing of the *Teilhard and Metamorphosis* international symposium, Arcosanti Events, Arizona, 22 September 1981) in *Humboldt Journal of Social Relations*, 8 (3): 151-167.

Birx, H. James: "Teilhard de Chardin: his centennial year" in *Zoolog*, 6 (6): 8.

Birx, H. James: "The biological determinants of behavior" in *The Humanist*, 32 (4): 42-43. A review of *In the shadow of man* by Jane Goodall and *The imperial animal* by L. Tiger and R. Fox.

Birx, H. James: "The creation/evolution controversy" in *Free Inquiry*, 1 (1): 24-26. A version of this article appears as part of chapter 19 in this volume.

Birx, H. James: "The Galapagos Islands" in *Collections*, 62 (1): 12-16.

Birx, H. James: "The great apes: a close encounter with extinction" in *Collections*, 60 (2): 16-25.

Birx, H. James: "The theory of evolution" in *Collections*, 60 (1): 22-23.

Birx, H. James: "The vanishing lemurs" in *Zoolog*, 5 (12): 3.

Birx, H. James: "Those glorious gorillas!" in *Zoolog*, 6 (4): 4-5. Also see, H. James Birx: "The Tropical Forest Gorilla Habitat" in *Zoolog*, 6 (9, 10): 3.

Birx, H. James, and Gary R. Clark: "Extraterrestrial life" in *Zoolog*, 6 (7): 4-5.

Birx, H. James, and Gary R. Clark: "The cosmic quest" in *Cosmic Search*, 4 (1): 29, 38. A version of this article appears in the epilogue of this volume.

Bitterman, M. E.: "The evolution of intelligence" in *Scientific American*, 212 (1): 92-100.

Bleibtreu, H. K.: *Evolutionary anthropology.* Boston, Allyn and Bacon, 1969.

Blum, Harold F.: *Time's arrow and evolution.* Princeton, New Jersey, Princeton University Press, 1968.

Blunt, Wilfrid: *The compleat naturalist: a life of linnaeus.* New York, Viking Press, 1971.

Bober, M. M.: *Karl Marx's interpretation of history.* Rev. 2nd ed. New York, W. W. Norton, 1965.

Boller, J., Paul F.: *American thought in transition: the impact of evolutionary naturalism, 1865-1900.* Chicago, Rand McNally, 1971.

Boslough, John: "The unfettered mind: Stephen Hawking encounters the dark edges of the universe" in *Science 81*, 2 (9): 66-73.

Bowler, Peter J.: "Herbert Spencer and 'evolution'" in *Journal of the History of Ideas*, 36 (2): 367.

Bowler, Peter J.: "The changing meaning of 'evolution'" in *Journal of the History of Ideas*, 36 (1): 95-114.

Brackman, Arnold C.: *A delicate arrangement: the strange case of Charles Darwin and Alfred Russel Wallace.* New York, Times Books, 1980. Reviewed by H. James Birx in *The American Rationalist*, 25 (5): 85-86. Also see, David Kohn: "On the origin of the principle of diversity" in *Science*, 213 (4512): 1105-1108.

Broad, William J.: "Creationists limit scope of evolution case" in *Science*, 211 (4488): 1331-1332.

Brodkorb, Pierce: "Origin and evolution of birds" in *Avian Biology*, vol. 1, chapter 2, pp. 19-55. Donald S. Farner, James R. King, and Kenneth C. Parkes, eds. New York, Academic Press, 1971.

Brodsky, Garry M.: "Dewey on experience and nature" in *The Monist*, 48 (3): 366-381.

Bronowski, Jacob: *The ascent of man*. Boston, Little/Brown, 1973.

Bronowski, Jacob: *The identity of man*. Garden City, New York, Natural History Press, 1965.

Bronowski, Jacob: *The origins of knowledge and imagination*. New Haven, Yale University Press, 1978.

Brosseau, Jr., George E., ed.: *Evolution: a book of readings*. Dubuque, Iowa, Wm. C. Brown, 1967.

Brown, Jerran L.: *The evolution of behavior*. New York, W. W. Norton, 1975.

Brush, Stephen G.: "Creationism/evolution: the case against 'equal time'" in *The Science Teacher*, 48 (4): 29-33.

Buvet, R., and C. Ponnamperuma, eds." *Molecular evolution, volume one: chemical evolution and the origin of life*. New York, American Elsevier, 1971.

Capouya, Emile, and Keitha Tompkins, eds.: *The essential Kropotkin*. New York, Liveright, 1975.

Carson, Hampton L.: "Evolutionary biology: its value to society" in *BioScience*, 22 (6): 349-352.

Carter, G. S.: *A hundred years of evolution*. New York, Macmillan, 1957.

Chagnon, Napoleon A., and William Irons, eds.: *Evolutionary biology and human social behavior: an anthropological perspective*. North Scituate, Massachusetts, Duxbury Press, 1979.

Chai, Chen Kang: *Genetic evolution*. Chicago, University of Chicago Press, 1976.

Chaisson, Eric: *Cosmic dawn: the origins of matter and life*. Boston, Little/Brown, 1981.

Chambers, Robert: *Vestiges of the natural history of creation* (1844). New York, Humanities Press, 1969.

Chancellor, John: *Charles Darwin*. New York, Taplinger, 1976.

Chedd, Graham: "Genetic gibberish in the code of life" in *Science 81*, 2, (9): 50-55.

Cheney, Margaret: *Tesla: man out of time*. Englewood Cliffs, New Jersey, Prentice-Hall, 1981.

Childe, V. Gordon: *Social evolution*. New York, Meridian Books, 1963.

Cho, Kah Kyung, and Lynn E. Rose: "Marvin Farber (1901-1980)" in *Philosophy and Phenomenological Research*, 42 (1): 1-4.

Christianson, Gale E.: *This wild abyss: the story of the men who made modern astronomy.* New York, Free Press, 1979. Reviewed by H.James Birx in *The American Rationalist,* 26 (3): 45.

Clark, Gary R., and H. James Birx: "A cosmic visitor" in *Space Dimensions,* 1 (4): 2.

Clark, Robert E. D.: *Darwin: before and after.* Chicago, Moody Press, 1967.

Cloud, Preston: "'Scientific creationism': a new inquisition brewing" in *The Humanist,* 37 (1): 6-15.

Cole, Fay-Cooper: "A witness at the Scopes trail" in *Scientific American,* 200 (1): 120-128, 130, 162.

Cole, Sonia: *Leakey's luck: the life of Louis Seymour Bazett Leakey (1903-1972).* New York, Harcourt Brace Jovanovich, 1975.

Coleman, W.: *Georges Cuvier, zoologist: a study in the history of evolution theory.* Cambridge, Harvard University Press, 1964.

Collin, R.: *Evolution.* New York, Hawthorn Books, 1962.

Colp, Jr., Ralph: "The contacts between Karl Marx and Charles Darwin" in *Journal of the History of Ideas,* 35 (2): 329-338.

Colp, Jr., Ralph: *To be an invalid: the illness of Charles Darwin.* Chicago, University of Chicago Press, 1977.

Colp, Jr., Ralph, and Margaret A. Fay: "Independent scientific discoveries and the 'Darwin-Marx' letter" in *Journal of the History of Ideas,* 40 (3): 479.

Colver, A. Wayne, and Robert D. Stevick: *Composition and research: problems in evolutionary theory.* New York, Bobbs-Merrill, 1963.

Condorcet, Marquis de: *Sketch for a historical picture of the progress of the human mind* (1795). Trans. by June Barraclough. New York, Noonday Press, 1955.

Coonen, Lester P.: "Aristotle's biology" in *BioScience,* 27 (11): 733-738.

Cornford, F. M.: *Principium Sapientiae: the origins of Greek philosophical thought.* W.K.C. Guthrie, ed. New York, Harper Torchbooks, 1965.

Cox, C. B., I. N. Healey, and P. D. Moore. *Biogeography: an ecological and evolutionary approach.* New York, John Wiley & Şons, 1973.

Crick, Francis: *Life itself: its origin and nature.* New York, Simon and Schuster, 1981.

Crick, Francis: *Of molecules and men.* Seattle, University of Washington Press, 1966.

Crick, Francis: "Seeding the universe" in *Science Digest,* 89 (10): 82-84, 115-117.

Daniel, Glyn E.: "The idea of man's antiquity" in *Scientific American,* 201, (5): 167-168, 170, 172-176, 232.

Darlington, C. D.: "A diagram of evolution" in *Nature,* 276: 447-454.

Darlington, C. D.: *Darwin's place in history.* Oxford, Basil Blackwell, 1960.

Darlington, C. D.: *The evolution of man and society.* New York, Clarion, 1971.

Darlington, C. D.: "The origin of Darwinism" in *Scientific American,* 200 (5): 60-66, 204.

Darwin, Charles: *The expression of the emotions in man and animals.* Chicago, Phoenix Books, 1965.

Darwin, Charles: *The illustrated origin of species*. Ed. by Richard E. Leakey. New York, Hill and Wang, 1979.

Darwin, Charles: *The origin of species by means of natural selection, or the preservation of favored races in the struggle for life* (1859) and *The descent of man, and selection in relation to sex* (1871). New York, Modern Library, 1936.

Darwin, Charles: *the red notebook*. Ed. by Sandra Herbert. Ithaca, New York, Cornell University Press, 1980.

Darwin, Charles: *The structure and distribution of coral reefs*. Berkeley, University of California Press, 1962.

Darwin, Charles: *The voyage of the Beagle*. Ed. by Leonard Engel. Garden City, New York, Doubleday, 1962.

Darwin, Charles, and Alfred Russel Wallace: "On the tendency of species to form varieities; and on the perpetuation of varieties and species by natural means of selection" (1859) in *Journal of the Linnaean Society*, 3:45-63.

Darwin, Francis, ed: *Charles Darwin's autobiography*. New York, Collier Books, 1961.

Darwin, Francis, ed.: *The life and letters of Charles Darwin*. 2 vols. New York, Basic Books, 1959.

Dawkins, Richard: *The selfish gene*. New York: Oxford University Press, 1978.

de Beer, Gavin: *Charles Darwin: a scientific biography*. Garden City, New York, Doubleday, 1965.

de Camp, L. Sprague: "The end of the monkey war" in *Scientific American*, 220 (2): 15-21, 132.

Deely, John N., and Raymond J. Nogar: *The problem of evolution: a study of the philosophical repercussions of evolutionary science*. New York, Appleton-Century-Crofts, 1973.

DeGrood, David H.: *Haeckel's theory of the unity of nature*. Boston, Christopher Publishing House, 1965.

Delson, Eric: "Paleoanthropology: pliocene and pleistocene human evolution" in *Paleobiology*, 7 (3): 298-305.

De Vries, Hugo: *Species and varieties: their origin by mutation*. Chicago, Open Court, 1912.

De Vries, Hugo: *The mutation theory* (1909). New York, Kraus, 1969.

Dewey, John: *Experience and nature* (1925).New York, Dover, 1958.

Dewey, John: *The influence of Darwin on philosophy, and other essays in contemporary thought* (1910). Bloomington, Indiana University Press, 1965.

Dick, Steven J.: "The origins of the extraterrestrial life debate and its relation to the scientific revolution" in *Journal of the History of Ideas*, 41 (1): 3-27.

Dickerson, Richard E.: "Chemical evolution and the origin of life" in *Scientific American*, 239 (3): 70-86, 242.

Diderot, Denis: *Interpretation of nature: selected writings* (1754). 2nd ed. Trans. by Jean Stewart and Jonathan Kemp. Ed. and Intro. by Jonathan Kemp. New York, International Publishers, 1963.

Dillon, L. S.: *Evolution: concepts and consequences*. Saint Louis, C. V. Mosby Company, 1973.

Dobzhansky, Theodosius: "Changing man" in *Science*, 155 (3761): 409-414.

Dobzhansky, Theodosius: "Creative evolution" in *Diogenes*, Number 60, Winter 1967, pp. 62-74.

Dobzhansky, Theodosius: *Evolution, genetics, and man*. New York, Wiley & Sons, 1955.

Dobzhansky, Theodosius: *Genetics and the origin of species*. Rev. 3rd. ed. New York, Columbia University Press, 1969.

Dobzhansky, Theodosius: *Genetics of the evolutionary process*. New York, Columbia University Press, 1970.

Dobzhansky, Theodosius: "Individuality, gene recombination and nonrepeatability of evolution" in *Australian Journal of Science*, 23: 71-74.

Dobzhansky, Theodosius: "Man and natural selection" in *American Scientist*, 40 (3): 285-299.

Dobzhansky, Theodosius: *Mankind evolving: the evolution of the human species*. New York, Bantam, 1970.

Dobzhansky, Theodosius: "Nothing in biology makes sense except in the light of evolution" in *The American Biology Teacher*, 35 (3): 125-129.

Dobzhansky, Theodosius: *The biological basis of human freedom*. New York, Columbia University, 1967.

Dobzhansky, Theodosius: *The biology of ultimate concern*. New York, New American Library, 1967.

Dobzhansky, Theodosius, M. K. Heckt, and W. C. Steere, eds.: *Evolutionary biology*. 3 vols. New York, Appleton-Century-Crofts, 1967-1969.

Dodson, Edward O., and Peter Dodson: *Evolution: process and product*. New York, D. Van Nostrand, 1976.

Dose, K., S. W. Fox, G. A. Deborin, and T. E. Pavlovskaya, eds.: *The origin of life and evolutionary biochemistry*. New York, Plenum, 1974.

Dowdeswell, W. H.: *The mechanism of evolution*. 2nd. ed. New York, Harper Torchbooks, 1960.

Drachman, Julian M.: *Studies in the literature of natural science*. New York, Macmillan, 1930.

Duffin, Kathleen E.: "Arthur O. Lovejoy and the emergence of novelty" in *Journal of the History of Ideas*, 41 (2): 267-281.

Eaton, Jr., T. H.: *Evolution*. New York, W. W. Norton, 1970.

Eckhardt, Robert C.: "Introduced plants and animals in the Galapagos Islands" in *BioScience*, 22 (10): 585-590.

Edwords, Frederick: "Why creationism should not be taught as science" in *Creation/Evolution*, 1: 2-23, and 3: 6-36. This excellent article discusses both the legal and educational issues.

Ehrle, Elwood B., and H. James Birx: "Organic evolution: selections from the literature" in *BioScience*, 20 (8, 10, 12, 14, 16, 18, 20, 24) 21 (2, 4).

Ehrlich, Paul, and Anne Ehrlich: *Extinction: the causes and consequences of the dissapparance of species*. New York, Random, House, 1981.

Ehrlich, Paul, and R. W. Holm: *Process of evolution*. New York, McGraw-Hill, 1963.

Ehrlich, Paul, R.W. Holm, and P.H. Raven, eds: *Papers on evolution.* Boston, Little, Brown & Co., 1969.

Eiseley, Loren C.: "Alfred Russel Wallace" in *Scientific American*, 200 (2): 70-82, 84, 172.

Eiseley, Loren C.: "Charles Lyell" in *Scientific American*, 201 (2): 90-106, 168.

Eiseley, Loren C.: *Darwin's century: evolution and the men who discovered it.* Garden City, New York, Anchor Books, 1961.

Eldredge, Niles, and Stephen Jay Gould: "Rates of evolution revisited" in *Paleobiology*, 2 (2): 174-179.

Emlen, J. Merritt: *Ecology: an evolutionary approach.* Reading, Massachusetts, Addison-Wesley, 1973.

Engel, A.E.J.: "Time and the earth" in *American Scientist*, 57 (4): 458-483.

Engels, Frederick: *Dialectics of nature* (1927). Trans. by Clemens Dutt. New York, International Publishers, 1963.

Engels, Frederick: *Herr Eugen Dühring's revolution in science* (1894). 3rd. ed. Trans. by Emile Burns. New York, International Publishers 1970.

Engels, Frederick: *Ludwig Feuerbach and the outcome of classical German philosophy* (1888). New York, International Publishers, 1967.

Engels, Frederick: *The origin of the family, private property, and the state: in the light of the researches of Lewis H. Morgan* (1891). New York, International Publishers, 1964.

Engels, Frederick: *The part played by labour in the transition from ape to man* (1876). Moscow, Foreign Languages, 1953.

Evans, Mary Alice: "Mimicry and the Darwinian heritage" in *Journal of the History of Ideas*, 26 (2): 211-220.

Farber, Marvin: *Basic issues of philosophy: experience, reality, and human values.* New York, Harper Torchbooks, 1968. Reviewed by Roy Wood Sellars in *Philosophy and Phenomenological Research*, 29 (4).

Farber, Marvin: *Naturalism and subjectivism.* Albany, State University of New York Press, 1968.

Farber, Marvin: "On subjectivism and the world-problem" in *Philosophy and Phenomenological Research*, 34 (1): 134-141.

Farber, Marvin: *Phenomenology and existence: toward a philosophy within nature.* New York, Harper Torchbooks, 1967. Reviewed by D.C. Mathur in *Philosophy and Phenomenological Research*, 29 (3): 462-463.

Farber, Marvin: *The aims of phenomenology: the motives, methods, and impact of Husserl's thought.* New York, Harper Torchbooks, 1966. Reviewed by Roy Wood Sellars in *Philosophy and Phenomenological Research*, 29 (1): 125-129.

Farber, Marvin: *The foundation of phenomenology: Edmund Husserl and the quest for a rigorous science of philosophy.* Rev. 3rd. ed. Albany, State University of New York Press, 1967.

Fay, Margaret A.: "Did Marx offer to dedicate 'Capital' to Darwin?: a reassessment of the evidence" in *Journal of the History of Ideas*, 39 (1): 133-146.

Feduccia, Alan: *The age of birds.* Cambridge, Harvard University Press, 1980.

Feuerbach, Ludwig: *Lectures on the essence of religion* (1848). Trans. by Ralph Manheim. New York, Harper & Row, 1967.

Feuerbach, Ludwig: *Principles of the philosophy of the future* (1843). Trans. by Manfred H. Vogel. New York, Bobbs-Merrill, 1966.

Feuerbach, Ludwig: *The essence of Christianity* (1841). Trans. by George Eliot. New York, Harper Torchbooks, 1957.

Fiske, John: *Outlines of cosmic philosophy based on the doctrine of evolution, with criticisms on the positive philosophy.* 4 vols. Intro. by Josiah Royce. Boston, Houghton/Mifflin, 1874.

Fiske, John: *The destiny of man viewed in the light of his origin.* Boston, Houghton/Mifflin, 1884.

Folsome, Clair Edwin: *The origin of life: a warm little pond.* San Francisco, W.H. Freeman, 1979.

Forest, Herman S., and Thomas Morrill: "Biological expansion: a perspective on evolution" in *The Monist,* 48 (2): 291-305.

Fox, S.W., and K. Dose: *Molecular evolution and the origin of life.* New York, W.H. Freeman, 1972.

Frankel, O.H., and Michael E. Soulé: *Conservation and evolution.* New York, Cambridge University Press, 1981.

Freeman, Derek: "The evolutionary theories of Charles Darwin and Herbert Spencer" in *Current Anthropology,* 15 (3): 211-237.

Fromm, Erich: *Marx's concept of man.* New York, Frederick Ungar, 1967. Contains Karl Marx's "Economic and philosophical manuscripts" (1844), trans. by T.B. Bottomore, pp. 90-196.

Gallant, Roy A.: *Charles Darwin: the making of a scientist.* Garden City, New York, Doubleday, 1972.

George, W.B.: *Biologist philosopher: a study of the life and writings of Alfred Russel Wallace.* New York, Abelard-Schuman, 1964.

Ghiselin, Michael T.: "Darwin and evolutionary psychology" in *Science,* 179 (4077): 964-968.

Gillispie, Charles Coulston: *Genesis and geology: a study in the relations of scientific thought, natural theology, and social opinion in Great Britain, 1790-1850.* New York, Harper Torchbooks, 1959.

Ginger, Ray: *Six days or forever?: Tennessee v. John Thomas Scopes.* New York, Signet Books, 1960.

Gish, Duane T.: *Evolution: the fossils say no!* San Diego, Institute for Creation Research, 1973.

Glaessner, M.F.: "Pre-Cambrian fossils" in *Biological Reviews of the Cambridge Philosophical Society,* 37 (4): 467-494.

Glass, Bentley, Owsei Temkin, and William L. Straus, Jr., eds.: *Forerunners of Darwin: 1745-1859.* Baltimore, John Hopkins Press, 1968.

Godfrey, Laurie R.: "Science and evolution in the public eye" in *The Skeptical Inquirer: The Zetetic,* 4 (1): 21-32.

Godfrey, Laurie R.: "The flood of antievolutionism: where is the science in 'scientific creationism'?" in *Natural History,* 90 (6): 4, 6, 9-10.

Goerke, Heinz: *Linnaeus*. Trans. by Denver Lindley. New York, Charles Scribner's Sons, 1973.

Gorman, James: "Creationists vs. evolution" in *Discover*, 2 (5): 32-33.

Gottlieb, L.D.: "The uses of place: Darwin and Melville on the Galapagos" in *BioScience*, 25 (3): 172-175.

Goudge, T.A.: "Pragmatism's contribution to an evolutionary view of mind" in *The Monist*, 57 (2): 133-150.

Goudge, T.A.: *The ascent of life: a philosophical study of the theory of evolution.* Toronto, University of Toronto Press, 1961.

Gould, Stephen Jay: "A visit to Dayton" in *Natural History*, 90 (10); 8, 12, 14, 18, 22.

Gould, Stephen Jay: *Ever since Darwin: reflections in natural history.* New York, W.W. Norton, 1979.

Gould, Stephen Jay: "Evolution as fact and theory" in *Discover*, 2 (5): 34-37, and "Essay: in praise of Charles Darwin" in *Discover*, 3 (2): 20-25.

Gould, Stephen Jay: "Generality and uniqueness in the history of life: an exploration with random models" In *BioScience*, 28 (4): 277-281.

Gould, Stephen Jay: "Is a new and general theory of evolution emerging?" in *Paleobiology* 6 (1): 119-130.

Gould, Stephen Jay: *Ontogeny and phylogeny.* Cambridge, Belknap Press, 1977.

Gould, Stephen Jay: *The mismeasure of man.* New York, W.W. Norton, 1981. Also refer to: James Gorman, "Scientist of the year: Stephen Jay Gould" in *Discover*, 3 (1): 56-58, 60, 62-63.

Gould, Stephen Jay: *The panda's thumb: more reflections in natural history.* New York, W.W. Norton, 1980.

Gould, Stephen Jay: "The promise of paleobiology as a nomothetic, evolutionary discipline" in *Paleobiology*, 6 (1): 96-118.

Gould, Stephen Jay, and Niles Eldredge: "Punctuated equilibria: the tempo and mode of evolution reconsidered" in *Paleobiology*, 3 (2): 115-151.

Grant, Susan T.: "Fabre and Darwin: a study of contrasts" in *BioScience*, 26 (6): 395-398.

Grant, V.: *The origin of adaptations.* New York, Columbia University Press, 1963.

Gray, Asa: *Darwiniana* (1876). Cambridge, Harvard University Press, 1960.

Green, John C.: *Darwin and the modern world view.* Baton Rouge, Louisiana State University Press, 1981.

Grene, Marjorie: "Aristotle and modern biology" in *Journal of the History of Ideas*, 33 (3): 395-424.

Grene, Marjorie: "Changing concepts of Darwinian evolution" in *The Monist*, 64 (2): 195-213.

Griffin, Donald R.: *The question of animal awareness: evolutionary continuity of mental experience.* New York, Rockefeller University Press, 1981.

Gruber, Howard E.: *Darwin on man: a psychological study of scientific creativity.* New York, E.P. Dutton, 1974.

Guest, William, and Frances E. Clayton: "Evolution-science versus creation-science" in *Evolution*, 35 (4): 822.

Gunter, P.A.Y.: "Bergson's theory of matter and modern cosmology" in *Journal of the History of Ideas*, 32 (4): 525-542.

Guthrie, W.K.C.: *The Greek philosophers from Thales to Aristotle*. New York, Harper Torchbooks, 1960.

Haber, F.C.: *Age of the world: Moses to Darwin*. Baltimore, John Hopkins Press, 1959.

Haeckel, Ernst: *Freedom in science and teaching* (1878). London, C. Kegan Paul, 1879.

Haeckel, Ernst: *Last words on evolution: a popular retrospect and summary* (1905).2nd ed. Trans. by Joseph McCabe. New York, Peter Eckler, 1905.

Haeckel, Ernst: *Monism as connecting religion and science: the confession of faith of man of science*. Trans. by J. Gilchrist. London, Adam & Charles Black, 1895.

Haeckel, Ernst: *The evolution of man: a popular scientific study* (1874). 2 vols. Trans. by Joseph McCabe. New York, G.P. Putnam's Sons, 1910.

Haeckel, Ernst: *The history of creation, or the development of the earth and its inhabitants by the action of natural causes: a popular exposition of the doctrine of evolution in general, and of that of Darwin, Goethe, and Lamarck in particular* (1873). 4th ed., 2 vols. Rev. trans. by E. Ray Lankseter. London, Kegan Paul/Trench/Trübner, 1892.

Haeckel, Ernst: *The riddle of the universe at the close of the nineteenth century* (1900). Trans. by Joseph McCabe. New York, Harper & Brothers, 1905.

Haeckel, Ernst: *The wonders of life: a popular study of biological philosophy* (1904). Trans. by Joseph McCabe. New York, Harper & Brothers, 1905.

Haldane, J.B.S.: *The causes of evolution*. Ithaca, New York, Cornell University Press, 1966.

Hamilton, Edith, and Huntington Cairns, eds: *The collected dialogues of Plato including the letters*. New York, Pantheon Books, 1961.

Hamilton, T.H.: *Process and pattern in evolution*. New York, Macmillan, 1967.

Hammond, Allen, and Lynn Margulis: "Creationism as science: farewell to Newton, Einstein, Darwin . . ." in *Science 81*, 2 (10): 55-57.

Hardin, Garrett: *Nature and man's fate*. New York, Rinehart & Company, 1959.

Harnack, William J.: "Carl Sagan: cosmic evolution vs. the creationist myth" in *The Humanist*, 41 (4): 5-11.

Harris, Marvin: *Cultural materialism: the struggle for a science of culture*. New York, Random House, 1979.

Harris, Marvin: *Culture, people, nature: an introduction to general anthropology*. 3rd ed. New York, Harper & Row, 1980.

Harris, Marvin: *The rise of anthropological theory: a history of theories of culture*. New York, Thomas Y. Crowell, 1968.

Harrison, James: "Erasmus Darwin's view of evolution" in *Journal of the History of Ideas*, 32 (2): 247-264.

Hartmann, William K: *Astronomy: the cosmic journey*. Belmont, California, Wasdsworth, 1978.

Hartshorne, Charles: *A natural theology for our time*. LaSalle, Illinois, Open Court, 1973.

Hartshorne, Charles, and Creighton Peden: *Whitehead's view of reality.* New York, Pilgrim Press, 1981.

Heffernan, William C.: "The singularity of our inhabited world: William Whewell and A. R. Wallace in dissent" in *Journal of the History of Ideas,* 39 (1): 81-100.

Henfrey, Norman, ed.: *Selected critical writings of George Santayana.* 2 vols. Cambridge, Cambridge University Press, 1968.

Henig, Robin Marantz: "Evolution called a 'religion,' creationism defended as a 'science'" in *BioScience,* 29 (9): 513-516.

Henig, Robin Marantz: "Exobiologists continue to search for life on other planets" in *BioScience,* 30 (1): 9-12.

Heppenheimer, T.A.: *Colonies in space.* New York, Warner Books, 1980.

Heppenheimer, T.A.: *Toward distant suns.* Harrisburg, Pennsylvania, Stackpole Books, 1979.

Higgins, Paul. J.: "The Galapagos iguanas: models of reptilian differentiation" in *BioScience,* 28 (8): 512-515.

Himmelfarb, Gertrude: *Darwin and the Darwinian revolution.* New York, W.W. Norton, 1968.

Hofstadter, Richard: *Social Darwinism in American thought.* Rev. ed. Boston, Beacon Press, 1971.

Holden, Constance: "The politics of paleoanthropology" in *Science,* 213 (4509): 737-740.

Holt, Niles R.: "Ernst Haeckel's monistic religion" in *Journal of the History of Ideas,* 32 (2): 265-280.

Horowitz, N.H.: "The search for extraterrestrial life" in *Science,* 151 (3712): 789-792.

Hoyle, Fred, and Chandra Wickramasinghe: *Diseases from space.* New York, Harper & Row, 1979.

Hoyle, Fred, and Chandra Wickramasinghe: *Lifecloud: the origin of life in the universe.* New York, Harper & Row, 1978.

Hull, D.L.: *Darwin and his critics.* Cambridge, Harvard University Press, 1973.

Huntley, William B.: "David Hume and Charles Darwin" in *Journal of the History of Ideas,"* 33 (3): 457-470.

Huxley, Julian S.: *Evolution in action.* New York, Mentor Books, 1957.

Huxley, Julian S.: *Evolution: the modern synthesis.* New York, John Wiley & Sons, 1964.

Huxley, Julian S.: *Man in the modern world.* New York, Mentor Books, 1948.

Huxley, Julian S.: *Religion without revelation.* New York, Mentor Books, 1957.

Huxley, Julian S., et al.: *A book that shook the world: anniversary essays on Charles Darwin's 'Origin of Species.'* Pittsburgh, University of Pittsburgh Press, 1961.

Huxley, Julian S., and H.B.D. Kettlewell: *Charles Darwin and his world.* New York, Viking Press, 1966.

Huxley, Thomas H.: *Darwiniana: essays.* New York, D. Appleton, 1895.

Huxley, Thomas H.: *Discourses biological and geological: essays.* New York, D. Appleton, 1894.

Huxley, Thomas H.: *Evidence as to man's place in nature* (1863). Ann Arbor, University of Michigan Press, 1959.

Huxley, Thomas H.: *Evolution and ethics: and other essays.* New York, D. Appleton, 1894.

Huxley, Thomas H.: *On the origin of species or, the causes of the phenomena of organic nature (1863).* Ann Arbor, University of Michigan Press, 1968.

Irvine, William: *Apes, angels, and victorians: the story of Darwin, Huxley, and evolution.* New York, McGraw-Hill, 1972.

Irvine, William: *Thomas Henry Huxley.* London, Longmans/Green, 1960.

James, William: *A pluralistic universe.* New York, Longmans/Green, 1925.

Jantsch, Erich: *Design for evolution: self-organization and planning in the life of human systems.* New York, George Braziller, 1975.

Jantsch, Erich: *The self-organizing universe: scientific and human implications of the emerging paradigm of evolution.* Elmsford, New York, Pergamon Press, 1979.

Johanson, Donald C., and T. D. White: "A systematic assessment of early African hominids" in *Science,* 202 (4378): 321-330.

Johanson, Donald C., and Maitland A. Edey: *Lucy: the beginnings of humankind.* New York, Simon and Schuster, 1981.

Jones, H.S.: *Life on other worlds.* New York, The New American Library, 1960.

Kant, Immanuel: *Universal natural history and theory of the heavens* (1755). Trans. by W. Hastie. Intro by Milton K. Munitz. Ann Arbor, University of Michigan Press, 1969.

Kardiner, Abram, and Edward Preble: *They studied man.* New York, Mentor Books, 1963.

Karp, Walter: *Charles Darwin and the origin of species.* New York, Harper & Row, 1968.

Keosian, J.: *The origin of life.* 2nd ed. New York, Reinhold, 1968.

Kerkut, G.A.: *Implications of evolution.* New York, Pergamon Press, 1968.

Kern, Edward P.H., and Donna Haupt: "Battle of the bones: a fresh dispute over the origins of man" in *Life,* 4 (12): 109-116, 118, 120.

Kettlewell, H.B.D.: "Darwin's missing evidence" in *Scientific American,* 200(3): 48-53, 184.

Kimura, Motoo: "The neutral theory of molecular evolution" in *Scientific American,* 241 (5): 98-100, 102, 104, 108, 110, 114, 117, 120, 124, 126, 206.

Kirk, G.S., and J.E. Raven: *The presocratic philosophers: a critical history with a selection of texts.* Cambridge, Cambridge University Press, 1966.

Kitts, David B.: "Paleontology and evolutionary theory" in *Evolution, 28 (3):* 458-472.

Kneževic, Božidar: *History, the anatomy of time: the final phase of sunlight.* Trans. by George V. Tomashevich in collaboration with Sherwood A. Wakeman. Preface by W. Warren Wagar. New York, Philosophical Library, 1980. Reviewed by H. James Birx in *The American Rationalist,* 25 (4): 69-70, and in *Philosophy and Phenomenological Research,* 42 (1): 137-139.

Komarov, V.L.: "Marx and Engels on biology" in *Marxism in modern thought,* pp. 190-234. Ed. by N.S. Bukarin. New York, Harcourt, 1935.

Kosin, I.L.: "Soviet genetics: biography, history, commentary" in *BioScience*, 24 (10): 583-589.

Kraus, John: *Our cosmic universe.* Foreword by Arno Penzias. Powell, Ohio, Cygnus-Quasar Books, 1980.

Kraus, John, and Gerard K. O'Neill (interview): "Space colonization and seti" in *Cosmic Search*, 1 (2): 16-23.

Kropotkin, Petr: *Mutual aid: a factor of evolution.* Boston, Extending Horizons Books, 1914. Also includes "The struggle for existence in human society" (1888) by Thomas H. Huxley, pp. 329-341.

Kuhn, Thomas S.: *The structure of scientific revolutions.* 2nd enlarged ed. Chicago, University Press, 1974.

Kurtz, Paul: *A secular humanist declaration.* Buffalo, New York, Prometheus Books, 1980. See "Evolution" pp. 21-22.

Lack, David: *Darwin's finches: an essay on the general biological theory on evolution.* New York, Harper Torchbooks, 1961.

LaMettrie, Julien Offray de: *Man à machine.* (1747). Trans. by G.C. Bussey. LaSalle, Illinois, Open Court, 1961.

Langdon-Davies, John: *Man and his universe.* London, Harper and Brothers, 1930.

Larson, James L.: "Goethe and Linneaus" in *Journal of the History of Ideas*, 28 (4): 590-596.

Leakey, Richard E.: *The making of mankind.* New York, E.P. Dutton, 1981.

Leakey, Richard E., and Roger Lewin: *Origins: what new discoveries reveal about the emergence of our species and its possible future.* New York, E.P. Dutton, 1977.

Leakey, Richard E., and Roger Lewin: *People of the lake: mankind and its beginnings.* Garden City, New York, Anchor Press, 1978.

Lear, John: *Recombinant DNA: the untold story.* New York, Crown, 1978.

Lederberg, J.: "Exobiology: approaches to life beyond the earth" in *Science*, 131 (3412): 1503-1508.

Lehrman, Nathaniel S.: "Human sociobiology: Wilson's fallacy" in *The Humanist*, 41 (4): 39-42, 58.

LeMahieu, D.L.: "Malthus and the theology of scarcity" in *Journal of the History of Ideas*, 40 (3): 467-474.

Lenin, V.I.: *Materialism and empirio-criticism: critical comments on a reactionary philosophy* (1920). New York, International Publishers, 1927.

Leroy, Pierre, ed.: *Letters from my friend Teilhard de Chardin: 1948-1955.* New York, Paulist Press, 1980. Trans. with a preface by Mary Lukas.

Levin, Samuel M.: "Malthus and the idea of progress" in *Journal of the History of Ideas*, 27 (1): 92-108.

Lewin, Roger: "Do jumping genes make evolutionary leaps?" in *Science*, 213 (4508): 634-636.

Lewin, Roger: "Lamarck will not lie down" in *Science*, 213 (4505): 316-321.

Lewin, Roger, and Sally Anne Thompson: *Darwin's forgotten world.* Los Angeles, Reed Books, 1978.

Lewontin, Richard C.: "Adaptation" in *Scientific American*, 239 (3): 212-218, 220, 222, 225, 228, 230, 242.

Lewontin, Richard C.: "Evolution/creation debate: a time for truth" in *BioScience*, 31(8):559.

Lewontin, Richard C., ed.: *Population biology and evolution*. New York, Syracuse University Press, 1968.

Lewontin, Richard C.: *The genetic basis of evolutionary change*. New York, Columbia University Press, 1974.

Lieberman, Janet J.: "Malthus: his life and work" in *The American Biology Teacher*, 35(3):130-131.

Loewenberg, B.J.: *Darwin, Wallace, and the theory of natural selection*. Cambridge, Arlington Books, 1959.

Lonegran, Bernard J.F.: *Insight: a study of human understanding*. London, Longmans/Green, 1964.

Lorenz, Konrad: *Evolution and modification of behavior*. Chicago, Chicago University Press, 1965.

Lorenz, Konrad: *On aggression*. Trans. by Marjorie Kerr Wilson. New York, Harcourt, Brace & World, 1966.

Lovejoy, Arthur O.: *The great chain of being: a study of the history of an idea*. New York, Harper Torchbooks, 1965.

Lovejoy, C. Owen: "The origin of man" in *Science*, 211(4480):341-350.

Lucretius: *On the nature of things*. Ed. by Wendell Clausen. Trans. by H.A.J. Munro. New York, Washington Square Press, 1965.

Lukas, Mary: "Teilhard and the Piltdown 'hoax'" in *America*, 144(20):424-427.

Lumsden, Charles J., and Edward O. Wilson: *Genes, mind, and culture: the coevolutionary process*. Cambridge, Harvard University Press, 1981. See "The channeling of social behavior" (review) by C. Robert Cloninger and Shozo Yokoyama in *Science*, 213(4509):749-751.

Lyell, Charles: *Principles of geology; or, the modern changes of the earth and its inhabitants considered as illustrations of geology* (1830-1833). Rev. 9th ed. New York, D. Appleton, 1865.

Lyell, Charles: *The geological evidences of the antiquity of man with remarks on theories of the origin of species by variation*. Rev. 3rd ed. London, John Murray, 1863.

McArthur, Robert H., and Edward O. Wilson: *The theory of island biogeography*. Princeton, Princeton University Press, 1967.

Macbeth, Norman: *Darwin retried: an appeal to reason*. New York, Delta, 1973.

MacLean, Paul D.: "A mind of three minds: evolution of the human brain" in *The Science Teacher*, 45(4):31-39.

Madden, Edward H., ed.: *The philosophical writings of Chancey Wright: representative selections*. New York, Liberal Arts Press, 1958.

Magee, Bryan: *Karl Popper*. New York, Viking Press, 1973. See pp. 40, 51-54, 57, 50-51, 64-65, 92.

Malthus, Thomas Robert: *An essay on the principle of population*. Ed. with a preface and introduction by Philip Appleman. New York, W.W. Norton, 1976.

Mandelbaum, Maurice: "A note on Thomas S. Kuhn's 'The Structure of Scientific Revolutions'" in *The Monist*, 60(4):445-452.

Marquand, Josephine: *Life: its nature, origins and distribution*. New York, W.W. Norton, 1971.

Mathur, D.C.: *Naturalistic philosophies of experiences: studies in James, Dewey and Farber against the background of Husserl's phenomenology*. St. Louis, Warren H. Green, 1971.

Matthews, L. Harrison: "Piltdown man: the missing links" in *New Scientist*, 90:280-282, 376, 450, 515-516, 578-579, 647-648, 710-711, 785, 861-862, and 91: 26-28.

May, Robert M.: "The evolution of ecological systems" in *Scientific American*, 239(3):160-164, 166, 168-172, 175, 242.

Mayer, William V.: "Evolution: yesterday, today, tomorrow" in *The Humanist*, 37(1): 16-22.

Mayr, Ernst: *Animal species and evolution*. Cambridge, Harvard University Press, 1963.

Mayr, Ernst: "Evolution" in *Scientific American*, 239(3):46-55, 242.

Mayr, Ernst, ed.: *The species problem*. Washington, D.C., American Association for the Advancement of Science (50), 1957.

Mayr, Ernst, and William B. Provine, eds.: *The evolutionary synthesis: perspectives on the unifaction of biology*. Cambridge, Harvard University Press, 1980.

McKeon, Richard, ed.: *The basic works of Aristotle*. New York, Random House, 1966.

Mead, Margaret: *Continuities in cultural evolution*. New Haven, Yale University Press, 1964.

Meggers, Betty J., ed.: *Evolution and anthropology: a centennial appraisal*. Washington, D.C., The Anthropological Society of Washington, 1959.

Mendel, Gregor Johann: *Experiments in plant hybridization* (1866). Foreword by Paul C. Mangelsdorf. Cambridge, Harvard University Press, 1967.

Metraux, Rhoda: "Margaret Mead: a biographical sketch" in *American Anthropologist*, 82(2):262-269. This issue is devoted to the memory of Margaret Mead (1901-1978).

Mikulak, Maxim W.: "Darwinism, Soviet genetics, and Marxism-Leninism" in *Journal of the History of Ideas*, 31(3):359-376.

Milhauser, M.: *Just before Darwin: Robert Chambers and the vestiges*. Middletown, Wesleyan University Press, 1959.

Miller, Martin A.: *Kropotkin*. Chicago, University of Chicago Press, 1976.

Miller, S.L., and H.C. Urey: "Organic compound synthesis on the primitive earth" in *Science*, 130(3370):245-251.

Miller, Stanley L., and Leslie E. Orgel: *The origins of life on the earth*. Englewood Cliffs, New Jersey, Prentice-Hall, 1974.

Milne, David H.: "How to debate with creationists—and 'win'" in *The American Biology Teacher*, 43(5):235-245, 266.

Montagu, Ashley: *Darwin, competition and cooperation*. New York, Henry Schuman, 1952.

Moody, Paul Amos: *Introduction to evolution.* 3rd ed. New York, Harper & Row, 1970.

Moog, F.: "Alfred Russel Wallace: evolution's forgotten man" in *The American Biology Teacher*, 22(7):414-418.

Moore, John A.: "Kuhn's 'The Structure of Scientific Revolutions' revisited" in *The American Biology Teacher*, 42(5):298-304.

Moore, P., and F. Jackson: *Life in the universe*, New York, Norton, 1962.

Moore, R.: *Charles Darwin.* New York, Alfred A. Knopf, 1955.

Moore, Tui De Roy: *Galapagos: islands lost in time.* New York, Viking Press, 1980.

Moorehead, Alan: *Darwin and the Beagle.* New York, Harper & Row, 1969.

Moorhead, P.S., and M.M. Kaplan, eds.: *Mathematical challenges to the Darwinian interpretation of evolution.* Philadelphia, Wistar Institute Press, 1967.

Morain, Lloyd L.: "Avicenna: Asian humanist forerunner" in *The Humanist*, 41(2):27-34.

Morgan, Lewis Henry: *Ancient society: or, researches in the line of human progress from savagery through barbarism to civilization* (1877). New York, Meridian Books, 1963.

Morrison, Philip, John Billingham, and John Wolfe, eds.: *The search for extraterrestrial intelligence.* New York, Dover, 1979.

Murphee, I.: "The evolutionary anthropologists: the progress of mankind" in *Proceedings of the American Philosophical Society* (1961), 105:265-300.

Nitecki, Matthew H., ed.: *Biotic crises in ecological and evolutionary time.* New York, Academic Press, 1981.

O'Brien, S.J., Elmer, ed.: *The essential Plotinus.* New York, Mentor Books, 1964.

Olby, R.C.: *Charles Darwin.* London, Oxford University Press, 1967.

Olson, E.C.: "Dialectics in evolutionary studies" in *Evolution*, 22(2):426-436.

Olson, E.C.: *The evolution of life.* New York, Mentor Books, 1966.

O'Neill, Gerard K.: "Habitats in space" in *The Science Teacher*, 44(6):22-26.

O'Neill, Gerard K.: *The high frontier: human colonies in space.* New York, Bantam, 1978.

O'Neill, Gerard K.: *2081: a hopeful view of the human future.* New York, Simon and Schuster, 1981.

O'Neill, John J.: *You and the universe: what science reveals.* New York, Ives Washburn, Inc., 1946. Concerning Nikola Tesla, see pp. 145-149.

Oparin, A.I.: *Genesis and evolutionary development of life.* Trans. by Eleanore Maass. New York, Academic Press, 1968.

Oparin, A.I.: *Life: its nature, origin and development.* New York, Academic Press, 1962.

Oparin. A.I.: *The chemical origin of life.* Springfield, Illinois, Charles C Thomas, 1964.

Oparin, A.I.: *The origin of life.* 2nd ed. Trans. by Sergius Morgulis. New York, Dover, 1965.

Opler, Morris: "Cause, process, and dynamics in the evolution of E.B. Tylor" in *Southwestern Journal of Anthropology*, 20:123-145.

Opler, Morris: "Integration, evolution, and Morgan" in *Current Anthropology*, 3:478-479.

Orgel, L.E.: *The origins of life: molecules and natural selection*. New York, John Wiley & Sons, 1973.

Osborn, H.F.: *From the Greeks to Darwin*. 2nd ed. New York, Scribner's, 1929.

Overden, M.W.: *Life in the universe: a scientific discussion*. New York, Doubleday, 1962.

Paley, W.: *Natural theology* (1802). Indianapolis, Bobbs-Merrill, 1963.

Paterson, Antionette Mann: *Francis Bacon and socialized science*. Springfield, Illinois, Charles C Thomas, 1973.

Paterson, Antionette Mann: *The infinite worlds of Giordano Bruno*. Springield, Illinois, Charles C Thomas, 1970.

Patterson, Colin: *Evolution*. London, British Museum (Natural History), 1978.

Persons, Stow, ed.: *Evolutionary thought in America*. New York, Archon Books, 1968.

Pfeiffer, John E.: *The emergence of man*. 2nd ed. New York, Harper & Row, 1972.

Phillips, D.C.: "Organicism in the late nineteenth and early twentieth centuries" in *Journal of the History of Ideas*, 31 (3): 413-432.

Pianka, E.R.: *Evolutionary ecology*. New York, Harper & Row, 1974.

Pitt-Rivers, A.L.F.: *The evolution of culture and other essays*. Ed. by J.L. Myers. Oxford, Clarendon Press, 1906.

Plaine, H.L., ed.: *Darwin, Marx, and Wagner*. Columbus, Ohio State University Press, 1962.

Platt, John: "The acceleration of evolution" in *The Futurist*, February 1981, 14-23.

Ponnamperuma, Cryil: "Chemical evolution and the origin of life" in *Nature*, 201 (4917): 337-340.

Ponnamperuma, Cryil: *The origins of life*. New York, E.P. Dutton, 1972.

Popper, Karl R.: *Conjectures and refutations: the growth of scientific knowledge*. New York, Harper Torchbooks, 1968. See pp. 38, 323, 340.

Popper, Karl R.: "Natural selection and the emergence of mind" in *Dialectica, International Review of Philosophy of Knowledge*, 32 (3-4): 339-355.

Popper, Karl R.: *Objective knowledge: an evolutionary approach*. Oxford, Clarendon Press, 1972. See chapter seven, "Evolution and the tree of knowledge," pp. 256-280.

Prenant, M.: *Biology and Marxism*. Trans. by C.D. Greaves. London, Lawrence and Wishart, 1943.

Primer, Irwin: "Erasmus Darwin's 'Temple of Nature': progress, evolution, and the Eleusiniam mysteries" in *Journal of the History of Ideas*, 25 (1): 58-76.

Pringle, J.W.S., ed.: *Essays on philosophical evolution*. New York, Macmillan, 1965.

Racle, Fred A.: *Introduction to evolution*. Englewood Cliffs, New Jersey, Prentice-Hall, 1979.

Rahner, Karl: *Hominization: the evolutionary origin of man as a theological problem*. New York, Herder and Herder, 1965.

Ralling, Christopher, ed.: *The voyage of Charles Darwin*. New York, Mayflower Books, 1979.

Randall, Jr., John Herman: "The changing impact of Darwin on philosophy" in *Journal of the History of Ideas*, 22 (4): 435-462.

Raychaudhuri, A.K.: *Theoretical cosmology*. Oxford, Clarendon Press, 1979.

Reed, Edward S.: "Darwin's evolutionary philosophy: the laws of change" in *Acta Biotheoretica*, 27, (3§4): 201-235.

Reed, Edward S.: "The lawfulness of natural selection" in *The American Naturalist*, 118 (1): 61-71.

Reed, S.C.: "The evolution of human intelligence: some reasons why it should be a continuing process" in *American Scientist*, 53 (3): 317-326.

Reese, Ronald Lane, Steven M. Evertt, and Edwin D. Craun: "The chronology of Archbishop James Ussher" in *Sky and Telescope*, 62 (5): 404-405.

Rensberger, Boyce: "Tinkering with life" in *Science 81*, 2 (9): 44-49.

Rensch, Bernhard: *Evolution above the species level*. New York, Columbia University Press, 1970.

Resek, C.: *Lewis Henry Morgan: American scholar*. Chicago, University of Chicago Press, 1960.

Richter, Jean Paul, ed.: *The notebooks of Leonardo da Vinci*. 2 vols. New York, Dover, 1970. See volume two, chapters 14, 15, and 16.

Ricklefs, Robert E.: "Phyletic gradualism vs. punctuated equilibrium: applicability of neontological data" in *Paleobiology*, 6 (3): 271-275.

Riddle, Oscar: *The unleashing of evolutionary thought*. New York, Vantage Press, 1954.

Ridley, Mark: "Who doubts evolution?" in *New Scientist*, 90 (1259): 830-832.

Riepe, Dale, ed.: *Phenomenology and natural existence: essays in honor of Marvin Farber*. Albany, State University of New York Press, 1973.

Rightmire, G. Philip: "Patterns in the evolution of Homo erectus" in *Paleobiology*, 7 (2): 241-246.

Rogers, James Allen: "Darwinism and social Darwinism" in *Journal of the History of Ideas*, 33 (2): 265-280.

Ross, H.H.: *A synthesis of evolutionary theory*. Englewood Cliffs, New Jersey, Prentice-Hall, 1962.

Ross, H.H.: *Understanding evolution*. Englewood Cliffs, New Jersey, Prentice-Hall, 1966.

Rostand, Jean: *Humanly possible: a biologist's notes on the future of mankind*. Trans. by Lowell Bair. New York, Saturday Review Press, 1973.

Ruse, Michael: "Charles Darwin and artificial selection" in *Journal of the History of Ideas*, 36 (2): 339-350.

Ruse, Michael: "Darwin's theory: an exercise in science" in *New Scientist*, 90 (1259): 828-830.

Ruse, Michael: *The Darwinian revolution: science red in tooth and claw.* Chicago, The University of Chicago Press, 1979.

Ryan, B.: *The evolution of man: some theological, philosophical and scientific considerations.* Westminister, The Newman Press, 1965.

Sagan, Carl: *Broca's brain: reflections on the romance of science.* New York, Ballantine, 1980.

Sagan, Carl: *Cosmos.* New York, Random House, 1980. Reviewed by H. James Birx in *The American Rationalist,* 25 (6): 102.

Sagan, Carl: "Is the early evolution of life related to the development of the earth's core?" in *Nature,* 206:448.

Sagan, Carl: *The cosmic connection: an extraterrestrial perspective.* New York, Anchor Books, 1980.

Sagan, Carl: *The dragons of Eden: speculations on the evolution of human intelligence.* New York, Ballantine, 1978.

Sagan, Carl: "The other world that beckons" in *The New Republic,* September 1978, 11-15.

Sagan, Carl: "The quest for extraterrestrial intelligence" in *Cosmic Search,* 1 (2): 2-8.

Sagan, Carl, and Frank Drake: "The search for extraterrestrial intelligence" in *Scientific American,* 232 (5): 80-89, 122.

Sagan, Carl, and I.S. Shklovskii: *Intelligent life in the universe.* San Francisco, Holden-Day, 1966.

Sahlins, Marshall D., and Elman R. Service, eds.: *Evolution and culture.* Ann Arbor, University of Michigan Press, 1960.

Sanford, Jr., William F.: "Dana and Darwinism" in *Journal of the History of Ideas,* 26 (4): 531-546.

Savage, Jay M.: *Evolution.* 2nd ed. New York, Holt, Rinehart and Winston, 1969.

Scheer, Bradley T.: "Evolution: help for the confused" in *BioScience,* 29 (4): 238-241.

Schoenwald, Richard L., ed.: *Nineteenth-century thought: the discovery of change.* Englewood Cliffs, New Jersey, Prentice Hall, 1965.

Schopenhauer, Arthur: *The world as will and representation.* 2 vols. Trans. by E.F.J. Payne. New York, Dover, 1966.

Schopf, J. William: "The evolution of the earliest cells" in *Scientific American,* 239 (3): 110-112, 114, 116-120, 126, 128-134, 137-138, 242.

Schopf, Thomas J.M.: "Evolving paleontological views on deterministic and stochastic approaches" in Paleobiology, 5 (3): 337-352.

Schubert-Soldern, R.: *Mechanism and vitalism: philosophical aspects of biology.* Notre Dame, University of Notre Dame Press, 1962.

Schwab, Mark F.: "Man's evolution: a conversation with Dr. C. Owen Lovejoy" in *Zoolog,* 5 (4): 3-4.

Schweber, S.S.: "The genesis of natural selection—1838: some further insights" in *BioScience,* 28 (5): 321-326.

Sears, Paul B.: *Charles Darwin: the naturalist as a cultural force.* New York, Charles Scribner's Sons, 1950.

Seeds, Michael A.: *Horizons: exploring the universe*. Belmont, California, Wadsworth, 1981.

Sellars, Roy Wood, ed.: *Philosophy for the future: the quest of modern materialism*. New York, Macmillan, 1949.

Service, Elman R.: *Primitive social organization: an evolutionary perspective*. 2nd ed. New York, Random House, 1971.

Service, Elman R.: *Profiles in ethnology*. Rev. ed. New York, Harper & Row, 1971.

Sibley, Jack R., and Pete A.Y. Gunter, eds.: *Process philosophy: basic writings*. Washington, D.C., University Press of America, 1978.

Silk, Joseph: *The Big Bang: the creation and evolution of the universe*. San Francisco, W.H. Freeman, 1980.

Simon, Walter M.: "Herbert Spencer and the 'Social Organism'" in *Journal of the History of Ideas*, 21 (2): 294-299.

Simpson, George Gaylord: *Life of the past*. New Haven, Yale University Press, 1953.

Simpson, George Gaylord: *Tempo and mode in evolution*. New York, Hafner, 1965.

Simpson, George Gaylord: *The geography of evolution*. Philadelphia, Chilton, 1965.

Simpson, George Gaylord: *The major features of evolution*. New York, Columbia University Press, 1953.

Simpson, George Gaylord: *The meaning of evolution*. Rev. ed. New York, Mentor Books, 1951.

Simpson, George Gaylord: *This view of life: the world of an evolutionist*. New York, Harcourt, Brace & World, 1964.

Singh, Bhagwan B.: *The self and the world in the philosophy of Josiah Royce*. Springfield, Illinois, Charles C Thomas, 1973.

Skinner, B.F.: *Beyond freedom and dignity*. New York, Alfred A. Knopf, 1971.

Skinner, B.F.: "Selection by consequences" in *Science*, 213 (4507): 501-504.

Skinner, Brian J., ed.: *Earth's history, structure and materials*. Los Altos, California, William Kaufman, 1981.

Skinner, Brian J., ed.: *Paleontology and paleoenvironments*. Los Altos, California, William Kaufman, 1981.

Skow, John: "Creationism as social movement: the genesis of equal time" in *Science 81*, 2 (10): 54, 57-60. Also refer to: Roger Lewin, "Creationism on the defensive in Arkansas" in *Science*, 215 (4528): 33-34.

Smith, John Maynard: "The evolution of behavior" in *Scientific American*, 239 (3): 176-178, 180, 182, 184-186, 189-192, 242.

Soleri, Paolo: *Arcology: the city in the image of man*. Cambridge, MIT Press, 1973.

Soleri, Paolo: *Fragments: the tiger paradigm-paradox*. San Francisco, Harper & Row, 1981.

Soleri, Paolo: *Sketchbooks*. Cambridge, MIT Press, 1971.

Soleri, Paolo: *The bridge between matter and spirit is matter becoming spirit*. Garden City, New York, Anchor Books, 1973.

Sorokin, Pitirim A.: *Modern historical and social philosophies.* New York, Dover, 1963.

Spencer, Herbert: *First principles* (1862). New York, De Witt Revolving Fund, 1958.

Spencer, Herbert: *Synthetic philosophy.* 10 vols. New York, D. Appleton, 1862-1893.

Stansfield, William D.: *The science of evolution.* New York, Macmillan, 1977.

Stebbins, G. Ledyard: *Processes of organic evolution.* Englewood Cliffs, New Jersey, Prentice-Hall, 1966.

Stebbins, G. Ledyard, and Francisco J. Ayala: "Is a new evolutionary synthesis necessary?" in *Science,* 213 (4511): 967-971.

Stern, B.J.: *Lewis Henry Morgan.* Chicago, University of Chicago Press, 1931.

Stern, Curt: *Principles of human genetics.* 3rd ed. San Francisco, Freeman, 1973.

Steward, Julian H.: "Cultural evolution" in *Scientific American* 194: 69-80.

Steward, Julian H.: *Theory of culture change: the methodology of multilinear evolution.* Urbana, University of Illinois Press, 1955.

Stirton, R.A.: *Time, life, and man: the fossil record.* New York, John Wiley & Sons, 1967.

Strickberger, Monroe W.: "Evolution and religion" in *BioScience,* 23 (7): 417-421.

Sullivan, Walter: *Black holes: the edge of space, the end of time.* New York, Warner Books, 1979.

Sullivan, Walter: *We are not alone: the search for intelligent life on other worlds.* Rev. ed. New York, McGraw-Hill, 1966.

Swezey, Kenneth M.: "Nikola Tesla" in *Science,* 127 (3307): 1147-1159.

Tax, Sol, ed.: *Evolution after Darwin.* 3 vols. Chicago, University of Chicago Press, 1960.

Teilhard de Chardin, Pierre: *Man's place in nature: the human zoological group.* New York, Harper & Row, 1966.

Teilhard de Chardin, Pierre: *The divine milieu.* New York, Harper Torchbooks, 1968.

Teilhard de Chardin, Pierre: *The phenomenon of man.* 2nd ed. New York, Harper Torchbooks, 1965.

Terborgh, John: "Preservation of natural diversity: the problem of extinction prone species" in *BioScience,* 24 (12): 715-722.

Thornton, Ian: *Darwin's islands: a natural history of the Galapagos.* New York, Natural History Press, 1971.

Tomashevich, George V.: "Božidar Knežević: a Yugoslav philosopher of history" in *The Slavic and East European Review,* 35 (85): 445-461.

Tomashevich, George V.: "Reflections on science and religion" in *Free Inquiry,* 1 (3): 34-36.

Tomovich, Vladislav A., ed.: *Definitions in sociology: convergence, conflict and alternative vocabularies.* St. Catherines, Ontario, Canada, Diliton Publications, Inc., 1979.

Toulmin, Stephen: "Historical inference in science: geology as a model for cosmology" in *The Monist,* 47 (1): 142-158.

Tylor, Edward Burnett: *Anthropology: an introduction to the study of man and civilization (1881)*. Ann Arbor, University of Michigan Press, 1960.

Tylor, Edward Burnett: *Primitive culture: researches into the development of mythology, philosophy, religion, language, art and custom* (1871). 2 vols. New York, Harper Torchbooks, 1958.

Tylor, Edward Burnett: *Researches into the early history of mankind and the development of civilization* (1865). Chicago, Phoenix Books, 1964.

Urey, H.C., and J.S. Lewis, Jr.: "Organic matter in carbonaceous chondrites" in *Science*, 152 (3718): 102-104.

Valentine, James W.: "The evolution of multicellular plants and animals" in *Scientific American*, 239 (3): 140-146, 148-149, 152-153, 156-158, 242.

Van Melsen, Andrew G.: *Evolution and philosophy*. Pittsburgh, Duquesne University Press, 1965.

Van Valen, L.: "Treeshrews, primates, and fossils" in *Evolution*, 19 (2): 137-151.

Van Valen, L., and R.E. Sloan: "The earliest primates" in *Science*, 150 (3697): 743-745.

Verschuur, Gerrit L.: "Is the cosmos alive?" in *Science Digest*, 80 (6): 31.

Volpe, E. Peter: *Understanding evolution*. Dubuque, Iowa, Wm. C. Brown, 1967.

Vorzimmer, Peter: "Darwin, Malthus, and the theory of natural selection" in *Journal of the History of Ideas*, 30 (4): 527-542.

Verdeveld, Gene, and Ruth Vredeveld: "The Galapagos: a laboratory for evolution" in *The American Biology Teacher*, 43 (4): 201-204.

Walker, Alan, and Richard E.F. Leakey: "The hominids of East Turkana" in *Scientific American*, 239 (2): 54-66, 148.

Wallace, Alfred Russel: *Contributions to the theory of natural selection: a series of essays*. New York, Macmillan, 1870.

Wallace, Alfred Russel: *Darwinism: an exposition on the theory of natural selection*. London, Macmillan, 1899.

Wallace, Alfred Russel: *Island life: or, the phenomena and causes of insular faunas and floras including a revision and attempted solution of the problem of geological climates*. New York, Harper & Brothers, 1881.

Wallace, Alfred Russel: *The world of life: a manifestation of creative power, directed mind and ultimate purpose* (1910). New York, Moffat/Yard, 1916.

Wartofsky, Marx W.: *Conceptual foundations of scientific thought: an introduction to the philosophy of science*. New York, Macmillan, 1968.

Wartofsky, Marx W.: *Feuerbach*. Cambridge, Cambridge University Press, 1977.

Washburn, Sherwood L.: "The evolution of man" in *Scientific American*, 239 (3): 194-198, 201-202, 204, 206, 208, 242.

Watson, James D.: *Molecular biology of the gene*. 2nd ed. New York, W.A. Benjamin, 1970.

Watson, James D.: *The double helix: a personal account of the discovery of the structure of DNA*. New York, Mentor Books, 1969.

Weinberg, Steven: *The first three minutes: a modern view of the origin of the universe*. New York, Basic Books, 1977.

Weller, J.M.: *The course of evolution*. New York, McGraw-Hill, 1969.

Wells, George A.: "Goethe and evolution" in *Journal of the History of Ideas*, 28 (4): 537-550.

Westphal, Merold: "Nietzsche and the phenomenological ideal" in *The Monist*, 60 (2): 278-288.

Wheelwright, Philip, ed.: *The Presocratics*. New York, Odyssey Press, 1966.

White, Leslie A.: "Diffusion versus evolution: an anti-evolutionist fallacy" in *American Anthropologist*, 47: 339-356.

White, Leslie A.: "Energy and the evolution of culture" in *American Anthropologist*, 45: 335-336.

White, Leslie A.: *The evolution of culture: the development of civilization to the fall of Rome*. New York, McGraw-Hill, 1959.

White, Leslie A.: *The science of culture: a study of man and civilization*. New York, Grove Press, 1949.

White, M.J.D.: *Modes of speciation*. San Francisco, Freeman, 1978.

Whitehead, Alfred North: *Adventures of ideas* (1933). New York, The Free Press, 1967.

Whitehead, Alfred North: *Modes of thought* (1938). New York, The Free Press, 1968.

Whitehead, Alfred North: *Process and reality: an essay in cosmology (1929)*. New York, The Free Press, 1969.

Whitehead, Alfred North: *Science and the modern world* (1925). New York, The Free Press, 1967.

Whitehead, Alfred North: *The concept of nature* (1920). Cambridge, Cambridge University Press, 1964.

Whitrow, G.J.: *The structure and evolution of the universe*. New York, Harper & Brothers, 1959.

Wichler, G.: *Charles Darwin: the founder of the theory of evolution and natural selection*. New York, Pergamon Press, 1961.

Wiener, Philip P.: *Evolution and the founders of pragmatism*. New York, Harper Torchbooks, 1965.

Williams, George C.: *Adaptation and natural selection: a critique of some current evolutionary thought*. Princeton, New Jersey, Princeton University Press, 1974.

Wilson, Daniel J.: "Arthur O. Lovejoy and the moral of 'The Great Chain of Being'" in *Journal of the History of Ideas*, 41 (2): 249-265.

Wilson, Edward O.: *On human nature*. Cambridge, Harvard University Press, 1978.

Wilson, Edward O.: *Sociobiology: the new synthesis*. Cambridge, Belknap Press, 1975. See also the abridged edition: Cambridge Belknap Press, 1980.

Wilson, John B.: "Darwin and the transcendentalists" in *Journal of the History of Ideas*, 26 (2): 286-290.

Wilson, Leonard G., ed.: *Sir Charles Lyell's scientific journal on the species question*. New Haven, Yale University Press, 1970.

Wilson, R.J., ed.: *Darwinism and the American intellectual: a book of readings*. Homewood, Illinois, The Dorsey Press, 1967.

Wissler, C.: "Doctrine of evolution and anthropology" in *American Anthropologist*, 15: 355-356.

Wolfe, Elaine Claire Daughetee: "Acceptance of the theory of evolution in America: Louis Agassiz vs. Asa Gray" in *The American Biology Teacher*, 37 (4): 244-247.

Wolfe, Linda D., and J. Patrick Gray: "Creationism and popular sociobiology as myths" in *The Humanist*, 41 (4): 43-48, 50.

Wright, Sewall: *Evolution and the genetics of populations.* Chicago, University of Chicago Press, 1968.

Wylie, Philip: "Cultural evolution: the fatal fallacy" in *BioScience*, 21 (13): 729-731.

Yang, Suh Y., and James L. Patton: "Genic variability and differentiation in the Galapagos finches" in *The Auk*, 98 (2): 230-242.

Young, Robert M.: "Darwin's metaphor: does nature select?" in *The Monist*, 55 (3): 442-503.

Young, R.S.: *Extraterrestrial biology.* New York, Holt, Rinehart & Winston, 1965.

Young, R.S.: "Life on other planets: some exponential speculations" in *Science*, 144 (3619): 613-615.

Young, R.S., and C. Ponnamperuma: "Life: origin and evolution" in *Science*, 143 (3604): 384-388.

Zeller, Eduard: *Outlines of the history of Greek philosophy.* New York, Meridian Books, 1965.

Zimmerman, P.A., J.W. Klotz, W.H. Rusch, and R.F. Surburg: *Darwin, evolution, and creation.* Saint Louis, Concordia, 1959.

Zimmerman, Robert L.: "On Nietzsche" in *Philosophy and Phenomenological Research*, 29 (2): 274-281.

Zirkle, C.: *Evolution, Marxian biology and the social sciences.* Philadelphia, University of Philadelphia Press, 1959.

INDEX

Dr. H. James Birx with a giant tortoise (*Geochelone elephantopus*) at the Charles Darwin Research Station on Santa Cruz Island of the Galapagos Archipelago, 4 July 1981. (Photo by Joyce Bittner)

ABOUT THE AUTHOR

Dr. H. James Birx is professor of anthropology and chairman of the anthropology/sociology department at Canisius College. He holds a B.S. in the natural sciences, M.S. in biology, M.A. in anthropology, and Ph.D. with distinction in philosophy from the State University of New York at Buffalo.

Professor Birx is the author of *Man's Place in the Universe: An Introduction to Scientific Philosophical Anthropology* (1977) and *Pierre Teilhard de Chardin's Philosophy of Evolution* (1972). His research areas have included botany, human craniometry, naturalist phenomenology, and the historical development of the theory of evolution. Extensive travels have taken him to Kauai, Stonehenge, Teotihuacan, and Machu Picchu; he has twice visited both the Galapagos Islands and Charles Darwin's Down House in Kent, England.

His scholarly articles and critical reviews have appeared in

BioScience, The Humanist, Free Inquiry, The Science Teacher, Serbian Studies, American Scientist, Creation/Evolution, American Biology Teacher, Humboldt Journal of Social Relations, and *Philosophy and Phenomenological Research.* His writings advocate scientific naturalism and rational humanism, embracing a cosmic perspective and the evolutionary framework. Emphasizing holism, their central theme is the place of humankind within this universe.

Dr. Birx is the editor of *The Tesla Journal,* editorial associate of *Free Inquiry,* and a contributing editor of *Cosmic Search.* During the past five years, as a faculty member of the Consortium of the Niagara Frontier, he has taught various courses in anthropology, sociology, and philosophy as well as participated in lecture/ seminar series at the Attica Correctional Facility. He has also offered classes and lectured at a number of colleges and universities in the United States, Canada, and Europe (including Brock University and the Freie Universität Berlin) and frequently appears as a guest on educational radio and television programs.

Professor H. James Birx is a U.S. Chancellor and Executive Board member of *Delta Tau Kappa* (the international social science honor society). He recently delivered papers at three international symposiums commemorating the intellectual achievements of Charles Darwin, Marvin Farber, and Pierre Teilhard de Chardin. In November of 1983, Dr. Birx will chair the *Ring* reinterpreted session of an interdisciplinary symposium "Wagner in Retrospect: A Centennial Reappraisal" at the University of Illinois at Chicago as well as present his paper "Philosophies of Evolution" at an interdisciplinary faculty seminar on evolution at the State University of New York at Buffalo (1983-1984 academic year).